VOLTAGE-SOURCED CONVERTERS IN POWER SYSTEMS

VOLTAGE-SOURCED CONVERTERS IN POWER SYSTEMS

Modeling, Control, and Applications

Amirnaser Yazdani
University of Western Ontario

Reza Iravani
University of Toronto

IEEE PRESS

A JOHN WILEY & SONS, INC., PUBLICATION

Published by John Wiley & Sons, Inc., Hoboken, New Jersey
Published simultaneously in Canada.

For general information on our other products and services or for technical support, please contact our
Customer Care Department within the United States at (800) 762-2974, outside the United States at (317)
572-3993 or fax (317) 572-4002.

Wiley also publishes its books in a variety of electronic formats. Some content that appears in print may
not be available in electronic formats. For more information about Wiley products, visit our web site at
www.wiley.com.

Library of Congress Cataloging-in-Publication Data:

Yazdani, Amirnaser, 1972–
    Voltage-sourced converters in power systems : modeling, control, and
applications / Amirnaser Yazdani, Reza Iravani.
        p. cm.
    ISBN 978-0-470-52156-4 (cloth)
    1. Electric current converters. 2. Electric power systems–Control. 3. Electric
power systems–Equipment and supplies. 4. Interconnected electric utility systems.
I. Iravani, Reza, 1955– II. Title.
    TK1007.Y39 2010
    621.31′3–dc22

                                                2009052122

Printed in the United States of America

10 9 8 7 6 5 4 3 2 1

*To Farzaneh and Arman,
and Suzan*

CONTENTS

PREFACE

The concept of electronic (static) power conversion has gained widespread acceptance in power system applications. As such, electronic power converters are increasingly employed for power conversion and conditioning, compensation, and active filtering. The gradual increase in the depth of penetration of distributed energy resource (DER) units in power systems and further acceptance of new trends and concepts, for example, microgirds, active distribution systems, and smart grids, also indicate a wider role for power-electronic converters in the electric power system.

While a fairly large number of books on various power-electronic converter configurations and their principles of operation do exist, there is a gap in terms of modeling, analysis, and control of power-electronic converters in the context of power systems. This book addresses this gap and concentrates on power conversion and conditioning applications and presents the analysis and control design methodologies for a specific class of high-power electronic converters, namely, the three-phase voltage-sourced converter (VSC). It provides systematic, comprehensive, unified, and detailed coverage of the relevant materials.

This book serves as a reference book for senior undergraduate and graduate students in power engineering programs, practicing engineers who deal with grid integration and operation of DER systems, design engineers, and researchers in the areas of electric power generation, transmission, distribution, and utilization. The book does not cover implementation details of controllers; however, it contains adequate details for system analysts and control designers and

- describes various functions that the VSC can perform in an electric power system,
- introduces different classes of applications of the VSC in electric power systems,
- provides a systematic approach to modeling a VSC-based system with respect to its class of application,
- presents a comprehensive and detailed control design approach for each class of applications, and
- illustrates the control design procedures and evaluates the performance, based on digital computer time-domain simulation studies.

The text is organized in 13 chapters. Chapter 1 provides a brief introduction to the most commonly used electronic switches and converter configurations in the power system. The rest of the book is divided into two parts. The first part, Chapters 2–10, provides theory and presents fundamental modeling and design methodologies. The second part, Chapters 11–13, covers applications of theory and design methodologies, through three selected application cases: the static compensator (STATCOM), the forced-commutated back-to-back HVDC converter system, and the variable-speed wind-power systems based on the doubly fed asynchronous generator. The second part could have included more application varieties. However, only three application cases have been presented to highlight the main concepts, within a limited number of pages. The PSCAD/EMTDC software package has been used to generate most of the time-domain simulation results in the text. We would like to emphasize that the main purpose of the numerical examples in this book is to highlight the concepts and design methodologies. As such, the numerical values of some parameters may not be fully consistent with the values typically adopted for specific applications.

The reader is expected to have, at least, an undergraduate-level background in electric circuits, electric machinery, electric power system fundamentals, and classical (linear) control. Familiarity with power electronics and the state-space representation of systems is a bonus but not a necessity. Relevant references are also cited throughout the book to help the reader trace back the developments to their original sources. While we have tried to be as comprehensive as possible, it is very likely that we have missed some important references due to the richness of the technical literature and the breadth of the subject matter. We would greatly appreciate any comments and feedback from the readers, for future modifications of the book.

AMIRNASER YAZDANI
REZA IRAVANI

London, Ontario, Canada
Toronto, Ontario, Canada
January 2010

ACKNOWLEDGMENTS

I am very grateful to my former Ph.D. supervisor Professor Reza Iravani (the University of Toronto). Without his encouragement and support this book would have never been envisaged. During the preparation of this book, I have benefited immensely from many colleagues and friends. In particular, I would like to thank Professors Tarlochan S. Sidhu and Serguei Primak (the University of Western Ontario) for their mentorship and support; Professor Rajni V. Patel (the University of Western Ontario) for his discussions and invaluable insight into the subject of control theory; Professor Richard Bonert (the University of Toronto) for his enlightening ideas on power electronics and electromechanical energy conversion; and the late Professor Shashi B. Dewan (DPS Inc. and University of Toronto) for providing me with the opportunity to further enrich the book concepts through my exposure to high-power electronic converter systems at Digital Predictive Systems (DPS) Inc. While teaching and revising the drafts of this book, I received invaluable feedback from my graduate students to whom I am thankful.

A. Y.

I would like to express my sincere thanks to the late Professor Shashi B. Dewan whose generous and unconditional support made this work possible. Many thanks to Professor R. Mohan Mathur who has always been a source of encouragement, and to my colleagues Dr. Milan Graovac, Mr. Xiaolin Wang, and Dr. Armen Baronijan for their invaluable discussions. And finally, thanks to all my former and current graduate students and postdoctoral fellows whose research work has immensely enriched the text.

R. I.

ACRONYMS

AC	Alternating current
CSC	Current-sourced converter
DC	Direct current
DCC	Diode-clamped converter
DER	Distributed energy resource
DFIG	Doubly-fed induction generator
DG	Distributed generation
DES	Distributed energy storage
FACTS	Flexible AC transmission systems
GTO	Gate-turn-off thyristor
HVDC	High-voltage DC
IGBT	Insulated-gate bipolar transistor
IGCT	Integrated gate-commutated thyristor
LHP	Left half plane
MIMO	Multi-input-multi-output
MOSFET	Metal-oxide-semiconductor field-effect transistor
NPC	Neutral-point clamped
PCC	Point of common coupling
PI	Proportional-integral
PLL	Phase-locked loop
PMSM	Permanent-magnet synchronous machine
PWM	Pulse-width modulation
pu	Per-unit
PV	Photovoltaic
RHP	Right half plane
SCR	Silicon-controlled rectifier
SISO	Single-input-single-output
SM	Synchronous machine
STATCOM	Static compensator
SVC	Static VAR compensator
UPS	Uninterruptible power supply
VCO	Voltage-controlled oscillator
VSC	Voltage-sourced converter

1 Electronic Power Conversion

1.1 INTRODUCTION

Historically, power-electronic converters have been predominantly employed in domestic, industrial, and information technology applications. However, due to advancements in power semiconductor and microelectronics technologies, their application in power systems has gained considerably more attention in the past two decades. Thus, power-electronic converters are increasingly utilized in power conditioning, compensation, and power filtering applications.

A power-electronic converter consists of a power circuit—which can be realized through a variety of configurations of power switches and passive components—and a control/protection system. The link between the two is through gating/switching signals and feedback control signals. This chapter briefly introduces power circuits of the most commonly used power-electronic converters for high-power applications. In the subsequent chapters, two specific configurations, that is, the two-level voltage-sourced converter (VSC) and the three-level neutral-point clamped (NPC) converter, are analyzed in more detail. This book focuses on the modeling and control aspects of the two-level VSC and the three-level NPC converter. However, the presented analysis techniques and the control design methodologies are conceptually also applicable to the other families of power-electronic converters introduced in this chapter.

1.2 POWER-ELECTRONIC CONVERTERS AND CONVERTER SYSTEMS

In this book we define a power-electronic (or static) converter as a multiport circuit that is composed of semiconductor (electronic) switches and can also include auxiliary components and apparatus, for example, capacitors, inductors, and transformers. The main function of a converter is to facilitate the exchange of energy between two (or more) subsystems, in a desired manner, based on prespecified performance specifications. The subsystems often have different attributes in terms of voltage/current waveforms, frequency, phase angle, and number of phases, and therefore cannot be directly

Voltage-Sourced Converters in Power Systems,   by Amirnaser Yazdani and Reza Iravani
Copyright © 2010 John Wiley & Sons, Inc.

interfaced with each other, that is, without power-electronic converters. For instance, a power-electronic converter is required to interface a wind turbine/generator unit, that is, an electromechanical subsystem that generates a variable-frequency/variable-voltage electricity, with the constant-frequency/constant-voltage utility grid, that is, another electromechanical subsystem.

In the technical literature, converters are commonly categorized based on the type of electrical subsystems, that is, AC or DC, that they interface. Thus,

- A DC-to-AC or DC/AC converter interfaces a DC subsystem to an AC subsystem.
- A DC-to-DC or DC/DC converter interfaces two DC subsystems.
- An AC-to-AC or AC/AC converter interfaces two AC subsystems.

Based on the foregoing classification, a DC/AC converter is equivalent to an AC/DC converter. Hence, in this book, the terms DC/AC converter and AC/DC converter are used interchangeably. The conventional diode-bridge rectifier is an example of a DC/AC converter. A DC/AC converter is called a *rectifier* if the flow of average power is from the AC side to the DC side. Alternatively, the converter is called an *inverter* if the average power flow is from the DC side to the AC side. Specific classes of DC/AC converters provide bidirectional power-transfer capability, that is, they can operate either as a rectifier or as an inverter. Other types, for example, the diode-bridge converter, can only operate as a rectifier.

DC/DC converter and AC/AC converter are also referred to as *DC converter* and *AC converter*, respectively. A DC converter can directly interface two DC subsystems, or it can employ an intermediate AC link. In the latter case, the converter is composed of two back-to-back DC/AC converters which are interfaced through their AC sides. Similarly, an AC converter can be direct, for example, the matrix converter, or it can employ an intermediate DC link. The latter type consists of two back-to-back DC/AC converters which are interfaced through their DC sides. This type is also known as *AC/DC/AC converter*, which is widely used in AC motor drives and variable-speed wind-power conversion units.

In this book, we define a *power-electronic converter system* (or a converter system) as a composition of one (or more) power-electronic converter(s) and a control/protection scheme. The link between the converter(s) and the control/protection scheme is established through gating signals issued for semiconductor switches, and also through feedback signals. Thus, the transfer of energy in a converter system is accomplished through appropriate switching of the semiconductor switches by the control scheme, based on the overall desired performance, the supervisory commands, and the feedback from a multitude of system variables.

This book concentrates on modeling and control of a specific class of converter systems, this is, the VSC systems. This class is introduced in Section 1.6.

1.3   APPLICATIONS OF ELECTRONIC CONVERTERS IN POWER SYSTEMS

For a long time, applications of high-power converter systems in electric power systems were limited to high-voltage DC (HVDC) transmission systems and, to a lesser extent, to the conventional static VAR compensator (SVC) and electronic excitation systems of synchronous machines. However, since the late 1980s, the applications in electric power systems, for generation, transmission, distribution, and delivery of electric power, have continuously gained more attention [1–6]. The main reasons are

- Rapid and ongoing developments in power electronics technology and the availability of various types of semiconductor switches for high-power applications.
- Ongoing advancements in microelectronics technology that have enabled realization of sophisticated signal processing and control strategies and the corresponding algorithms for a wide range of applications.
- Restructuring trends in the electric utility sector that necessitate the use of power-electronic-based equipment to deal with issues such as power line congestion.
- Continuous growth in energy demand that has resulted in close-to-the-limit utilization of the electric power utility infrastructure, calling for the employment of electronic power apparatus for stability enhancement.
- The shift toward further utilization of green energy, in response to the global warming phenomenon, and environmental concerns associated with centralized power generation. The trend has gained momentum due to recent technological developments and has resulted in economic and technical viability of alternative energy resources and, in particular, renewable energy resources. Such energy resources are often interfaced with the electric power system through power-electronic converters.

In addition, development of new operational concepts and strategies, for example, microgrids, active networks, and smart grids [7], also indicates that the role and importance of power electronics in electric power systems will significantly grow. The envisioned future roles of power-electronic converter systems in power systems include

- Enhancement of efficiency and reliability of the existing power generation, transmission, distribution, and delivery infrastructure.
- Integration of large-scale renewable energy resources and storage systems in electric power grids.
- Integration of distributed energy resources, both distributed generation and distributed storage units, primarily, at subtransmission and distribution voltage levels.

- Maximization of the depth of penetration of renewable distributed energy resources.

Power-electronic converter systems are employed in electric power systems for

- *Active Filtering:* The main function of a power-electronic-based active filter is to synthesize and inject (or absorb) specific current or voltage components, to enhance power quality in the host power system. A comprehensive treatment of the concepts and controls of active power filters is given in Ref. [8].
- *Compensation:* The function of a power-electronic (static) compensator, in either a transmission or a distribution line, is to increase the power-transfer capability of the line, to maximize the efficiency of the power transfer, to enhance voltage and angle stability, to improve power quality, or to fulfill a combination of the foregoing objectives. Various static compensation techniques have been extensively discussed in the technical literature under the general umbrella of flexible AC transmission systems (FACTS) and custom-power controllers [1–6]. The FACTS controllers include, but are not limited to, the static synchronous compensator (STATCOM), the static synchronous series compensator (SSSC), the intertie power flow controller (IPFC), the unified power flow controller (UPFC), and the semiconductor-controlled phase shifter.
- *Power Conditioning:* The main function of an electronic power conditioner is to enable power exchange between two electrical (or electromechanical) subsystems in a controlled manner. The power conditioner often has to ensure that specific requirements of subsystems, for example, the frequency, voltage magnitude, power factor, and velocity of the rotating machines, are met. Examples of electronic power conditioning systems include but are not limited to
  1. the back-to-back HVDC system that interfaces two AC subsystems that can be synchronous, asynchronous, or even of different frequencies [9];
  2. the HVDC rectifier/inverter system that transfers electrical power through a DC tie line between two electrically remote AC subsystems [10, 11];
  3. the AC/DC/AC converter system that transfers the AC power from a variable-frequency wind-power unit to the utility grid; and
  4. the DC/AC converter system that transfers the DC power from a DC distributed energy resource (DER) unit, for example, a photovoltaic (PV) solar array, a fuel cell, or a battery storage unit, to the utility grid [12, 13].

1.4   POWER-ELECTRONIC SWITCHES

Power-electronic semiconductor switches (or electronic switches) are the main building blocks of power-electronic converters. A power-electronic switch is a semiconductor device that can permit and/or interrupt the flow of current through a branch of the host circuit, by the application of a gating signal.[1] This is in contrast to the operation of a mechanical switch in which the on/off transition is achieved through a mechanical process, for example, the movement of a mechanical arm. A mechanical switch

- is slow and thus not intended for repetitive switching;
- essentially includes moving parts and therefore is subject to loss of lifetime during each switching action and thus, compared to an electronic switch, provides a limited number of on/off operations; and
- introduces relatively low power loss during conduction, such that it can be practically considered as a closer representation to an ideal switch.

By contrast, an electronic switch

- is fast and intended for continuous switching;
- includes no moving part and thus is not subject to loss of lifetime during turn-on and turn-off processes; and
- introduces switching and conduction power losses.

The above-mentioned characteristics of the mechanical and electronic switches indicate that for some applications a combination of mechanical and electronic switches can provide an optimum solution in terms of switching speed and power loss. However, the trend in the development of power semiconductor switches [14, 15] points toward ever-increasing utilization of electronic switches. The effort to increase the maximum permissible switching frequency and to minimize switching and conduction losses is the subject of major research and development programs of the power semiconductor switch industry.

1.4.1 Switch Classification

The characteristics of a power-electronic converter mainly depend on the type of its semiconductor switches. It is therefore warranted to briefly review different switch types. Further details regarding the operation and characteristics of the most commonly used switches can be found in Refs. [16, 17].

1.4.1.1 Uncontrollable Switches   The power diode is a two-layer semiconductor device and the only uncontrollable switch. It is uncontrollable since the current conduction and interruption instants are determined by the host electrical circuit. Power

---

[1]The only exception is diode that conducts current based on the conditions of the host circuit and not in response to a gating signal.

diodes are extensively used in power-electronic converter circuits as stand-alone components, and/or as integral parts of other switches.

1.4.1.2   *Semicontrollable Switches*

The most widely used semicontrollable electronic switch is the thyristor or the silicon-controlled rectifier (SCR). The thyristor is a four-layer semiconductor device that is half- or semicontrollable, since only the instant at which its current conduction starts can be determined by a gating signal, provided that the device is properly voltage biased. However, the current interruption instant of the thyristor is determined by the host electrical circuit. The thyristor has been, and even currently is, the switch of choice for HVDC converters, although in recent years fully controllable switches have also been considered and utilized for HVDC applications.

1.4.1.3   *Fully Controllable Switches*

The current conduction and interruption instants of a fully controllable switch can be determined by means of a gating command. Most widely used fully controllable switches include

- *Metal-Oxide-Semiconductor Field-Effect Transistor (MOSFET):* The MOSFET is a three-layer semiconductor device. Compared to other fully controllable power switches, current and voltage ratings of power MOSFETs are fairly limited. Consequently, the application of power MOSFETs is confined to relatively lower power converters where a high switching frequency is the main requirement.

- *Insulated-Gate Bipolar Transistor (IGBT):* The IGBT is also a three-layer semiconductor device. The power IGBT has significantly evolved since the early 1990s, in terms of the switching frequency, the current rating, and the voltage rating. At present, it is used for a broad spectrum of applications in electric power systems.

- *Gate-Turn-Off Thyristor (GTO):* The GTO is structurally a four-layer semiconductor device and can be turned on and off by external gating signals. The GTO requires a relatively large, negative current pulse to turn off. This requirement calls for an elaborate and lossy drive scheme. Among the fully controllable switches, the GTO used to be the switch of choice for high-power applications in the late 1980s and early 1990s. However, it has lost significant ground to the IGBT in the last several years.

- *Integrated Gate-Commutated Thyristor (IGCT):* The IGCT conceptually and structurally is a GTO switch with mitigated turn-off drive requirements. In addition, the IGCT has a lower on-state voltage drop and can also be switched faster compared to the GTO. In recent years, the IGCT has gained considerable attention for high-power converters due to its voltage/current handling capabilities.

In terms of voltage/current handling capability, the semicontrollable and fully controllable switches are classified as follows:

- *Unidirectional Switch:* A unidirectional switch can conduct current in only one direction. Hence, the switch turns off and assumes a reverse voltage when its current crosses zero and attempts to go negative. A unidirectional switch can be bipolar (symmetrical) or unipolar (asymmetrical). A bipolar switch can withstand a relatively large reverse voltage. The thyristor is an example of a bipolar, unidirectional switch. A unipolar switch, however, has a relatively small reverse breakdown voltage; thus, a voltage exceeding the switch reverse breakdown voltage results in a reverse in-rush current that can damage the switch. Therefore, to prevent the reverse breakdown and the consequent damage, a diode can be connected in antiparallel with the unipolar switch that also makes the switch *reverse conducting*. The GTO and the IGCT are commercially available in both unipolar and bipolar types. The current-sourced converter (CSC), described in Section 1.5.2, requires bipolar, unidirectional switches.

- *Reverse-Conducting Switch:* A reverse-conducting switch is realized when a unidirectional switch, whether unipolar or bipolar, is connected in antiparallel with a diode. Hence, a reverse-conducting switch can be regarded as a unipolar switch whose reverse breakdown voltage is approximately equal to the forward voltage drop of a diode. Thus, a reverse-conducting switch starts to conduct in the opposite direction if it is reverse biased by only a few volts. The IGBT and the power MOSFET are examples of reverse-conducting switches. Reverse-conducting IGCT switches are also commercially available. In this book, we refer to a fully controllable reverse-conducting switch also as a *switch cell*, generically illustrated in Figure 1.1(a). The VSC, defined later in this chapter, requires reverse-conducting switches (switch cells). Figure 1.1(b) shows two common symbolic representations of a switch cell in which the gate control terminal is not shown.

- *Bidirectional Switch:* A bidirectional switch can conduct and interrupt the current in both directions. Essentially, a bidirectional switch is also a bipolar switch since in the off state it must withstand both forward and reverse voltage biases. An example of a (semicontrollable) bidirectional switch are two thyristors that are connected in antiparallel. It should be pointed out that, to date, there

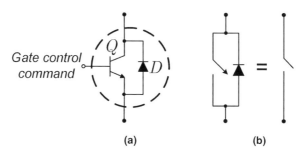

(a)   (b)

FIGURE 1.1   (a) Generic schematic diagram of a switch cell. (b) Symbolic representations of a switch cell.

exists no fully controllable bidirectional single-device switch technology. Hence, such a switch must be realized through antiparallel connection of two bipolar unidirectional switches. Fully controllable bidirectional switches are required for matrix converters [18].

1.4.2 Switch Characteristics

In the context of electronic power conversion, semiconductor switches are almost exclusively used in the switching mode, that is, the switch is either in the on state or in the off state. The steady-state and switching properties of an electronic switch are conventionally illustrated and characterized by, respectively, the switch current/voltage waveforms and the characteristic curves in the current-versus-voltage (v–i) plane. For system studies and control design purposes, especially for high-power converters where the switching frequencies are typically low, simplified switch models are often adopted. Such models retain the device features relevant to the study, while considerably reduce the modeling, analytical, and computational burden. However, depending on the objectives of a specific investigation, the accuracy of waveforms and results can be enhanced if more elaborate switch models are employed. For example, if the switching loss of a converter is of interest, the diode reverse recovery and the transistor tailing current effects [16] must be included in the model of switches.

In this book, the on- and off-state characteristics of an electronic switch are approximated by corresponding straight lines in the v–i plane. Thus, transient switching processes such as the reverse recovery, the tailing current, and so on are ignored, and transition from one state to the other is generally assumed to be instantaneous. However, to demonstrate the methodology, in Section 2.6 we employ more detailed models of switches to estimate the power loss of a DC/AC voltage-sourced converter.

1.5 CLASSIFICATION OF CONVERTERS

There are a variety of approaches to classification of power-electronic converters. This section introduces two categorization methods relevant to high-power applications.

1.5.1 Classification Based on Commutation Process

One widely used approach to the categorization of converters is based on the commutation process, defined as the transfer of current from branch i to branch j of a circuit, when the switch of branch i turns off while that of branch j turns on. Based on this definition, the following two classes of converters are identified in the technical literature:

- *Line-Commutated Converter:* For a line-commutated (naturally-commutated) converter, the electrical AC system dictates the commutation process. Thus, the commutation process is initiated by the reversal of the AC voltage polarity. The conventional six-pulse thyristor-bridge converter, widely used in HVDC transmission systems, is an example of a line-commutated converter [19]. The line-commutated converter is also known as the naturally-commuted converter.

- *Forced-Commutated Converter:* For a forced-commutated converter, the transfer of current from one switch to another one is a controlled process. Thus, in this type of converter, either the switches must be fully controllable, that is, they must have the *gate-turn-off capability*, or the turn-off process must be accomplished by auxiliary turn-off circuitry, for example, an auxiliary switch or a capacitor. A forced-commutated converter that utilizes switches with the gate-turn-off capability is also known as a *self-commutated converter*. Self-commutated converters are of great interest for power systems applications and are the main focus of this book.

It should be noted that in specific power-electronic converter configurations, switches may not be subjected to the current commutation process, in which case the converter is referred to as a *converter without commutation*. For example, the two antiparallel thyristors in the conventional SVC can be regarded as a converter without commutation.

1.5.2   Classification Based on Terminal Voltage and Current Waveforms

DC/AC converters can also be classified based on voltage and current waveforms at their DC ports. Thus, a *current-sourced converter (CSC)* is a converter in which the DC-side current retains the same polarity, and therefore, the direction of average power flow through the converter is determined by the polarity of the DC-side voltage. The DC side of a CSC is typically connected in series with a relatively large inductor that maintains the current continuity and is more representative of a current source. For example, the conventional, six-pulse, thyristor-bridge rectifier is a CSC. In a *voltage-sourced converter (VSC)*, however, the DC-side voltage retains the same polarity, and the direction of the converter average power flow is determined by the polarity of the DC-side current. The DC-side terminals of a VSC are typically connected in parallel with a relatively large capacitor that resembles a voltage source.

Compared to the VSC, the forced-commutated CSC has not been as widely used for power system applications. The reason is that a CSC requires bipolar electronic switches. However, the power semiconductor industry has not yet fully established a widespread commercial supply of fast, fully controllable bipolar switches. Although bipolar versions of the GTO and the IGCT are commercially available, they are limited in terms of switching speed and are mainly tailored for very high-power electronic converters. Unlike the CSC, a VSC requires reverse-conducting switches or switch cells. The switch cells are commercially available as the IGBT or the reverse-conducting IGCT. Prior to the dominance of the IGBT and the IGCT, each switch of the VSC was realized through antiparallel connection of a GTO with a diode.

1.6   VOLTAGE-SOURCED CONVERTER (VSC)

The focus of this book is on modeling and control of the VSC and VSC-based systems. In the next section, the most common VSC configurations are briefly introduced.

1.7   BASIC CONFIGURATIONS

Figure 1.2 shows the basic circuit diagram of a half-bridge, single-phase, two-level VSC. The half-bridge VSC consists of an upper switch cell and a lower switch cell. Each switch cell is composed of a fully controllable, unidirectional switch in antiparallel connection with a diode. As explained in Section 1.4.1.3, this switch configuration constitutes a reverse-conducting switch that is readily available, for example, in the form of commercial IGBT and IGCT. The DC system that maintains the net voltage of the split capacitor can be a DC source, a battery unit, or a more elaborate configuration such as the DC side of an AC/DC converter. The half-bridge VSC of Figure 1.2 is called a two-level converter since the switched AC-side voltage, at any instant, is either at the voltage of node p or at the voltage of node n, depending on which switch cell is on. The fundamental component of the AC-side voltage is usually controlled based on a pulse-width modulation (PWM) technique [16, 20].

If two half-bridge VSCs are connected in parallel through their DC sides, the full-bridge single-phase VSC of Figure 1.3 is realized. Thus, as shown in Figure 1.3, the AC system can be interfaced with the AC-side terminals of the two half-bridge converters. One advantage is that, for a given DC voltage, the synthesized AC voltage of the full-bridge VSC is twice as large in comparison with the half-bridge VSC, which corresponds to a more efficient utilization of the DC voltage and switch cells. The full-bridge VSC of Figure 1.3 is also known as the *H-bridge converter*.

Figure 1.4 illustrates the schematic diagram of a three-phase two-level VSC. The three-phase VSC is also an extension of the half-bridge VSC of Figure 1.2. In power system applications, the three-phase VSC is interfaced with the AC system, typically, through a three-phase transformer, based on a three-wire connection. In case a four-wire interface is required, either the VSC must permit access to the midpoint of its split DC-side capacitor, through the fourth wire (or the neutral wire), or it must

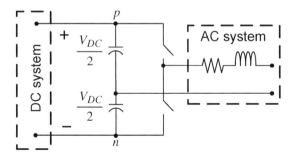

FIGURE 1.2   Schematic diagrams of the half-bridge, single-phase, two-level VSC.

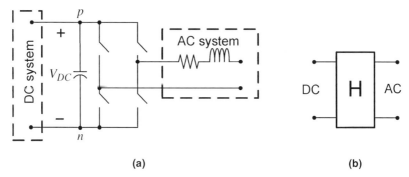

FIGURE 1.3   (a) Schematic diagram of the full-bridge, single-phase, two-level VSC (or an H-bridge converter). (b) Symbolic representation of the H-bridge converter.

be augmented with an additional half-bridge converter, to represent the fourth leg identical to the other three legs, whose AC terminal is connected to the fourth wire. Various PWM and space-vector modulation techniques for switching the three-phase two-level VSC are described in Ref. [20].

The principles of operation of the half-bridge VSC and the three-phase VSC are discussed in Chapters 2 and 5, respectively.

1.7.1   Multimodule VSC Systems

In high-voltage, high-power VSCs, the switch cell of Figure 1.2(b), which is composed of a fully controllable, unidirectional switch and a diode, may not be able to handle the voltage/current requirements. To overcome this limitation, the switch cells are connected in series and/or in parallel and form a composite switch structure which is called a *valve*. Figure 1.5 shows two valve configurations composed of parallel- and series-connected identical switch cells. In most applications, the existing power semiconductor switches meet current handling requirements. However, in

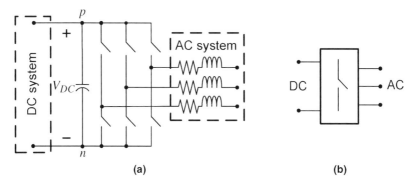

FIGURE 1.4   (a) Schematic diagram of the three-wire, three-phase, two-level VSC. (b) The symbolic representation of the three-phase VSC.

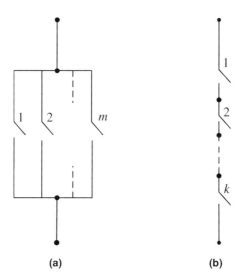

(a)   (b)

FIGURE 1.5   Symbolic representations of a valve composed of (a) m parallel-connected switch cells and (b) k series-connected switch cells.

most applications, series-connected switch cells are inevitably required to satisfy the voltage requirements. Due to various practical limitations including unacceptable form factor, unequal off-state voltage distribution, and simultaneous-gating require-ments, the number of series-connected switch cells within a valve is limited. Thus, a two-level VSC unit cannot be constructed for any voltage level, and an upper voltage limit applies.

The maximum permissible voltage limit of a VSC system can be increased by series connection of identical, three-phase, two-level VSC modules, to form a multimodule VSC [21]. Figure 1.6 illustrates a schematic diagram of an n-module VSC in which n identical two-level VSC modules are connected in series and parallel, respectively, at their AC and DC ports. Thus, the VSC modules share the same DC-bus capacitor. Figure 1.7 illustrates an alternative configuration of an n-module VSC in which the two-level VSC modules are connected in series at both the AC and DC sides. In both configurations of Figures 1.6 and 1.7, the AC-side voltages of the VSC modules are added up by the corresponding open-winding transformers, to achieve the desired voltage level (and waveform) for connection to the AC system. One of the salient features of the n-module VSC configurations of Figures 1.6 and 1.7 is their modularity, as all the VSC modules and transformers are identical. Modularity is a desired feature that reduces manufacturing costs, facilitates maintenance, and permits provisions for spare parts.

The multimodule converter of Figure 1.7 can be further enhanced to acquire the AC-side voltage harmonic reduction capability of a multipulse configuration while its modularity is preserved. This is achieved through appropriate phase shift in switching patterns of the constituent VSC modules, such that a prespecified set of voltage harmonics are canceled/minimized when added up by the open-winding

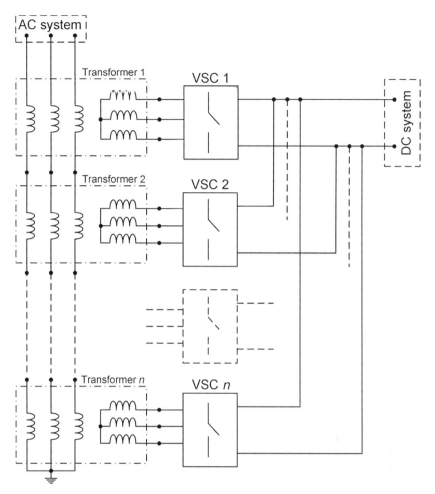

FIGURE 1.6 Schematic diagram of a multimodule VSC composed of *n* two-level VSC modules paralleled from their DC sides.

transformers. The harmonic minimization enables operation of the multimodule converter at low switching frequencies, which, in turn, results in lower switching losses; it also mitigates the need for low-frequency harmonic filters at the converter AC side [22].

The concept of multipulse conversion is another technique employed to minimize low-frequency harmonic components of the synthesized AC voltage of a VSC [23], and thus to minimize the associated filtering requirements. Figure 1.8 shows a schematic diagram of a 12-pulse, thyristor-bridge CSC system that has been extensively used in conventional HVDC transmission applications. As Figure 1.8 shows, the 12-pulse operation of the CSC requires a 30-degree phase shift between the AC-side voltages of the two CSCs. This phase shift is realized by means of two transformers

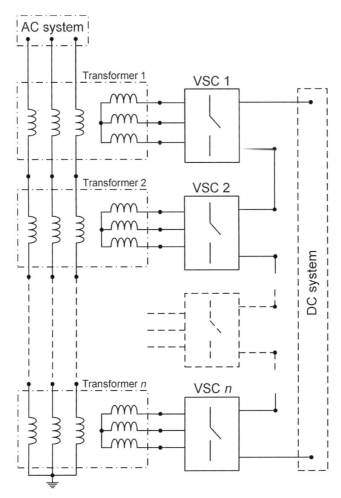

FIGURE 1.7   Schematic diagram of a multimodule VSC composed of n two-level VSC modules series with respect to their DC sides.

of different winding configurations. Therefore, the modularity is not fully preserved in the configuration of Figure 1.8.

1.7.2   Multilevel VSC Systems

Another option for a VSC system to accommodate the voltage requirements of a high-power application is to utilize a multilevel voltage synthesis strategy. Conceptually, the multilevel VSC configurations [17] can be divided into

- the H-bridge-based multilevel VSC;
- the capacitor-clamped multilevel VSC; and
- the diode-clamped multilevel VSC.

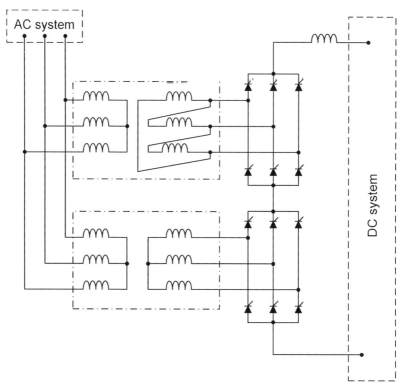

FIGURE 1.8   Schematic diagram of a 12-pulse, line-commutated, thyristor-bridge-based CSC.

The H-bridge-based multilevel VSC, also called the *cascaded H-bridge multi-level VSC*, is constructed through series connection of the H-bridge modules of Figure 1.3(b). Figure 1.9 shows a schematic diagram of a three-phase, wye-connected, H-bridge-based, multilevel VSC. For this configuration, especially if real-power exchange is involved, the DC-bus voltage of each H-bridge module must be independently supplied and regulated by an auxiliary converter system. This renders the H-bridge-based VSC of Figure 1.9 practically unattractive for general-purpose applications; rather, the H-bridge-based converter is more suitable for specific applications, for example, the STATCOM, where only reactive-power exchange is the objective.

One salient feature of the configuration of Figure 1.9 is that it permits independent control of the three legs of the converter. If a (grounded) neutral conductor is provided, the converter can also provide independent control over the zero-sequence components of the three-phase current, in addition to the positive-sequence and the negative-sequence components. It should be noted that the three-wire, three-phase VSC configurations of Figures 1.4 and 1.7 can be controlled to respond only to the positive-sequence and the negative-sequence components.

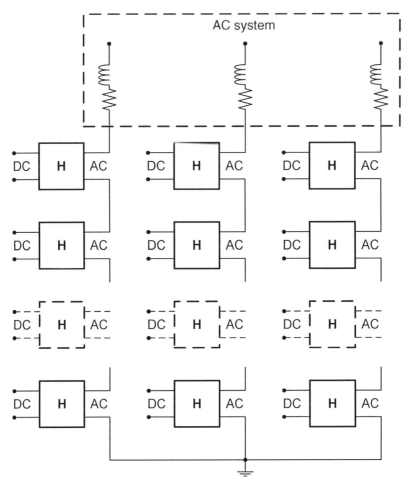

FIGURE 1.9   Schematic diagram of a wye-connected, H-bridge-based, multilevel VSC.

The capacitor-clamped multilevel VSC is another class of multilevel converter configurations. This type of converter is characterized by a large number of relatively large-size capacitors. One of the technical challenges associated with this type of converter is the regulation of its capacitors' voltages. Consequently, the application of the capacitor-clamped multilevel VSC in power systems has not been widely sought in practice and will not be discussed in this book.

The diode-clamped, multilevel, voltage-sourced converter (DCC) is a generalization of the two-level VSC of Figure 1.4(a). This configuration largely avoids the drawbacks of the other two multilevel converter configurations and is considered a promising configuration for power system applications. Figure 1.10 illustrates a conceptual diagram of an n-level DCC in which each leg of the converter is symbolically represented by a fictitious n-tuple-through switch, and the DC bus consists

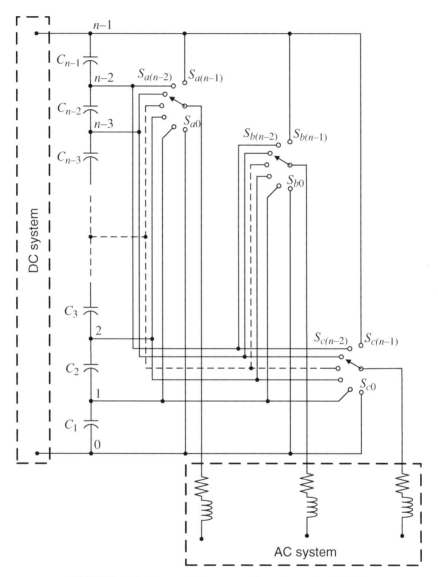

FIGURE 1.10   Conceptual representation of an n-level DCC.

of $n - 1$ nominally identical capacitors C_1 to C_{n-1}. Based on the switching strategy
devised for the n-level DCC, each switch in Figure 1.10 connects the corresponding AC-side terminal to one of the nodes 0 to $n - 1$ at the converter DC side. Thus,
the AC-side terminal voltage can assume one of the n discrete voltage values of
the DC-side nodes. Figures 1.11 and 1.12 illustrate circuit realizations for the three-
level DCC and the five-level DCC, respectively. The three-level DCC is also known
as the *neutral-point diode-clamped (NPC) converter*, which is widely accepted for

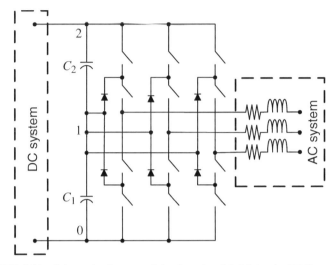

FIGURE 1.11   Schematic diagram of the three-level DCC (or the NPC converter).

high-power applications. Compared to a two-level VSC of the same rating, the three-level DCC can offer a less distorted synthesized AC voltage, lower switching losses, and reduced switch stress levels.

One main technical requirement for proper operation of an n-level DCC is to maintain the voltages of its DC-bus capacitors at their prescribed (usually equal) levels and to prevent voltage drifts during the steady-state and dynamic regimes. In a DCC, net DC-bus voltage regulation cannot guarantee proper operation of the converter system since the individual capacitor voltages may drift or even entirely collapse. This is in contrast to the case of the three-wire two-level VSC for which net DC voltage regulation guarantees proper operation. Provision of a DC-side voltage equalizing scheme is essential for the operation of the DCC.

Conceptually, there are two approaches to deal with the DC capacitor voltage drift phenomenon in a DCC. The first approach is to utilize an auxiliary power circuitry. The auxiliary circuitry can be a set of independent power supplies for capacitors, or it can be a dedicated electronic converter—of a considerably smaller capacity—that injects current into the capacitors and regulates their voltages. The approach is, however, not appealing for power systems applications due to its cost and complexity.

The second approach to equalize the voltages of the DC-side capacitors of a DCC is to enhance the converter control strategy, to modify the switching patterns of the switches such that the capacitors' voltages are regulated at their corresponding desired values. Although this approach calls for a more elaborate control strategy, it offers an economically viable and technically elegant solution to the problem.

It should be pointed out that the number of levels of the multilevel DCC can be higher than three. Therefore, the DCC can accommodate noticeably higher AC and

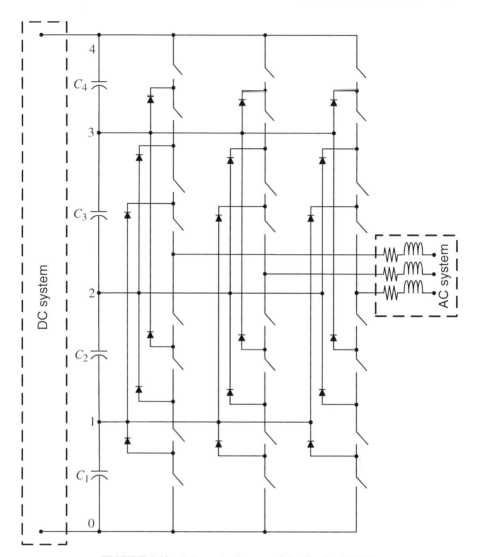

FIGURE 1.12   Schematic diagram of the five-level DCC.

DC voltages, compared to the two-level VSC. However, there is also a limit to the maximum attainable voltage level of a multilevel DCC. Thus, the application of a multilevel DCC in a high-voltage system, for example, an HVDC system, practically may require a multimodule structure similar to that of Figure 1.7, with each module being a multilevel DCC itself. Chapter 6 provides modeling and analysis techniques for the three-level DCC.

1.8 SCOPE OF THE BOOK

In a power-electronic converter system, the functions required for active filtering, compensation, and power conditioning are enabled through the proper operation of the converter control/protection scheme, which finally determines the switching instants of the converter switches. The remainder of this book discusses the mathematical modeling, transient and steady-state behavior, and control design methodologies for a number of VSC-based systems. To limit the number of pages, the methodologies are presented only for the two-level VSC and the three-level DCC. This provides the reader with a comprehensive understanding of the principles of operation, operational characteristics, and control design considerations for the two basic, yet most commonly used, VSC-based configurations. Thus, no attempt has been made in this book to present methodologies for every VSC-based configuration. However, with the understanding gained through this book, the reader should be able to extend and apply similar techniques to different VSC-based systems. Although the book primarily focuses on VSC applications in power conditioning systems, the developments are largely applicable also to compensation systems.

PART I
Fundamentals

2 DC/AC Half-Bridge Converter

2.1 INTRODUCTION

This chapter investigates the DC/AC half-bridge converter as a building block for the multiphase, in particular three-phase, DC/AC voltage-sourced converter (VSC). In Chapter 5, based on the insight obtained through this chapter, we extend the model of the half-bridge converter to study the three-phase DC/AC VSC. This chapter presents the steady-state and dynamic models of the half-bridge converter to the extent that we will be able to directly exploit the results for the multiphase DC/AC VSC.

2.2 CONVERTER STRUCTURE

Figure 2.1 shows a schematic diagram of a DC/AC half-bridge converter [16].[1] The half-bridge converter is composed of two switch cells. The upper and lower switch cells are numbered 1 and 4, respectively. Each switch cell is realized by antiparallel connection of a fully controllable unidirectional switch and a diode. As discussed in Section 1.4.1.3, such a switch cell is also known as a *reverse-conducting switch* and is commercially available as IGBT or IGCT.[2] For ease of reference, throughout this chapter we refer to the fully controllable switch as the *transistor*, for which we have adopted the circuit symbol of the conventional bipolar junction transistor. Thus, the upper switch cell consists of the transistor Q_1 and the diode D_1. Similarly, the lower switch cell is composed of the transistor Q_4 and the diode D_4. As Figure 2.1 shows, in each transistor the positive current is defined as the current flowing from the collector to the emitter. The positive current in the diode is defined as the current flowing from the anode to the cathode. The currents through the upper and lower switch cells are denoted by i_p and i_n, respectively, as shown in Figure 2.1. Thus, $i_p = i_{Q1} - i_{D1}$ and $i_n = -(i_{Q4} - i_{D4})$.

Nodes p and n in Figure 2.1 signify the *DC-side terminals* (or the DC side) of the half-bridge converter. Similarly, we specify the *AC-side terminal* (or the AC side) of the half-bridge converter by node t. From the DC side, the half-bridge converter

---

[1]The configuration shown in Figure 2.1 is also known as a four-quadrant chopper [19].
[2]IGBT, IGCT, and GTO stand, respectively, for *insulated-gate bipolar transistor*, *integrated gate-commutated thyristor*, and *gate-turn-off thyristor* (see Section 1.4.1.3.).

Voltage-Sourced Converters in Power Systems, by Amirnaser Yazdani and Reza Iravani
Copyright © 2010 John Wiley & Sons, Inc.

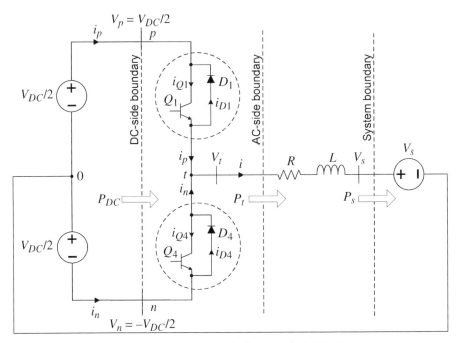

FIGURE 2.1   Simplified power circuit diagram of a half-bridge converter.

of Figure 2.1 is connected to two identical DC voltage sources, each with a voltage of $V_{DC}/2$. The common point of the voltage sources is labeled as node 0. We refer to this node as the *DC-side midpoint* and choose it as the voltage reference node.

From the AC side, the half-bridge converter is interfaced with the voltage source V_s, which we refer to as the *AC-side voltage source*. The negative terminal of the AC-side voltage source is connected to the DC-side midpoint.[3] The connection between the AC-side terminal and the AC-side voltage source is established through an interface reactor represented by a series RL branch. The AC-side terminal voltage, V_t, is a switched waveform and contains voltage ripple.[4] Thus, the interface reactor acts as a filter and ensures a low-ripple AC-side current. L and R, respectively, represent the inductance and the internal resistance of the interface reactor. In some cases, the load or the AC-side source embeds the interface reactor, and no external RL branch is provided. For example, in an electric drive system, the inductance of the machine stator is utilized as the interface reactor between the converter and the machine.

---

[3] In subsequent chapters, we remove the connection path between the AC-side voltage source and the DC-side midpoint, when we extend the DC/AC half-bridge converter to the three-phase VSC; then, the negative terminal of the AC-side voltage source assumes a voltage, V_n, with reference to the DC-side midpoint.

[4] Conventionally, the term *ripple* is defined for a DC waveform as the whole waveform minus its average component. In this book, we define the ripple of a—not necessarily DC—waveform as the time-domain superposition of unwanted harmonics of the waveform.

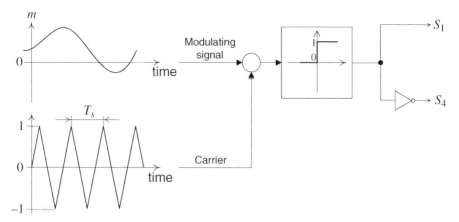

FIGURE 2.2   Schematic diagram of the mechanism to generate PWM gating pulses for Q_1 and Q_4.

In Figure 2.1, P_{DC} represents the (instantaneous) power at the DC side, P_t denotes the power at the AC side, and P_s signifies the power delivered to the AC-side voltage source. The positive direction of the power flow is defined from the DC voltage source(s) toward the AC-side voltage source, as indicated in Figure 2.1.

2.3   PRINCIPLES OF OPERATION

2.3.1   Pulse-Width Modulation (PWM)

The half-bridge converter operates based on the alternate switching of Q_1 and Q_4. The turn-on/off commands of Q_1 and Q_4 are issued through a pulse-width modulation (PWM) strategy. The PWM can be carried out following numerous techniques [24, 25]. However, the most common PWM strategy compares a high-frequency periodic triangular waveform, the carrier signal,[5] with a slow-varying waveform known as the modulating signal. The carrier signal has a periodic waveform with period T_s and swings between -1 and 1. The intersections of the carrier and the modulating signals determine the switching instants of Q_1 and Q_4. The PWM process is illustrated in Figure 2.2, where the switching function of a switch is defined as

$$s(t) = \begin{cases} 1, & \text{if the switch is commanded to conduct,} \\ 0, & \text{if the switch is turned off.} \end{cases}$$

Thus, as Figure 2.2 shows, when the modulating signal is larger than the carrier signal, a turn-on command is issued for Q_1, and the turn-on command of Q_4 is canceled.

---

[5]The carrier signal can also be a periodic sawtooth waveform. However, a triangular carrier signal is often used for a high-power converter.

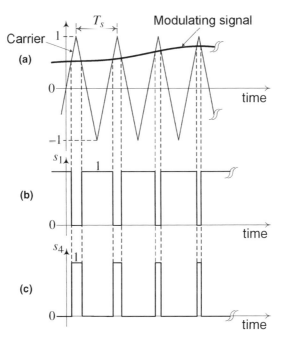

FIGURE 2.3   Signals based on PWM switching strategy: (a) carrier and modulating signals; (b) switching function of the switch Q_1; and (c) switching function of the switch Q_4.

Once the modulating signal is smaller than the carrier signal, the turn-on command for Q_1 is blocked while a turn-on command is issued for Q_4. It should be noted that a switch does not necessarily conduct if it is commanded to turn on; the switch conducts only if the turn-on command is provided and the current direction conforms to the switch characteristics. For example, in response to a turn-on command, an IGBT can conduct only if the current flow is from the collector to the emitter. Figure 2.3(b) and (c) illustrates the waveforms of the switching functions of Q_1 and Q_4, based on the PWM strategy of Figure 2.2. We note that $s_1(t) + s_4(t) \equiv 1$, as Figure 2.3 shows.

2.3.2   Converter Waveforms

In this section, we study the switching nature of the half-bridge converter of Figure 2.1, based on the switching scheme introduced in Section 2.3.1. To avoid unnecessary details, we make the following simplifying assumptions:

- Each transistor or diode acts as a short circuit in its conduction state.
- Each transistor or diode switch acts as an open circuit in its blocking state.
- The transistors have no turn-off tailing current.
- The diodes have no turn-off reverse recovery current.

- Transitions from a conduction state to a blocking state, and vice versa, take place instantly.
- The AC-side current i is a ripple-free DC quantity.

In the following subsections, we examine the converter switching waveforms on the basis of the abovementioned assumptions. Since the converter operation is different for positive and negative AC-side currents, we study each case separately.

2.3.2.1 Converter Waveforms for Positive AC-Side Current
Consider the half-bridge converter of Figure 2.1 with a positive AC-side current i. Assume that $s_1 = 0$ and thus Q_1 is blocked. Consequently, i cannot flow through D_1, since i_{D1} cannot be negative. For the same reason, Q_4 does not carry i, although $s_4 = 1$. Therefore, i flows through D_4 and $V_t = V_n = -V_{DC}/2$. Now consider a time instant at which $s_1 = 1$ and $s_4 = 0$. In this case, Q_1 conducts while Q_4 is blocked. When Q_1 is on, we have $V_t = V_p = V_{DC}/2$, and D_4 is reverse biased. Therefore, i flows through Q_1.

The waveforms of the half-bridge converter for positive i are illustrated in Figure 2.4(a)–(h). As it follows from on the foregoing discussion, Q_4 and D_1 play no role in the operation of the converter when i is positive. The fraction of the switching period T_s during which $s_1 = 1$ is called the *duty ratio* and denoted by d. Under the simplifying assumptions of Section 2.3.2, d is also equal to the fraction of the switching period during which $V_t = V_p = V_{DC}/2$; however, the latter is not the case if switching transients are taken into account.[6]

2.3.2.2 Converter Waveforms for Negative AC-Side Current
It follows from a similar analysis as presented for the case of positive i that Q_1 and D_4 do not take part in the converter operation when i is negative. In this case, when $s_4 = 1$, Q_4 conducts and $V_t = V_n = -V_{DC}/2$. Alternatively, when $s_4 = 0$, the AC-side current passes through D_1 and $V_t = V_p = V_{DC}/2$. The duty ratio, d, is defined in the same way as in the case of the positive AC-side current. Figure 2.5(a)–(h) illustrates the waveforms of the half-bridge converter for negative i.

2.4 CONVERTER SWITCHED MODEL

To employ the half-bridge converter as a component of a larger system, we need to identify the characteristics of the converter as observed from its terminals. The switched model of the half-bridge converter introduces the relationships among the converter terminal voltages and currents. A comparison between Figures 2.4 and 2.5 indicates that in a switch cell the waveform of the current through the transistor, or that through the diode, depends on the direction of the converter AC-side current. However, since $i_p = i_{Q1} - i_{D1}$ and $i_n = -i_{Q4} + i_{D4}$, the waveform of the switch cell

---

[6]We will show this in Section 2.6.

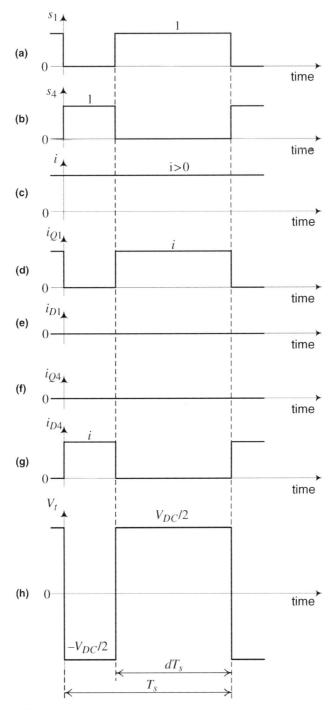

FIGURE 2.4 Half-bridge converter waveforms for positive AC-side current.

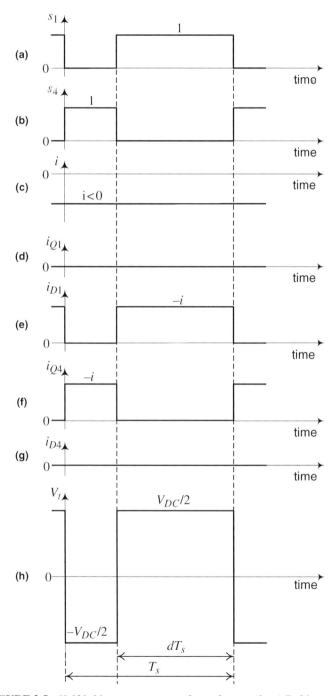

FIGURE 2.5   Half-bridge converter waveforms for negative AC-side current.

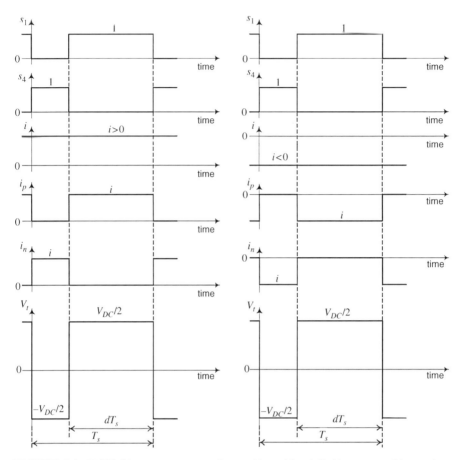

FIGURE 2.6 Half-bridge converter waveforms: (a) positive AC-side current; (b) negative AC-side current.

current is independent of the polarity of i. More importantly, the waveform of the AC-side terminal voltage V_t is independent of the polarity of i and is uniquely determined by the switching functions. Hence, from the terminal viewpoint, the operation of the half-bridge converter can be described as follows.

When $s_1 = 1$, the upper switch cell is closed and the lower one is open; therefore, $V_t = V_p = V_{DC}/2$, $i_p = i$, and $i_n = 0$. Alternatively, when $s_4 = 1$, the lower switch cell is closed but the upper one is open; consequently, $V_t = V_n = -V_{DC}/2$, $i_p = 0$, and $i_n = i$. This holds for both $i > 0$ and $i < 0$, as illustrated in Figure 2.6.

It follows from the foregoing discussion that the half-bridge converter of Figure 2.1 can be mathematically characterized by

$$s_1(t) + s_4(t) \equiv 1, \tag{2.1}$$

$$V_t(t) = (V_{DC}/2)\,s_1(t) - (V_{DC}/2)\,s_4(t), \tag{2.2}$$

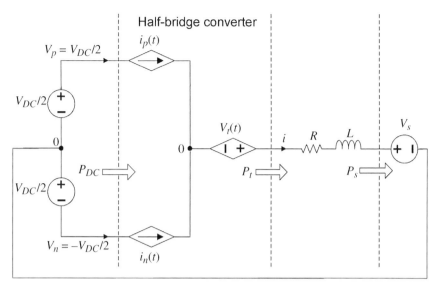

FIGURE 2.7   Switched equivalent circuit of the half-bridge converter of Figure 2.1.

$$i_p(t) = is_1(t), \tag{2.3}$$

$$i_n(t) = is_4(t). \tag{2.4}$$

Equations (2.1)–(2.4) describe the relationships between the half-bridge converter terminal voltages/currents and the switching functions. Figure 2.7 illustrates a switched equivalent circuit for the half-bridge converter of Figure 2.1, based on (2.1)–(2.4).

P_{DC}, P_t, and P_s are calculated as

$$P_{DC}(t) = V_p i_p + V_n i_n = \frac{V_{DC}}{2}\,[s_1(t) - s_4(t)]\,i, \tag{2.5}$$

$$P_t(t) = V_t(t)i = \frac{V_{DC}}{2}\,[s_1(t) - s_4(t)]\,i, \tag{2.6}$$

$$P_s(t) = V_s i. \tag{2.7}$$

The converter power loss is

$$P_{loss} = P_{DC} - P_t. \tag{2.8}$$

It then follows from (2.5) and (2.6) that $P_{loss} \equiv 0$, that is, the ideal half-bridge converter is lossless.

2.5   CONVERTER AVERAGED MODEL

Equations (2.1)–(2.7) provide a switched model for the half-bridge converter of Figure 2.1. The switched model accurately describes the steady-state and dynamic behavior of the converter. The accuracy can be enhanced if more elaborate models are adopted for the circuit components, for example, switches. Thus, given the switching functions for the transistors, the instantaneous values of the current and voltage variables can be computed by means of the switched model. Therefore, the computed variables include high-frequency components, for example, due to the switching process, as well as slow transients. However, the relationships between the modulating signal—which is the main control variable—and the current/voltage variables are not easily understood from the switched model. Moreover, for dynamic analysis and control design purposes, knowledge about the high-frequency details of variables is often not necessary, as the compensators and filters in a closed-loop control system typically exhibit low-pass characteristics and do not react to high-frequency components. For these reasons, we are often interested in the dynamics of the average values of variables, rather than in the dynamics of the instantaneous values. An averaged model also enables us to describe the converter dynamics as a function of the modulating signal.

Consider the switched equivalent circuit of the half-bridge converter, as shown in Figure 2.7. The AC-side current, i, satisfies

$$L\frac{d}{dt}i(t) + Ri(t) = V_t(t) - V_s. \tag{2.9}$$

Since $V_t(t)$ is a periodic function with period T_s, it can be described by the following Fourier series:

$$V_t(t) = \frac{1}{T_s}\int_0^{T_s} V_t(\tau)d\tau + \sum_{h=1}^{h=+\infty}[a_h\cos(h\omega_s t) + b_h\sin(h\omega_s t)], \tag{2.10}$$

where h is the harmonic order, $\omega_s = \frac{2\pi}{T_s}$, and a_h and b_h are given by

$$a_h = \frac{2}{T_s}\int_0^{T_s} V_t(\tau)\cos(h\omega_s\tau)d\tau, \tag{2.11}$$

$$b_h = \frac{2}{T_s}\int_0^{T_s} V_t(\tau)\sin(h\omega_s\tau)d\tau. \tag{2.12}$$

Substituting for $V_t(t)$ from (2.10) in (2.9), we obtain

$$L\frac{di}{dt} + Ri = \left(\frac{1}{T_s}\int_0^{T_s} V_t(\tau)d\tau - V_s\right) + \sum_{h=1}^{h=+\infty}[a_h\cos(h\omega_s t) + b_h\sin(h\omega_s t)]. \tag{2.13}$$

Equation (2.13) describes a low-pass filter with the output i. The input to the filter consists of two components, the constant (DC) component $\frac{1}{T_s} \int_0^{T_s} V_t(\tau)d\tau - V_s$ and the periodic component $\sum_{h=1}^{h=+\infty} (a_h \cos(h\omega_s t) + b_h \sin(h\omega_s t))$. Equation (2.13) is linear. Therefore, based on the superposition principle, the response of the filter to the composite input can be regarded as the summation of its responses to individual input components. This can be expressed as

$$L\frac{d\bar{i}}{dt} + R\bar{i} = \frac{1}{T_s} \int_0^{T_s} V_t(\tau)d\tau - V_s, \tag{2.14}$$

$$L\frac{d\tilde{i}}{dt} + R\tilde{i} = \sum_{h=1}^{h=+\infty} [a_h \cos(h\omega_s t) + b_h \sin(h\omega_s t)], \tag{2.15}$$

$$i(t) = \bar{i}(t) + \tilde{i}(t), \tag{2.16}$$

where $\bar{i}(t)$ and $\tilde{i}(t)$ are, respectively, the responses of the filter to the DC (low-frequency) component and the periodic (high-frequency) component of the filter input. We can also refer to $\tilde{i}(t)$ as the ripple. According to (2.15), if ω_s is adequately larger than R/L, then the periodic component of the input has a negligible contribution to the entire output, the ripple is small, and we can assume that $i(t) \approx \bar{i}(t)$. Thus, the dynamics of the converter system are primarily described by (2.14).

To extend the foregoing methodology to cases where the average of a variable is itself a function of time, that is, it changes from one switching cycle to the next, the averaging operator is defined as

$$\bar{x}(t) = \frac{1}{T_s} \int_{t-T_s}^{t} x(\tau)d\tau, \tag{2.17}$$

where $x(t)$ is a variable and the overbar denotes its average.[7] Thus, (2.14) can also be derived by applying the averaging operator (2.17) to both sides of (2.9). This concept is known as *averaging* in nonlinear systems theory [28, 29] and the power electronics literature [30, 31].

As explained in Section 2.3.1, the periodic switched waveforms are generated by a PWM process. Thus, it can be concluded from Figure 2.3 that if the modulating waveform is not constant but varies in time, the switched waveforms s_1 and s_4 will not retain their precisely periodic forms. Moreover, the average of the switched waveforms varies from one switching cycle to another. The definition of the average, based on (2.17), allows one to also include such switched waveforms in the averaging process. A prerequisite for validity of (2.17) is that the frequency of the carrier waveform should be sufficiently, for example, 10 times, larger than that of the modulating waveform.

---

[7] Also known as the *moving average*.

Applying the averaging operator (2.17) to $s_1(t)$ and $s_4(t)$, in view of Figure 2.6, one deduces

$$\bar{s}_1(t) = d,$$
$$\bar{s}_4(t) = 1 - d. \tag{2.18}$$

Figure 2.8 shows that if the carrier frequency is adequately higher than that of the modulating signal, $\bar{i}$ and $\overline{V}_{DC}$ can be assumed as constant values over one switching cycle [16, 26, 27]. Thus, taking averages of both sides of (2.2)–(2.8) and substituting for $\bar{s}_1(t)$ and $\bar{s}_4(t)$ from (2.18) in the results, we obtain

$$\overline{V}_t = \frac{V_{DC}}{2}(2d - 1), \tag{2.19}$$

$$\bar{i}_p = di, \tag{2.20}$$

$$\bar{i}_n = (1 - d)i, \tag{2.21}$$

$$\overline{P}_{DC} = \frac{V_{DC}}{2}(2d - 1)i, \tag{2.22}$$

$$\overline{P}_t = \frac{V_{DC}}{2}(2d - 1)i, \tag{2.23}$$

$$\overline{P}_s = V_s i, \tag{2.24}$$

$$\overline{P}_{loss} = \overline{P}_{DC} - \overline{P}_t \equiv 0. \tag{2.25}$$

The duty ratio d can assume any value between 0 and 1. If the PWM strategy of Figure 2.2 is adopted, $m = 2d - 1$ describes the relationship between the magnitude of the modulating signal and the duty ratio. This is highlighted in Figure 2.8, which illustrates that d changes from 0 to 1 as m is changed from -1 to 1. It is implicitly assumed that m is constant over the switching period.

Substituting for $d = (m + 1)/2$ in (2.19)–(2.23), we obtain

$$\overline{V}_t = m\frac{V_{DC}}{2}, \tag{2.26}$$

$$\bar{i}_p = \left(\frac{1 + m}{2}\right)i, \tag{2.27}$$

$$\bar{i}_n = \left(\frac{1 - m}{2}\right)i, \tag{2.28}$$

$$\overline{P}_{DC} = m\frac{V_{DC}}{2}i, \tag{2.29}$$

$$\overline{P}_t = m\frac{V_{DC}}{2}i. \tag{2.30}$$

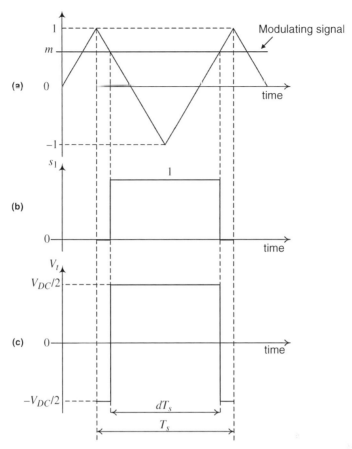

FIGURE 2.8   Generation of the switching signal with a desired duty ratio: if m is changed from -1 to 1, d changes linearly from 0 to 1.

The advantage of the change of variable $d = (m + 1)/2$ becomes evident in (2.26); if m is changed from -1 to 1, the averaged AC-side terminal voltage $\overline{V}_t$ changes linearly from $-V_{DC}/2$ to $V_{DC}/2$, with $m = 0$ corresponding to the zero averaged voltage. Figure 2.9 illustrates the averaged equivalent circuit of the half-bridge converter of Figure 2.1. In the following example, we study and compare the dynamic responses of the half-bridge DC/AC converter deduced from the switched and averaged models.

EXAMPLE 2.1   Dynamic Response of Half-Bridge Converter

Consider the half-bridge converter of Figure 2.1. The system parameters are $L = 690$ μH, $R = 5$ mΩ, $V_{DC}/2 = 600$ V, $V_s = 400$ V, $m = 0.68$, and $f_s = 1620$ Hz (or equivalently $T_s = 617$ μs). Initially, the converter is in a steady state. Then, m is changed from 0.68 to 0.685, V_s is changed from 400 to 415 V,

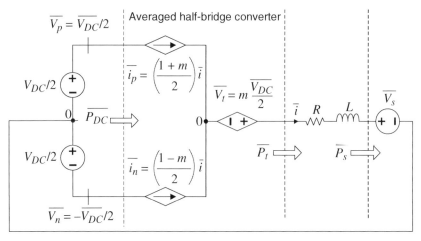

FIGURE 2.9   Averaged equivalent circuit of the half-bridge converter of Figure 2.1.

and $V_{DC}/2$ is changed from 600 to 605 V, respectively, at $t = 0.2$ s, $t = 0.7$ s, and $t = 1.5$ s.

Figures 2.10 and 2.11 illustrate the patterns of variation of the converter currents and powers, respectively. In each figure, columns (a) and (b) show the waveforms obtained from the switched model and the averaged model, re-

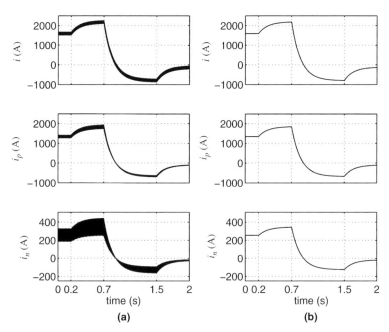

FIGURE 2.10   Transient behavior of the half-bridge converter AC- and DC-side currents, Example 2.1: (a) switched model; (b) averaged model.

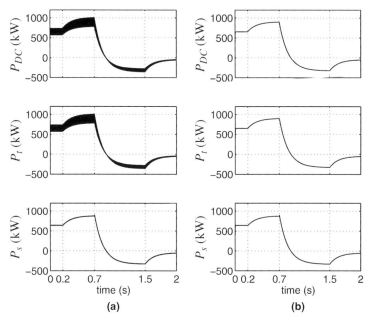

FIGURE 2.11   Transient behavior of the half-bridge converter powers, Example 2.1: (a) switched model; (b) averaged model.

spectively. As Figures 2.10 and 2.11 indicate, the averaged model accurately predicts the pattern of the converter dynamic behavior while it does not include the details of the switched waveforms, that is, the high-frequency components of the waveforms. To further highlight the accuracy of the averaged model, we have superimposed in Figure 2.12 the waveforms of i obtained from the switched and averaged models. Figure 2.12 illustrates that the waveform obtained from the averaged model is an average of that provided by the switched model.

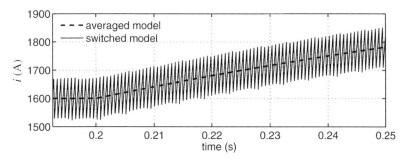

FIGURE 2.12   Current waveforms obtained from the switched and averaged models of Example 2.1.

It should be noted that although P_{DC} and P_t are equal, P_s is slightly less (in absolute value) than both of them (Fig. 2.11). This difference is due to the power loss in R. Figure 2.11 also illustrates that powers can be both positive and negative since the half-bridge converter of Figure 2.1 is a bidirectional power processor.

2.6   NONIDEAL HALF-BRIDGE CONVERTER

In the previous section, we analyzed the DC/AC half-bridge converter. In our analysis, we employed idealized models for the switches. Based on simplifying assumptions, we also presented an averaged model for the half-bridge converter. In this section, we extend the averaged model to represent a half-bridge converter whose switches are nonideal. Thus, we employ more elaborate models for transistors and diodes [16] and present a procedure to include the impacts of the intrinsic voltage drops, resistances, and switching transients of the transistors and diodes. The methodology presented here is a generalization of that presented in Ref. [32]. In this section, we assume the following:

- In its conducting state, each electronic switch is modeled by a voltage drop in series with a resistance.
- In its blocking state, each electronic switch is modeled by an open circuit.
- The turn-on process of each transistor is instantaneous, but its turn-off process is subject to the tailing current phenomenon.
- The turn-on process of each diode is instantaneous, but its turn-off process is subject to the reverse current recovery phenomenon.

Figure 2.13 shows the schematic diagram of a half-bridge converter whose transistors and diodes are nonideal. Thus, for each switch, V_d and r_{on} represent the corresponding on-state voltage drop and resistance, respectively. For the sake of clarity, compared to the ideal half-bridge converter of Figure 2.1, we signify the AC-side terminal voltage and the DC-side currents of the nonideal half-bridge converter by V_t', i_p', and i_n', respectively. Moreover, we denote the DC-side and the AC-side powers by P_1 and P_2, respectively. The following two subsections present and investigate the converter switching waveforms.

2.6.1   Analysis of Nonideal Half-Bridge Converter: Positive AC-Side Current

Let us consider one switching cycle of the operation of the half-bridge converter of Figure 2.13; the switching cycle extends from $t = 0$ to $t = T_s$, where T_s is the switching period. Figure 2.14(a) and (b), respectively, illustrates the waveforms corresponding to the switching functions of Q_1 and Q_4. Let us assume that the AC-side

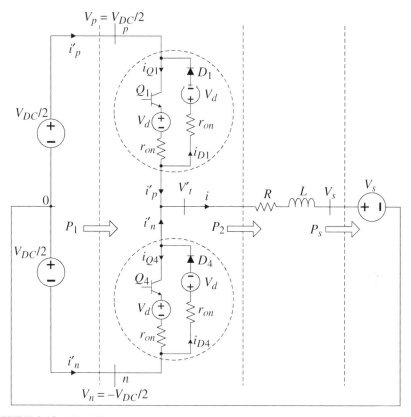

FIGURE 2.13 Simplified power circuit diagram of a half-bridge converter with nonideal switches.

current i is positive and remains relatively constant during the switching cycle. At $t = 0^-$, Q_1 is off and D_4 conducts. At $t = 0^+$, Q_1 is commanded to turn on and, consequently, i_{Q1} increases. However, since $i_{Q1} + i_{D4} = i$, i_{D4} decreases. Once i_{Q1} is equal to i, i_{D4} becomes zero and the diode reverse recovery starts (Fig. 2.14(d) and (e)). During the reverse recovery process, D_4 still conducts and

$$V_t' = V_n - r_{on} i_{D4} - V_d. \tag{2.31}$$

Since $|V_n| \gg r_{on} i_{D4} + V_d$, $V_t' \approx V_n = -V_{DC}/2$ during the reverse recovery interval, as Figure 2.14(c) illustrates. The reverse recovery of D_4 lasts for t_{rr} until the whole reverse recovery charge of Q_{rr} is removed from D_4. As Figure 2.14(d) and (e) shows, during the reverse recovery process, i_{D4} is negative and i_{Q1} is larger than i.

At $t = t_{rr}$, the reverse recovery charge is entirely removed and D_4 stops conduction. Hence, i_{D4} becomes zero and i_{Q1} becomes equal to i. Since the gating command of Q_1 is still in place at $t = t_{rr}$, Q_1 enters the saturation mode. From $t = t_{rr}$ to $t = dT_s$,

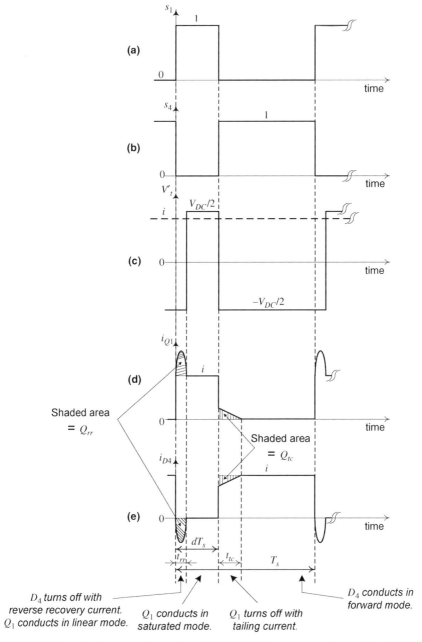

FIGURE 2.14 Switching waveforms of nonideal half-bridge converter for positive AC-side current.

Q_1 carries i (Fig. 2.14(d)), and

$$V_t' = V_p - r_{on}i - V_d. \tag{2.32}$$

Note that the term $r_{on}i + V_d$ is considerably smaller than V_p and therefore $V_t' \approx V_p = V_{DC}/2$, as Figure 2.14(c) indicates. At $t = dT_s$, the gating command of Q_1 is removed and i_{Q1} rapidly drops to its tailing current level (Fig. 2.14(d)). Therefore, D_4 starts conducting, and i_{D4} increases rapidly (Fig. 2.14(e)). The tailing current process lasts for t_{tc} until the whole tailing current charge of Q_{tc} is removed from the transistor. During the tailing current process, the following equation holds:

$$
\begin{aligned}
V_t' &= V_n - r_{on}i_{D4} - V_d \\
&= V_n - r_{on}(i - i_{Q1}) - V_d, \tag{2.33}
\end{aligned}
$$

and $V_t' \approx -V_{DC}/2$ (Fig. 2.14(c)). At $t = t_{rr} + dT_s + t_{tc}$, i_{Q1} becomes zero and $i_{D4} = i$, as Figure 2.14(e) illustrates. From $t = dT_s + t_{tc}$ to T_s, the whole AC-side current is carried by D_4, and we have

$$V_t' = V_n - r_{on}i - V_d, \tag{2.34}$$

and $V_t' \approx -V_{DC}/2$, as Figure 2.14(c) shows.

The average of the AC-side terminal voltage is

$$
\begin{aligned}
\overline{V_t'} &= \frac{1}{T_s} \int_0^{T_s} V_t'(\tau)d\tau \\
&= \frac{1}{T_s} \left(\int_0^{t_{rr}} V_t'd\tau + \int_{t_{rr}}^{dT_s} V_t'd\tau + \int_{dT_s}^{dT_s+t_{tc}} V_t'd\tau + \int_{dT_s+t_{tc}}^{T_s} V_t'd\tau \right), \tag{2.35}
\end{aligned}
$$

where V_t', for each time interval, is obtained from (2.31) to (2.34), respectively. Substituting for V_t' in (2.35), based on (2.31)–(2.34), knowing that $\int_0^{t_{rr}} i_{D4}(\tau)d\tau = -Q_{rr}$ and $\int_{dT_s}^{dT_s+t_{tc}} i_{Q1}d\tau = Q_{tc}$, and rearranging the result, we conclude that

$$
\begin{aligned}
\overline{V_t'} &= m\frac{V_{DC}}{2} - V_e - r_e i \\
&= \overline{V_t} - V_e - r_e i, \quad \text{for} \quad i > 0, \tag{2.36}
\end{aligned}
$$

where $\overline{V_t} = mV_{DC}/2$ based on (2.26), and

$$V_e = V_d - \left(\frac{Q_{rr} + Q_{tc}}{T_s} \right) r_{on} + V_{DC}\left(\frac{t_{rr}}{T_s} \right), \tag{2.37}$$

$$r_e = \left(1 - \frac{t_{rr}}{T_s} \right) r_{on}. \tag{2.38}$$

Equation (2.36) indicates that the average terminal voltage $\overline{V_t'}$ can be controlled by m. However, a comparison between (2.36) and (2.26) reveals that, compared to the ideal converter, the AC-side terminal voltage of the nonideal converter $\overline{V_t'}$ includes two parasitic terms in addition to $m(V_{DC}/2)$: the effective voltage offset V_e and the effective resistive voltage drop $r_e i$.

Since $P_1 = (V_{DC}/2)(i_p' - i_n')$, $i_p' = i_{Q1}$, and $i_n' = i_{D4}$, the averaged DC-side power is

$$\overline{P}_1 = \frac{1}{T_s} \int_0^{T_s} P_1(\tau) d\tau = \frac{V_{DC}}{2T_s} \int_0^{T_s} [i_{Q1}(\tau) - i_{D4}(\tau)] d\tau. \tag{2.39}$$

The right-hand side integral of (2.39) can be expanded as

$$\overline{P}_1 = \left(\frac{V_{DC}}{2T_s}\right) \left\{ \int_0^{t_{rr}} [i_{Q1}(\tau) - i_{D4}(\tau)] d\tau + \int_{t_{rr}}^{dT_s} [i_{Q1}(\tau) - i_{D4}(\tau)] d\tau \right\}$$

$$+ \left(\frac{V_{DC}}{2T_s}\right) \left\{ \int_{dT_s}^{dT_s + t_{tc}} [i_{Q1}(\tau) - i_{D4}(\tau)] d\tau + \int_{dT_s + t_{tc}}^{T_s} [i_{Q1}(\tau) - i_{D4}(\tau)] d\tau \right\}. \tag{2.40}$$

Each integral in (2.40) is equal to the area captured between the waveform $i_{Q1}(t) - i_{D4}(t)$ and the time axis, within the integral limits. As Figure 2.14(d) and (e) suggests, the integrals are calculated as

$$\int_0^{t_{rr}} [i_{Q1}(\tau) - i_{D4}(\tau)] d\tau = (it_{rr} + Q_{rr}) - (-Q_{rr}) = it_{rr} + 2Q_{rr}, \tag{2.41}$$

$$\int_{t_{rr}}^{dT_s} [i_{Q1}(\tau) - i_{D4}(\tau)] d\tau = (idT_s - it_{rr}) - (0) = idT_s - it_{rr}, \tag{2.42}$$

$$\int_{dT_s}^{dT_s + t_{tc}} [i_{Q1}(\tau) - i_{D4}(\tau)] d\tau = (Q_{tc}) - (it_{tc} - Q_{tc}) = 2Q_{tc} - it_{tc}, \tag{2.43}$$

$$\int_{dT_s + t_{tc}}^{T_s} [i_{Q1}(\tau) - i_{D4}(\tau)] d\tau = (0) - (iT_s - idT_s - it_{tc}) = -iT_s + idT_s + it_{tc}. \tag{2.44}$$

Substituting for the integrals (2.41)–(2.44) in (2.40), we obtain

$$\overline{P}_1 = \underbrace{m \frac{V_{DC}}{2} i}_{\overline{P}_{DC}} + V_{DC} \left(\frac{Q_{rr} + Q_{tc}}{T_s}\right) = \overline{P}_{DC} + V_{DC} \left(\frac{Q_{rr} + Q_{tc}}{T_s}\right), \quad \text{for} \quad i > 0.$$

$$\tag{2.45}$$

Based on (2.36), we obtain the average of the AC-side terminal power as

$$\overline{P}_2 = \overline{V_t'i} = m\underbrace{\frac{V_{DC}}{2}i}_{\overline{P}_1} - V_ei - r_ei^2 = \overline{P}_t - V_ei - r_ei^2, \quad \text{for} \quad i > 0. \tag{2.46}$$

As (2.29) and (2.30) indicate, $\overline{P}_{DC} = \overline{P}_t = m(V_{DC}/2)i$. Therefore, based on (2.45) and (2.46), the power loss $\overline{P}_{loss} = \overline{P}_1 - \overline{P}_2$ is

$$\overline{P}_{loss} = V_{DC}\left(\frac{Q_{rr} + Q_{tc}}{T_s}\right) + V_ei + r_ei^2, \quad \text{for} \quad i > 0, \tag{2.47}$$

which is nonzero, as expected.

2.6.2   Analysis of Nonideal Converter: Negative AC-Side Current

For the case where the AC-side current i is negative, the half-bridge converter of Figure 2.13 is analyzed in a way similar to that for the positive AC-side current case. In this case, i_{D1} and i_{Q4} are involved in the converter operation, and based on the directions of Figure 2.13, $i_p' = -i_{D1}$ and $i_n' = -i_{Q4}$.

Figure 2.15(a)–(e) illustrates the converter switching waveforms. From $t = 0$ to $t = t_{tc}$, Q_4 is turning off with the tailing current. From $t = t_{tc}$ to $t = dT_s$, Q_4 is off, and D_1 carries the current. From $t = dT_s$ to $t = dT_s + t_{rr}$, Q_4 is turning on and D_1 is undergoing the reverse recovery process. From $t = dT_s + t_{rr}$ to $t = T_s$, D_1 is off and Q_4 carries the current. The following equations describe the AC-side terminal voltage over each aforementioned time interval:

$$V_t' = V_p + r_{on}i_{D1} + V_d = V_p + r_{on}(-i - i_{Q4}) + V_d \quad \text{over} \quad 0 < t < t_{tc}, \tag{2.48}$$

$$V_t' = V_p - r_{on}i + V_d \quad \text{over} \quad t_{tc} < t < dT_s, \tag{2.49}$$

$$V_t' = V_p + r_{on}i_{D1} + V_d \quad \text{over} \quad dT_s < t < dT_s + t_{rr}, \tag{2.50}$$

$$V_t' = V_n - r_{on}i + V_d \quad \text{over} \quad dT_s + t_{rr} < t < T_s. \tag{2.51}$$

Based on (2.48)–(2.51), and since $\int_0^{t_{tc}} i_{Q4}(\tau)d\tau = Q_{tc}$ and $\int_{dT_s}^{dT_s+t_{rr}} i_{D1}d\tau = -Q_{rr}$, the average of the AC-side terminal voltage is

$$\overline{V_t'} = \overline{V}_t + V_e - r_ei, \quad \text{for} \quad i < 0, \tag{2.52}$$

where $\overline{V}_t = mV_{DC}/2$, and V_e and r_e are defined by (2.37) and (2.38), respectively.

We calculate the averaged DC-side and AC-side powers following the same steps as those taken for the case of positive AC-side current. Thus,

$$\overline{P}_1 = \overline{P}_{DC} + V_{DC}\left(\frac{Q_{rr} + Q_{tc}}{T_s}\right), \quad \text{for} \quad i < 0, \tag{2.53}$$

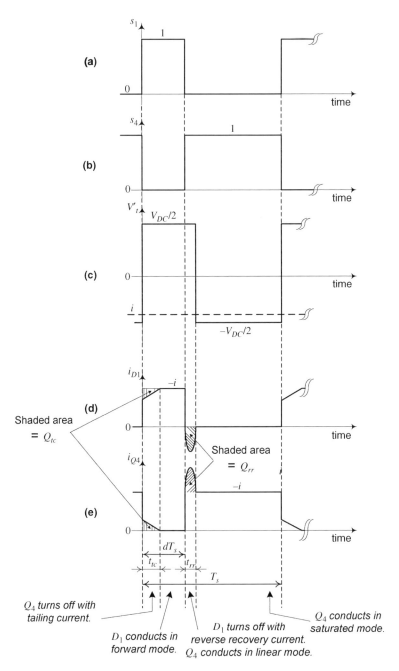

FIGURE 2.15   Switching waveforms of nonideal half-bridge converter for negative AC-side current.

which is identical to (2.45), and

$$\overline{P}_2 = \overline{P}_t + V_e i - r_e i^2, \quad \text{for} \quad i < 0. \tag{2.54}$$

Based on (2.29) and (2.30), $\overline{P}_{DC} = \overline{P}_t - m(V_{DC}/2)i$. Therefore, based on (2.53) and (2.54), the converter power loss is

$$\overline{P}_{loss} = \overline{P}_1 - \overline{P}_2 = V_{DC} \left(\frac{Q_{rr} + Q_{tc}}{T_s} \right) - V_e i + r_e i^2, \quad \text{for} \quad i < 0. \tag{2.55}$$

2.6.3 Averaged Model of Nonideal Half-Bridge Converter

Equations (2.36) and (2.52) provide expressions for the averaged values of the AC-side terminal voltage of the nonideal half-bridge converter for positive and negative AC-side currents, respectively. The two equations can be expressed in a unified form as

$$\overline{V}_t' = \overline{V}_t - \frac{i}{|i|} V_e - r_e i, \quad i \neq 0, \tag{2.56}$$

where $|\cdot|$ denotes the absolute value function, and V_e and r_e are given by (2.37) and (2.38), respectively. Similarly, (2.46) and (2.54) can be combined to express the averaged AC-side terminal power of the nonideal converter as

$$\overline{P}_2 = \overline{P}_t - V_e |i| - r_e i^2. \tag{2.57}$$

The averaged DC-side power of the nonideal converter is described by (2.45) and (2.53) as

$$\overline{P}_1 = \overline{P}_{DC} + V_{DC} \left(\frac{Q_{rr} + Q_{tc}}{T_s} \right). \tag{2.58}$$

As (2.56)–(2.58) indicate, the expressions of the terminal voltage and powers of the nonideal converter possess additional terms compared to their counterparts in the ideal converter, that is, $\overline{V}_t$, $\overline{P}_t$, and $\overline{P}_{DC}$. These additional terms represent the impacts of nonideal characteristics of the switches on the averaged model. Thus, to develop an averaged model for the nonideal converter, one can augment the ideal averaged model of Figure 2.9 with additional components that represent the nonidealities. The augmented circuit model is illustrated in Figure 2.16.

Figure 2.16 shows that the averaged AC-side terminal voltage of the nonideal converter is a superposition of three components: (i) the averaged AC-side terminal voltage of the ideal converter, that is, $\overline{V}_t$; (ii) a current-dependent voltage source; and (iii) a resistive voltage drop. $\overline{V}_t$ can be linearly controlled by m, according to Figure 2.9. However, the current-dependent voltage $(i/|i|)V_e$ is added to or subtracted from $\overline{V}_t$, depending on the polarity of i. The resistive voltage drop is represented by

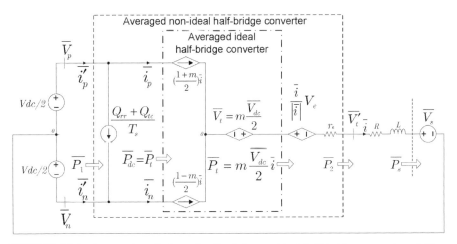

FIGURE 2.16 Averaged equivalent circuit of the nonideal half-bridge converter of Figure 2.13.

the resistor r_e, which, for analysis or control design purposes, may be considered as a part of the resistance of the interface reactor. Figure 2.16 also shows that the DC side of the nonideal half-bridge converter can be modeled by an independent current source that is paralleled with the DC side of the corresponding ideal converter. The independent current source represents the term $V_{DC}(Q_{rr} + Q_{tc})/T_s$ of (2.58). Based on (2.57) and (2.58), and since $\overline{P}_{DC} = \overline{P}_t$, the converter power loss is

$$\overline{P}_{loss} = V_{DC}\left(\frac{Q_{rr} + Q_{tc}}{T_s}\right) + V_e|i| + r_e i^2. \tag{2.59}$$

As Figure 2.16 illustrates, the DC-side terminal currents of the nonideal converter are

$$\overline{i'_p} = \left(\frac{1 + m}{2}\right)i + \frac{Q_{rr} + Q_{tc}}{T_s}, \tag{2.60}$$

$$\overline{i'_n} = \left(\frac{1 - m}{2}\right)i - \frac{Q_{rr} + Q_{tc}}{T_s}. \tag{2.61}$$

The averaged circuit model of Figure 2.16 can be simplified to that of Figure 2.17, based on the following considerations.

As Figure 2.16 shows, the effective resistance r_e is in series and can be lumped with the resistance R. On the other hand, t_{rr} is typically much smaller than T_s and, based on (2.38), r_e can be approximated by r_{on}. Therefore, r_e can be omitted from the averaged model of Figure 2.16 and combined with the resistance of the interface reactor. Figure 2.16 also suggests that the internal (averaged) AC voltage of the nonideal converter can be approximately expressed by $\overline{V}_t = mV_{DC}/2$; that is, the impact of the current-dependent voltage $(i/|i|)V_e$ can be ignored. The justifications are as follows: (i) in a properly designed converter, $\overline{V}_t = mV_{DC}/2$ is typically much larger than V_e

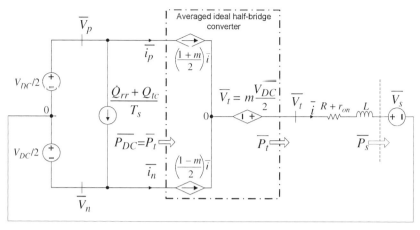

FIGURE 2.17   Simplified averaged equivalent circuit for the nonideal half-bridge converter of Figure 2.13.

and (ii) practically, m is determined by a closed-loop control scheme whose function is to regulate i.

It should be noted that the above-mentioned approximations render the simplified averaged circuit model of Figure 2.17 adequately accurate for dynamic analysis and control design tasks. However, for estimation of the converter power loss, (2.59) should still be used to achieve a higher accuracy. Example 2.2 provides a sense of typical numerical values and their relative significance.

EXAMPLE 2.2   Half-Bridge Converter with Nonideal Elements

Consider the half-bridge converter of Example 2.1, with the parameters $r_{on} = 0.88$ mΩ, $V_d = 0.94$ V, $t_{rr} = t_{tc} = 0.96$ μs, and $Q_{rr} = Q_{tc} = 135$ μC.[8] On the basis of (2.37) and (2.38), we calculate $V_e = 2.8$ V and $r_e = 0.879$ mΩ. Assuming $m = 0.68$, we calculate $mV_{DC}/2 = 408$ V, which is about 145 times V_e.

Let us assume that the AC-side current is regulated at $i = 1300$ A. Then, based on (2.59) we find the converter power loss to be about 5650 W, of which $V_{DC}(Q_{rr} + Q_{tc})/T_s = 525$ W, that is, less than 10%, is the no-load power loss and does not depend on i. However, the remaining 90%, that is, $V_e|i| + r_e i^2 = 5125$ W, is a quadratic function of i. Therefore, to achieve a higher efficiency under the rated load condition, a lower AC-side current at a higher DC-side voltage should be adopted.

---

[8] A 1200 V/1200 A IGBT switch, FF1200R12KE3 from EUPEC, is considered.

3 Control of Half-Bridge Converter

3.1 INTRODUCTION

Chapter 2 introduced the four-quadrant DC/AC half-bridge converter and developed its dynamic model based on the averaging method. The half-bridge converter serves as a building block for the three-phase DC/AC voltage-sourced converter (VSC). Therefore, a clear understanding of the dynamics and control of the half-bridge converter is essential to a comprehensive treatment of the dynamics and control of a three-phase VSC system.

This chapter employs the averaged dynamic model developed in the previous chapter to study the control of the half-bridge converter. We introduce the structure of the controller, review the frequency-response approach to the controller design, and identify the controller requirements to ensure quality command following and disturbance rejection properties. Furthermore, we demonstrate the effectiveness of the feed-forward compensation technique in mitigating the dynamic couplings between the half-bridge converter and the AC system, improving the start-up transient of the converter system, and enhancing the disturbance rejection capability of the converter system. The insight gained from this chapter will be used in subsequent chapters for the control of the three-phase VSC.

3.2 AC-SIDE CONTROL MODEL OF HALF-BRIDGE CONVERTER

Hereinafter, for the sake of compactness, the overbar is dropped from the averaged variables throughout the rest of the book. However, we shall remember that the averaging operator is applied to variables, over one switching cycle, unless otherwise stated.

Consider the half-bridge converter of Figure 2.13, duplicated as Figure 3.1 for ease of reference. Based on the converter averaged equivalent circuit (Fig. 2.17), the dynamics of the AC-side current i are described by

$$L\frac{di}{dt} + (R + r_{on})i = V_t - V_s, \tag{3.1}$$

Voltage-Sourced Converters in Power Systems, by Amirnaser Yazdani and Reza Iravani
Copyright © 2010 John Wiley & Sons, Inc.

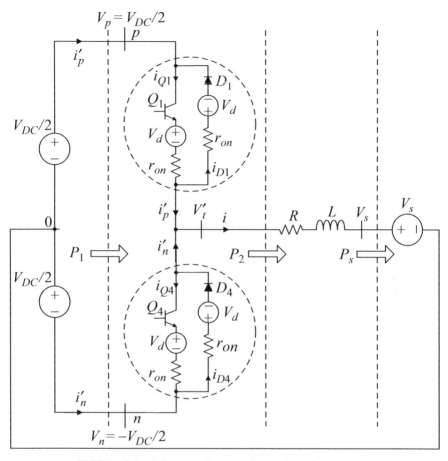

FIGURE 3.1   Schematic diagram of the half-bridge converter.

where

$$V_t = \frac{V_{DC}}{2}m. \tag{3.2}$$

Equation (3.1) represents a system in which i is the state variable, V_t is the control input, and V_s is the disturbance input. The output of the system can, for example, be the power exchanged with the AC-side voltage source, that is, $P_s = V_s i$. Based on (3.2), the control input V_t is proportional to, and can be controlled by, the modulating signal m. Figure 3.2 shows a control block diagram of the system described by (3.1) and (3.2), for which the next section presents a closed-loop control structure to regulate i at its reference value.

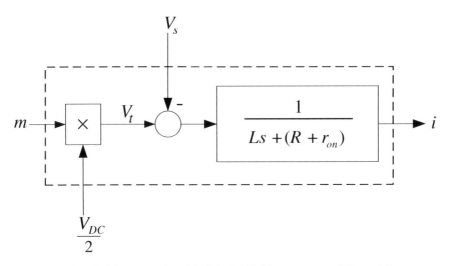

FIGURE 3.2   Control model of the half-bridge converter of Figure 3.1.

3.3   CONTROL OF HALF-BRIDGE CONVERTER

Consider the open-loop system of Figure 3.2 in which the current i is the output and the control objective is to regulate i at a prespecified reference value. This can be achieved via the closed-loop system of Figure 3.3 in which the reference command i_{ref} is compared with i, and the error signal e is generated. Then, the compensator $K(s)$ processes e and provides the control signal u. Then, u is divided by $V_{DC}/2$ to compensate for the converter voltage gain, as described by (3.2). If the DC-side voltage is constant, the numerical value of $V_{DC}/2$ is known in advance and the gain compensation becomes trivial. However, if the DC-side voltage is subject to variations, V_{DC} must be measured, for example, by a Hall-effect voltage transducer, to enable the gain compensation that can be regarded as a feed-forward compensation. The output of the controller must be limited prior to being delivered to the converter pulse-width modulation (PWM) signal generator, to ensure that $|m| \leq 1$.

Depending on the type of the reference signal and the desired performance, different types of compensator may be used for the control. For example, if i_{ref} is a

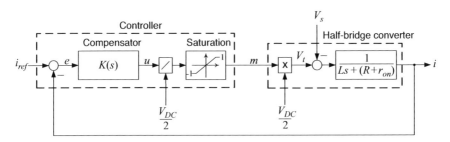

FIGURE 3.3   Control block diagram of the closed-loop half-bridge converter system.

step function and V_s is a DC voltage, a proportional-integral (PI) compensator of the generic form $K(s) = (k_p s + k_i)/s$ is sufficient for the control. The integral term of the compensator guarantees that i tracks i_{ref}, with zero steady-state error, in spite of the disturbance V_s.

As understood from Figure 3.3, if $K(s) = (k_p s + k_i)/s$, the control system loop gain is

$$\ell(s) = \left(\frac{k_p}{Ls}\right) \left(\frac{s + \frac{k_i}{k_p}}{s + \frac{R+r_{on}}{L}}\right). \tag{3.3}$$

Based on the block diagrams of Figures 3.2 and 3.3, the open-loop half-bridge converter has a stable pole at $p = -(R + r_{on})/L$. Typically, this pole is fairly close to the origin and corresponds to a slow natural response. To improve the open-loop frequency response, the pole can be canceled by the zero of the PI compensator. Thus, choosing $k_i/k_p = (R + r_{on})/L$ and $k_p/L = 1/\tau_i$, where τ_i is the desired time constant of the closed-loop system, one obtains the closed-loop transfer function

$$G_i(s) = \frac{i(s)}{i_{ref}(s)} = \frac{1}{\tau_i s + 1}, \tag{3.4}$$

which is a first-order transfer function with the unity gain. τ_i should be made small for a fast current-control response, but adequately large such that $1/\tau_i$, that is, the bandwidth of the closed-loop control system, is considerably smaller, for example, 10 times smaller, than the switching frequency of the half-bridge converter (expressed in rad/s). Depending on the requirements of a specific application and the converter switching frequency, τ_i is typically selected in the range of 0.5–5 ms.

Example 3.1 illustrates the performance of the current-controlled half-bridge converter in tracking a step command.

EXAMPLE 3.1   Closed-Loop Response of the Half-Bridge Converter

Consider the half-bridge converter of Figure 3.1 with parameters $L = 690\ \mu H$, $R = 5\ m\Omega$, $r_{on} = 0.88\ m\Omega$, $V_d = 1.0\ V$, $V_{DC}/2 = 600\ V$, $V_s = 400\ V$, and $f_s = 1620\ Hz$. To achieve a closed-loop time constant of 5 ms, the compensator parameters are chosen as $k_p = 0.138\ \Omega$ and $k_i = 1.176\ \Omega/s$.

Initially, the half-bridge converter system is in a steady state and $i = 0$. The current command, i_{ref}, is first changed from 0 to 1000 A, at $t = 0.1$ s, and then changed from 1000 to -1000 A, at $t = 0.2$ s. These correspond to changes in the AC-side power from 0 to 400 kW and from 400 to -400 kW, respectively. Figure 3.4(a)–(d) shows the half-bridge converter response to the step changes in i_{ref}. Figure 3.4(a) shows the compensator response to the command changes. The response of the compensator output u is equivalent to that of the AC-side terminal voltage V_t, required for the command tracking. u is then translated into the modulating waveform m (Fig. 3.4(b)), which is rendered to the converter PWM scheme. As expected, the step response of i is a first-order exponential

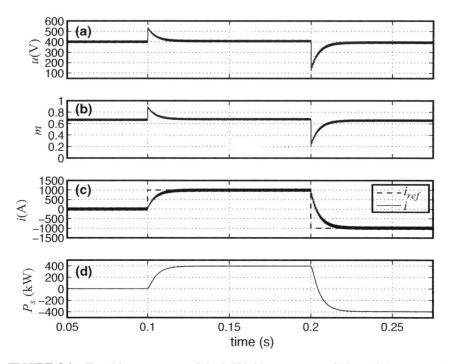

FIGURE 3.4 Closed-loop response of the half-bridge converter of Figure 3.1 to current reference changes; Example 3.1.

signal that settles at its final level of 1000 A in less than 25 ms (Fig. 3.4(c)). Since V_s is constant, the power delivered to the AC-side voltage source P_s has the same pattern of variations as that of i (Fig. 3.4(d)). Thus, P_s can be rapidly controlled through the current control process.

The closed-loop response illustrated in Figure 3.4 was obtained following a steady-state regime. In practice, however, the converter system starts from a zero-state condition. Figure 3.5 illustrates the start-up response of the half-bridge converter system, based on the following sequence of events. Initially, i_{ref} is set to zero, and V_s is equal to 400 V. However, the gating signals of both Q_1 and Q_2 are blocked and, therefore, i is zero. At $t = 0.1$ s, the gating signals are unblocked and the PWM is exercised.

Figure 3.5(a) shows that subsequent to the activation of the control, i undergoes a large undershoot before coming back to zero. This response, certainly not desirable due to the additional stress it imposes on the converter, can be explained by the block diagram of Figure 3.3. When the gating signals are activated, the half-bridge converter generates an AC-side terminal voltage that is equal to the output of the compensator, u. Since the system is initially in the zero-state condition, u starts from zero (Fig. 3.5(b)), and so does V_t. However, V_s has a positive value at all times. Consequently, i becomes negative immediately after the controller activation, until the compensator regulates it back to zero.

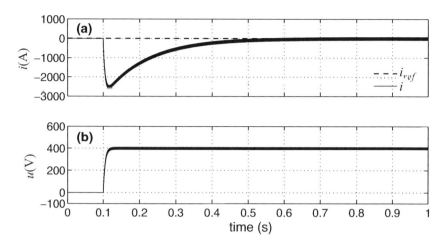

FIGURE 3.5   Start-up response of the half-bridge converter of Figure 3.1; Example 3.1.

3.4   FEED-FORWARD COMPENSATION

3.4.1   Impact on Start-Up Transient

The undesirable start-up transient of Figure 3.5 can be avoided if the control scheme of Figure 3.3 is augmented with a feed-forward compensation scheme, as shown in Figure 3.6. The feed-forward scheme augments the compensator output with a measure of V_s. This measure, denoted by $\breve{V}_s$, can be obtained from a voltage transducer whose (dynamic) gain is $G_{ff}(s)$, where $G_{ff}(0) = 1$. Therefore, at the start-up instant when the compensator output is zero, the AC-side terminal voltage to be generated starts from a value equal to V_s, and the AC-side current remains at zero.

Example 3.2 illustrates the effectiveness of the feed-forward compensation in eliminating the start-up transient observed in Figure 3.5.

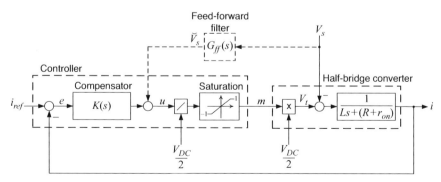

FIGURE 3.6   Control block diagram of the half-bridge converter with feed-forward compensation.

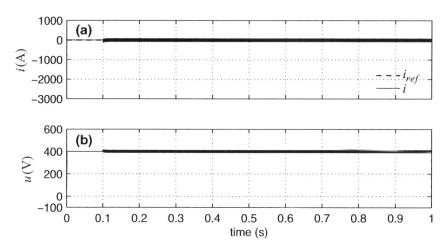

FIGURE 3.7   Start-up response of the half-bridge converter of Figure 3.1 with feed-forward compensation; Example 3.2.

EXAMPLE 3.2   Start-Up Transient of the Half-Bridge Converter with Feed-Forward Compensation

Figure 3.7 illustrates the start-up transient of the half-bridge converter of Example 3.1 for which the feed-forward compensation, with $G_{ff}(s) = 1$, is employed. Figure 3.7(a) shows that the output current i stays at zero following the unblocking of the gating pulses. The reason is that, as Figure 3.7(b) shows, the initial value of u is equal to V_s through the feed-forward compensation signal path. Therefore, $V_t = V_s$, the voltage across the series RL interface reactor equals zero, and thus no current excursion is experienced.

3.4.2   Impact on Dynamic Coupling Between Converter System and AC System

The benefit of the feed-forward compensation is not restricted to improving the converter start-up transients. The feed-forward compensation can also decouple dynamics of the converter system from those of the AC system with which the converter system is interfaced. For example, consider the half-bridge converter of Figure 3.1 in which the ideal AC-side voltage source V_s is more realistically represented by an AC system, as shown in Figure 3.8. The AC system can be described by the state-space equations

$$\dot{\underline{x}} = f(\underline{x}, \underline{r}, i, t),$$
$$V_s = g(\underline{x}, \underline{r}, i, t), \tag{3.5}$$

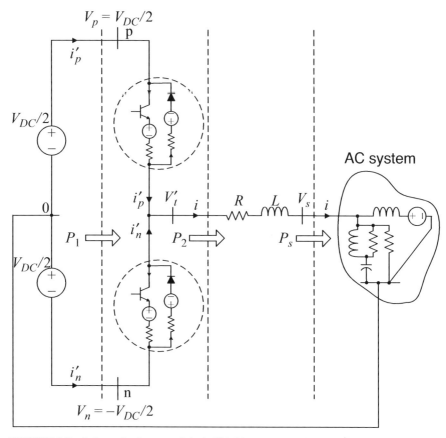

FIGURE 3.8   Schematic diagram of the half-bridge converter interfaced with an AC system.

where x and $\underline{r}$ denote vectors of the state variables and the AC system voltage/current sources, respectively; $f(\,\cdot\,)$ and $g(\,\cdot\,)$ are, in general, nonlinear functions of their arguments.

Let us assume that the control objective is to regulate i at i_{ref}, according to the control block diagram of Figure 3.9. Without the feed-forward compensation, that is, $G_{ff}(s) \equiv 0$, the compensator design must take into account dynamics of the AC system described by (3.5). The dynamics may be uncertain, time varying, nonlinear, and of a high order. Therefore, the compensator design would be laborious and may not be straightforward. Moreover, the impact of the AC system transients on the control cannot be readily mitigated. However, if the feed-forward compensation is employed, a prior knowledge of the AC system dynamics is not essential for the controller design. The reason is that the dynamics of the half-bridge converter system become effectively decoupled from those of the AC system if $G_{ff}(s) \approx 1$ in the range of frequencies over which the AC system modes have considerable energy. As Figure 3.9 illustrates, if $\check{V}_s \approx V_s$, the voltage difference across the series RL interface reactor is determined exclusively by the PI compensator.

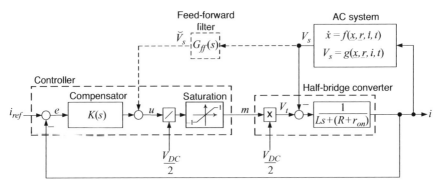

FIGURE 3.9   Control block diagram of the half-bridge converter of Figure 3.8.

Example 3.3 highlights the effectiveness of the feed-forward compensation in mitigating dynamic interactions between the AC system and the half-bridge converter system.

EXAMPLE 3.3   Decoupling from AC System Dynamics by Feed-Forward Compensation

Consider the half-bridge converter of Figure 3.10 with parameters $L = 450 \, \mu H$, $R = 5 \, m\Omega$, $r_{on} = 0.88 \, m\Omega$, $V_d = 1.0$ V, $V_{DC}/2 = 600$ V, and $f_s = 3780$ Hz. The AC system parameters are $V_{th} = 400$ V, $L_f = 250 \, \mu H$, $C_f = 2500 \, \mu F$, and $R_f = 50 \, \Omega$. The compensator parameters are set as $k_p = 0.15 \, \Omega$ and $k_i = 1.96 \, \Omega/s$.

Figure 3.11 shows the response of the closed-loop system to a 1000-A step change in i_{ref} when no feed-forward compensation is employed. Figure 3.11(a) indicates that subsequent to the command change, V_s experiences low-frequency oscillations due to poorly damped oscillatory modes of the AC system. The oscillations lead to a poor transient response of i (Fig. 3.11(b)).

Figure 3.12 shows the response of the system to the same current command, however, when the feed-forward compensator is in service. The transfer function of the feed-forward filter is $G_{ff}(s) = 1/(8 \times 10^{-6}s + 1)$. Figure 3.12(b) shows that in spite of the oscillations of V_s, observed in Figure 3.12(a), the response of i to i_{ref} is a first-order exponential function with a time constant of 3 ms.

A comparison between Figures 3.12(a) and 3.11(a) highlights the impact of the feed-forward compensation on V_s. Figure 3.12(a) shows that the oscillations of V_s decay relatively slowly when the feed-forward compensation is employed. The reason is that, under the feed-forward compensation, the converter system is equivalent to an independent current source as viewed by the AC system. Consequently, damping of the AC system modes corresponding to the eigenvalues $\lambda_{1,2} \simeq -4 \pm 1256j$ is left only to the resistance R_f. However, when the feed-forward compensation is disabled, the dynamics of the AC

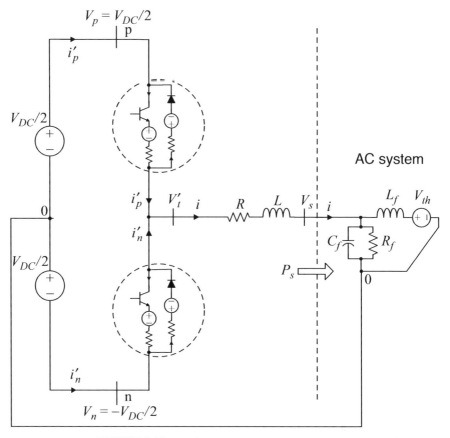

FIGURE 3.10   Half-bridge converter of Example 3.3.

system interact through the feedback loop with those of the compensator and the converter. Thus, the modes corresponding to the AC system are relocated with respect to their initial location and assume different natural frequencies and damping ratios. It is thus expected that due to the converter loss and the damping of the compensator modes, the closed-loop modes have higher damping ratios than the open-loop modes, as Figure 3.11(a) suggests.

3.4.3   Impact on Disturbance Rejection Capability

The feed-forward compensation can also enhance the disturbance rejection capability of the closed-loop converter system. Example 3.4 illustrates that, if the feed-forward filter $G_{ff}(s)$ in Figure 3.6 has an adequately large bandwidth, the transient impact of V_s on i in the converter system of Figure 3.1 is noticeably mitigated.

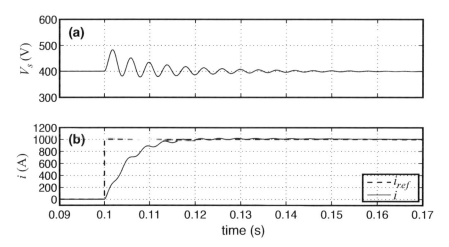

FIGURE 3.11 Transient response of the converter system with no feed-forward compensation; Example 3.3.

EXAMPLE 3.4   Disturbance Rejection of the Half-Bridge Converter with Feed-Forward Compensation

Consider the half-bridge converter of Figure 3.1 with parameters $L = 690$ μH, $R = 5$ mΩ, $r_{on} = 0.88$ mΩ, $V_d = 1.0$ V, $V_{DC}/2 = 600$ V, $V_s = 400$ V, and $f_s = 1620$ Hz. The compensator parameters are $k_p = 0.138$ Ω and $k_i = 1.176$ Ω/s. The closed-loop half-bridge converter system is initially in a steady state and $i = 1000$ A. At $t = 0.1$ and $t = 0.2$ s, V_s is subjected to step changes from

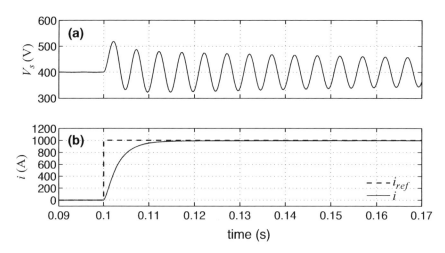

FIGURE 3.12 Transient response of the converter system with the feed-forward compensation in service; Example 3.3.

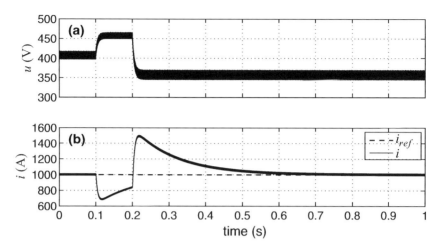

FIGURE 3.13   Disturbance rejection of the system when no feed-forward compensation is employed; Example 3.4.

400 to 450 V and from 450 to 350 V, respectively, while i_{ref} is kept constant at 1000 A.

Figure 3.13 shows the response of the half-bridge converter to the disturbances when the feed-forward scheme is not in service and the control effort u is merely determined based on the reaction of the PI compensator to the error signal. Thus, since the error signal does not change until there is a discrepancy between i_{ref} and i, the corrective response of u to the disturbances is fairly slow (Fig. 3.13(a)), and consequently, i diverts from i_{ref} due to the changes in V_s, as Figure 3.13(b) illustrates. In the steady state, i settles at i_{ref} (Fig. 3.13(b)). Figure 3.14 shows the response of the half-bridge converter to the same disturbances when the feed-forward scheme is enabled. The transfer function of the feed-forward filter is $G_{ff}(s) = 1/(0.0005s + 1)$. In this case, when V_s changes, u reacts rapidly via the feed-forward signal path (Fig. 3.14(a)). Therefore, the impact of the disturbances on i is counteracted and effectively mitigated, as Figure 3.14(b) shows. Similar to the previous case, the error becomes zero in the steady state.

3.5   SINUSOIDAL COMMAND FOLLOWING

In subsequent chapters, the half-bridge converter is employed as the main building block of three-phase VSC systems, in which the sinusoidal command following is a typical requirement. As such, it is warranted to evaluate the capability of the half-bridge converter in tracking a sinusoidal current command.

Consider the half-bridge converter of Figure 3.1 with the closed-loop control block diagram of Figure 3.6. Assume that V_s is a sinusoidal function of time with the angular

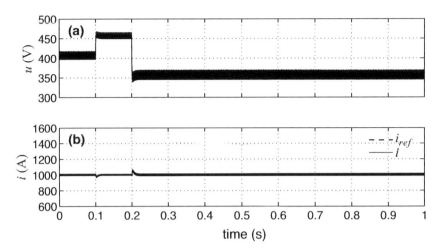

FIGURE 3.14 Disturbance rejection of the system with feed-forward compensation in service; Example 3.4.

frequency ω_0, corresponding to a power line frequency of 50 or 60 Hz. Also assume that there is a need to track a sinusoidal current command characterized by

$$i_{ref}(t) = \widehat{I}\cos(\omega_0 t + \phi)\text{unit}(t), \tag{3.6}$$

where $\widehat{I}$ and ϕ are the amplitude and the initial phase angle of the sinusoidal command, respectively, and unit(t) is the unit step function. Then, if the closed-loop transfer function of the converter system is given by (3.4), the steady-state response of i to the command has the form

$$i(t) = \frac{\widehat{I}}{\sqrt{1 + (\tau_i \omega_0)^2}}\cos(\omega_0 t + \phi + \delta), \tag{3.7}$$

where the phase shift δ is given by

$$\delta = -\tan^{-1}(\tau_i \omega_0). \tag{3.8}$$

Equations (3.7) and (3.8) suggest that if the PI compensator of Section 3.3 is employed, i tracks i_{ref} with errors in both the amplitude and the phase angle. According to (3.7), the amplitude of i is inversely proportional to $\sqrt{1 + (\tau_i \omega_0)^2}$ and, therefore, smaller than $\widehat{I}$. In addition, as (3.8) indicates, i lags i_{ref} by an angle that can be significant depending on the product $\tau_i \omega_0$.

The capability of the closed-loop system to faithfully track a sinusoidal command depends on the system closed-loop bandwidth. The bandwidth of the closed-loop system of (3.4) is equal to $1/\tau_i$. Therefore, a sinusoidal command can be followed with negligible attenuation or phase delay if τ_i is adequately small. However, selection of a very small closed-loop time constant may not be possible due to practical limita-tions/requirements. For example, while a closed-loop current controller with a time constant of $\tau_i = 2$ ms is considered as being reasonably fast for most high-power con-verter systems, it tracks a 60 Hz sinusoidal command with an amplitude attenuation of 20% and a phase delay of 37°

Example 3.5 illustrates the sinusoidal command following performance of the half-bridge converter system if a PI compensator is used.

EXAMPLE 3.5   Sinusoidal Command Following with PI Compensator

Consider the half-bridge converter of Figure 3.1, in conjunction with the control scheme of Figure 3.6, with the following parameters: $L = 690$ µH, $R = 5$ mΩ, $r_{on} = 0.88$ mΩ, $V_d = 1.0$ V, $V_{DC}/2 = 600$ V, and $f_s = 3420$ Hz. The compensator parameters are $k_p = 0.345$ Ω and $k_i = 2.94$ Ω/s, which correspond to $\tau_i = 2$ms. The transfer function of the feed-forward filter is $G_{ff}(s) = 1/(8 \times 10^{-6}s + 1)$.

Let us assume that $V_s = 400\cos(377t - \frac{\pi}{2})$ V and that we intend to deliver 200 kW to the AC system, at unity power factor. Thus, the current command $i_{ref} = 1000\cos(377t - \frac{\pi}{2})$ A must be tracked by the closed-loop converter sys-tem. Figure 3.15 shows the system closed-loop response to i_{ref} when a PI compensator is employed. As Figure 3.15 illustrates, $i(t)$ is about 37° delayed with respect to $i_{ref}(t)$ and $V_s(t)$. Moreover, the amplitude of i is only about 800 A. Consequently, rather than 200 kW at unity power factor, 128 kW and 96 kVAr are delivered to the AC-side source.

To further investigate the mechanism of the time-varying command tracking, consider the closed-loop control system of Figure 3.16 with the transfer function

$$\frac{i(s)}{i_{ref}(s)} = G_i(s) = \frac{\ell(s)}{1 + \ell(s)}, \tag{3.9}$$

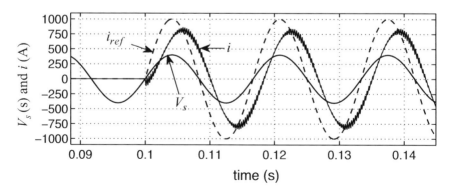

FIGURE 3.15   Steady-state error in phase and amplitude of the current when PI compensator is employed; Example 3.5.

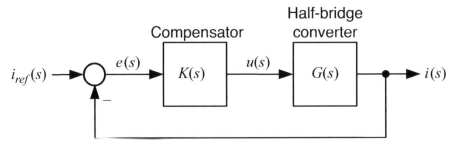

FIGURE 3.16   A simplified control block diagram of the half-bridge converter.

where the loop gain $\ell(s)$ is defined as

$$\ell(s) = K(s)G(s). \tag{3.10}$$

The frequency response of the closed-loop system is

$$G_i(s)|_{s=j\omega} = \frac{\ell(j\omega)}{1 + \ell(j\omega)}, \tag{3.11}$$

which can be expressed in polar coordinates as

$$G_i(j\omega) = |G_i(j\omega)|e^{j\delta}, \tag{3.12}$$

where $|G_i(j\omega)|$ and δ denote the magnitude and the phase of $G_i(j\omega)$, respectively. Based on the definition of the frequency response, the steady-state response of the closed-loop system to a sinusoidal command (with frequency ω_0) is scaled by $|G_i(j\omega_0)|$ and phase shifted by δ, with respect to the command. If the sinusoidal command is to be tracked with zero steady-state error, $|G_i(j\omega_0)|$ must be equal to unity, and δ must be zero. Based on (3.11), this is fulfilled if $|\ell(j\omega_0)|$, that is, the loop-gain magnitude, is infinity at the frequency of the command signal. For example, if two of the poles of $K(s)$ are located at $s = \pm j\omega_0$, then $|\ell(j\omega_0)| = +\infty$, and the sinusoidal command is tracked with zero steady-state errors. In general, to follow a command with zero steady-state error, the unstable poles of the Laplace transform of the command must be included in the compensator. This not only ensures command tracking with zero steady-state error but also eliminates all disturbances of the same type, in the steady state.[1] In a commonly faced scenario, to track a step (DC) command with zero steady-state error, and to reject constant disturbances, $K(s)$ includes an integral term, that is, it possesses a pole at $s = 0$; the PI compensator is a special form of such a compensator.

As an alternative method, the sinusoidal command tracking can be achieved if $K(s)$ is designed in such a way that the bandwidth of the closed-loop system is

[1]This is known as the *internal-model principle* [33–36].

adequately larger than the frequency of the command signal. In this approach, no attempt is usually made to include the unstable poles of the command signal in the compensator. Consequently, the tracking will not be perfect and a steady-state error, although small, is inevitable. The design procedure is almost the same in both methods and illustrated by Example 3.6

EXAMPLE 3.6   Sinusoidal Command Following with a Modified Compensator

Consider the half-bridge converter of Example 3.5 and the control block diagram of Figure 3.6. Let us assume that i_{ref} is required to be tracked with zero steady-state error and that a closed-loop bandwidth of about 3500 rad/s (i.e., about 9 times ω_0) is desired.

To satisfy the zero steady-state error requirement, we include a pair of complex-conjugate poles in the compensator, at $s = \pm 377 j$ rad/s. Thus, a candidate compensator is $K(s) = (s^2 + (377)^2)^{-1} H(s)$, where $H(s) = h(N(s)/D(s))$ is a rational fraction of the polynomials $N(s)$ and $D(s)$, and h is a constant.[2] The compensator zeros, and the other poles (if required) must be located in the s plane such that the closed-loop system is stable, a reasonable phase margin is achieved, and the switching ripple content of the control signal u is low. The compensator can be designed based on either the root-locus method or the frequency-response approach. For this example, we adopt the frequency-response method (also known as the *loop shaping*).

If $H(j\omega) = 1$, that is, $K(j\omega) = [-\omega^2 + (377)^2]^{-1}$, the magnitude and phase plots of $\ell(j\omega) = K(j\omega)G(j\omega) = K(j\omega)[jL\omega + (R + r_{on})]^{-1}$ assume the shapes shown by dashed lines in Figure 3.17. It is observed that at very low frequencies $\ell(j\omega)$ has a constant magnitude, and the phase delay is insignificant. However, due to the open-loop pole $s = -(R + r_{on})/L$, the magnitude starts to roll off at about $\omega = 8.52$ rad/s. The pole also results in a phase drop with a slope of $-45°$/dec, such that the phase settles at $-90°$ for frequencies larger than 85 rad/s. At $\omega = 377$ rad/s, that is, the resonance frequency of the complex-conjugate poles, the loop-gain magnitude peaks to infinity but continues to roll off with a slope of -60 dB/dec. The resonance also results in a $-180°$ phase delay, such that the loop-gain phase drops to $-270°$, for frequencies larger than 377 rad/s.

To achieve a stable closed-loop system, one must ensure that the loop-gain phase at the gain crossover frequency is larger than $-180°$, by a value that is referred to as the *phase margin*. The gain crossover frequency, denoted by ω_c, is the frequency at which the loop-gain magnitude becomes unity (0 dB) [37]. On the other hand, the gain crossover frequency and the -3 dB bandwidth of the closed-loop system, denoted by ω_b, are closely correlated, such that, in general, ω_b satisfies the inequality $\omega_c < \omega_b < 2\omega_c$ and can be approximated as $\omega_b \approx 1.5\omega_c$. Therefore, ω_c is imposed if a certain value is required for ω_b. In this

---

[2]$N(s)$ and $D(s)$ are arranged such that the coefficients of their highest order terms are unity.

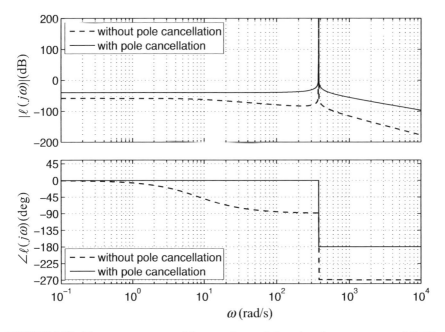

FIGURE 3.17 Frequency response of the open-loop gain based on the compensator of (3.13); Example 3.6.

example, $\omega_b = 3500$ rad/s and, therefore, ω_c needs to be placed at about 2333 rad/s. However, at this frequency the loop-gain phase is $-270°$, corresponding to an unstable closed-loop system.

The excessively low loop-gain phase described above is partly due to the pole of $G(s)$ at $s = -8.52$ rad/s. This pole introduces a $90°$ phase delay for frequencies larger than 85 rad/s. Therefore, to improve the loop-gain phase, we cancel the pole by a compensator zero, placed at $s = -8.52$ rad/s. This pole/zero cancellation is permitted since the pole is on the left half plane (LHP). Thus, the modified compensator is

$$K(s) = \frac{s + 8.52}{s^2 + (377)^2} H(s). \qquad (3.13)$$

The solid lines in Figure 3.17 illustrate the magnitude and phase plots of $\ell(j\omega) = K(j\omega)G(j\omega)$ for the modified compensator (3.13), with $H(j\omega) = 1$. It is noted that, using the modified compensator, the phase is zero and the magnitude is constant up to about $\omega = 377$ rad/s; thereafter, the phase drops to $-180°$ and remains constant, while the magnitude rolls off with a slope of -40 dB/dec. This frequency response suggests that for the gain crossover frequency of $\omega_c = 2333$ rad/s the phase margin is still insufficient (it is zero). To achieve a reasonably large phase margin, the loop-gain phase at ω_c must be supplemented, for example, by means of a lead filter [38]. The lead filter is of the general form

$$F_{lead}(s) = \frac{s + (p_1/\alpha)}{s + p_1},$$ (3.14)

where p_1 is the filter pole and $\alpha > 1$ is a real constant. The maximum phase of the lead filter is

$$\delta_m = \sin^{-1}\left(\frac{\alpha - 1}{\alpha + 1}\right),$$ (3.15)

which occurs at the frequency

$$\omega_m = \frac{p_1}{\sqrt{\alpha}}.$$ (3.16)

To add the maximum possible value to the loop-gain phase, one can choose ω_m to be equal to ω_c.

In this example, let us assume that a phase margin of about $45°$ is desired. Since the loop-gain phase at $\omega > 377$ rad/s is $-180°$, we choose $\delta_m = 45°$. This, based on (3.15), yields $\alpha = 5.83$, which is the pole to zero ratio of the lead filter. Then, if $\omega_m = \omega_c = 2333$ rad/s, it follows from (3.16) that $p_1 = 5633$ rad/s and $z_1 = p_1/\alpha = 966$ rad/s. Thus, the compensator (3.13) is modified to

$$K(s) = h\left(\frac{s + 8.52}{s^2 + (377)^2}\right)\left(\frac{s + 966}{s + 5633}\right).$$ (3.17)

Finally, the constant gain h is determined as $h = 8680\ \Omega/s$, based on $\ell(j\omega_c) + 1 = 0$ (or $|\ell(j\omega_c)| = 1$).

The dashed lines in Figure 3.18 show the magnitude and phase plots of $\ell(j\omega)$ when the compensator (3.17) is employed and illustrate that at ω_c the loop-gain phase is about $-135°$, corresponding to a phase margin of about $45°$. It should be noted that the loop-gain magnitude is constant over a wide range of frequencies, up to $\omega = 377$ rad/s. To ensure that the loop gain exhibits a large magnitude at low frequencies, we should also include a lag filter in the compensator.[3] The loop-gain magnitude is increased at low frequencies by about 32 dB if the following lag filter is introduced:

$$F_{lag}(s) = \frac{s + 2}{s + 0.05}.$$ (3.18)

[3] An integral term in the compensator makes this possible. However, the integral term imposes a phase shift of $-90°$ at all frequencies, including the crossover frequency, and thus has a destabilizing impact on the closed-loop system.

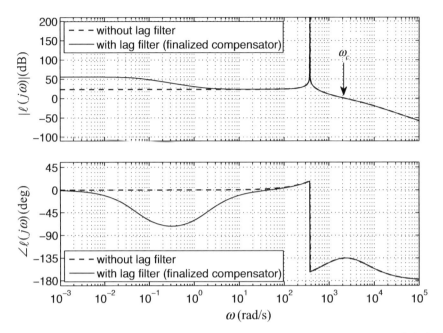

FIGURE 3.18   Open-loop frequency responses based on the compensators (3.17) and (3.19); Example 3.6.

The lag filter (3.18) has the property that $F_{lag}(j\omega) \approx 1$ for frequencies larger than about 100 rad/s. Therefore, it does not change the phase or magnitude of the loop gain around the crossover frequency $\omega_c = 2330$ rad/s. Thus, the phase margin, bandwidth, and command tracking capability of our design remain unchanged in spite of the introduction of the lag filter. The final compensator is thus expressed as

$$K(s) = 8680 \left(\frac{s + 8.52}{s^2 + (377)^2} \right) \left(\frac{s + 966}{s + 5633} \right) \left(\frac{s + 2}{s + 0.05} \right) \quad [\Omega]. \quad (3.19)$$

The solid lines in Figure 3.18 illustrate the magnitude and phase plots of $\ell(j\omega)$ when the compensator (3.19) is employed. It is noted that for frequencies around the crossover frequency, the loop gain exhibits the same behavior under the compensators (3.19) and (3.17). Two remaining tasks of the controller design are to examine (i) the adequacy of the gain margin and (ii) the loop-gain magnitude at the converter switching frequency.[4]

---

[4]To be more precise, we should consider the switching frequency side bands since the modulating waveform is a sinusoid. However, in our case the side bands are close to the switching frequency, and we can consider the switching frequency instead.

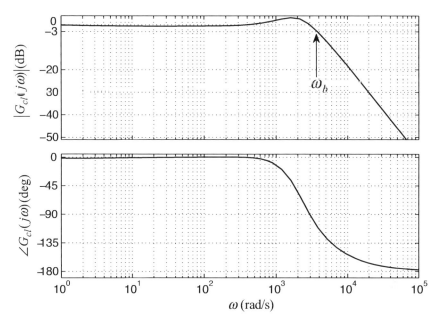

FIGURE 3.19   Closed-loop frequency response of the half-bridge converter; Example 3.6.

As Figure 3.18 indicates, as the loop-gain phase approaches $-180°$ at high frequencies, the loop-gain magnitude drops to very small values. Thus, the phase crossover frequency and the gain margin are infinity in this closed-loop system. In addition, the converter switching frequency is equal to 3420 Hz, corresponding to 21488 rad/s, at which the loop-gain magnitude is about -30 dB. Furthermore, it can be verified that the switching ripple content of the control signal u is about 2.5 times smaller than that of the error signal e.

The foregoing compensator design procedure is comprehensive but cumbersome. Moreover, (3.19) is a relatively high-order compensator and may be difficult to implement. As discussed in Chapter 7, two such compensators will be required for a three-phase VSC system that is controlled in the $\alpha\beta$-frame. Therefore, a special case of (3.19) is proposed in the literature that is analogous to a conventional PI compensator, and thus has only two parameters to tune. This class of compensators is referred to as a *stationary-frame generalized integrator* [39] or a *proportional-plus-resonant compensator* (P+Resonant) [40], and is applicable to both single-phase and three-phase converter systems.

Figure 3.19 shows the closed-loop frequency response of the control system of Figure 3.9. It can be observed that the -3 dB bandwidth of the closed-loop system is about $\omega_b = 3820$ rad/s. Moreover, at $\omega = 377$ rad/s the magnitude and phase of the closed-loop transfer function are unity and zero, respectively. Figure 3.20 shows the closed-loop time response of the half-bridge converter to $i_{ref} = 1000\cos(377t - \frac{\pi}{2})$ A, when the compensator (3.19) is employed.

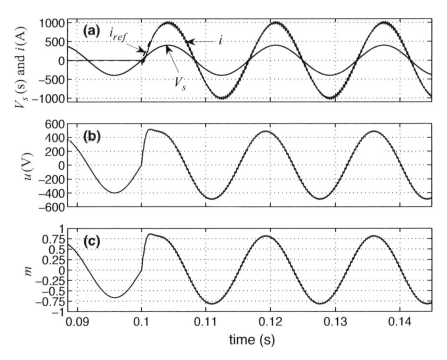

FIGURE 3.20   Response of the half-bridge converter to a sinusoidal command, based on the compensator of (3.19); Example 3.6.

Figure 3.20(a) illustrates that $i(t)$ rapidly reaches and tracks i_{ref}, without amplitude or phase-angle errors. Figure 3.20(b) and (c) shows the waveforms of the compensator output (control signal) and the modulating signal, respectively. It is observed that the switching ripples of these signals are small.

A comparison of the corresponding results in Examples 3.5 and 3.6 shows that the compensator structure is significantly more complex for sinusoidal command tracking than that for the DC command tracking. As noted in Example 3.5, a PI compensator is sufficient to ensure DC command following with a satisfactory performance. However, as indicated by Example 3.6, a more elaborate compensator is required to track a sinusoidal command, with a high degree of fidelity. Furthermore, the control loop must be designed for a much wider bandwidth in case of sinusoidal command following.

In three-phase VSC systems, we are often interested in tracking a sinusoidal command, rapidly and with small steady-state errors. We also need to stipulate rapid changes in the amplitude and/or the phase of the commands. Therefore, the control design is noticeably simplified if we can transform the problem of sinusoidal command tracking to a DC command tracking problem. The reference-frame theory and the techniques introduced in Chapter 4 are instrumental to that end.

4 Space Phasors and Two-Dimensional Frames

4.1 INTRODUCTION

Chapter 3 investigated control of the half-bridge converter. The half-bridge converter is the main building block of the three-phase voltage-sourced converter (VSC), and the three-phase VSC control deals with simultaneous control of three half-bridge converters. As discussed in Chapter 3, while only a proportional-integral (PI) compensator can enable a half-bridge converter system to track a DC command, the compensator must be of higher order and bandwidth if a sinusoidal command is to be tracked. In a three-phase VSC system, we are invariably interested in tracking sinusoidal voltage or current commands. Therefore, the compensator design task inherently faces the same hardships as discussed in Chapter 3 for a half-bridge converter system tracking a sinusoidal command. The *αβ-frame* and the *dq-frame*, the two main classes of *two-dimensional frames*, are introduced in this chapter to simplify the analysis and control[1].

The *αβ*-frame enables one to transform the problem of controlling a system of three half-bridge converters to an equivalent problem of controlling two equivalent subsystems. Moreover, the concept of instantaneous reactive power can be defined in the *αβ*-frame [41]. The *dq*-frame possesses the same merits as the *αβ*-frame, in addition to the following:

- If the control is exercised in the *dq*-frame, a sinusoidal command tracking problem is transformed to an equivalent DC command tracking problem. Hence, PI compensators can be used for the control.

- In *abc*-frame, models of specific types of electric machine exhibit time-varying, mutually coupled inductances. If the model is expressed in *dq*-frame, the time-varying inductances are transformed to (equivalent) constant parameters.

- Conventionally, components of large power systems are formulated and analyzed in *dq*-frame [42]. Therefore, representation of VSC systems in the *dq*-frame

---

[1] In the literature, the *αβ*-frame and the *dq*-frame are also called the stationary frame and the rotating frame, respectively.

Voltage-Sourced Converters in Power Systems, by Amirnaser Yazdani and Reza Iravani
Copyright © 2010 John Wiley & Sons, Inc.

enables analysis and design tasks based on methodologies that are commonly employed for power systems, in a unified framework.

In this chapter, first, the space phasor is introduced as a generalization of the conventional phasor. In addition, appropriate procedures are presented to (i) express a balanced three-phase function by an equivalent space phasor, (ii) stipulate dynamic changes in the amplitude and phase angle of a three-phase signal, and (iii) formulate a compact, equivalent, space-phasor representation for a balanced three-phase system. Then, the $\alpha\beta$-frame and the dq-frame are introduced as immediate by-products of the space-phasor concept. Finally, generic control schemes are introduced for the control of a three-phase VSC system in the $\alpha\beta$-frame and the dq-frame.

4.2   SPACE-PHASOR REPRESENTATION OF A BALANCED THREE-PHASE FUNCTION

4.2.1   Definition of Space Phasor

Consider the following balanced, three-phase, sinusoidal function [2]

$$f_a(t) = \widehat{f} \cos{(\omega t + \theta_0)},$$

$$f_b(t) = \widehat{f} \cos\left(\omega t + \theta_0 - \frac{2\pi}{3}\right),$$

$$f_c(t) = \widehat{f} \cos\left(\omega t + \theta_0 - \frac{4\pi}{3}\right), \tag{4.1}$$

where $\widehat{f}$, θ_0, and ω are the amplitude, the initial phase angle, and the angular frequency of the function, respectively. For the sinusoidal function (4.1), the space phasor is defined as

$$\overrightarrow{f}(t) = \frac{2}{3}\left[e^{j0} f_a(t) + e^{j\frac{2\pi}{3}} f_b(t) + e^{j\frac{4\pi}{3}} f_c(t)\right]. \tag{4.2}$$

Substituting for f_{abc} from (4.1) in (4.2), and using the identities $\cos\theta = \frac{1}{2}(e^{j\theta} + e^{-j\theta})$ and $e^{j0} + e^{j\frac{2\pi}{3}} + e^{j\frac{4\pi}{3}} \equiv 0$, one obtains

$$\overrightarrow{f}(t) = (\widehat{f}e^{j\theta_0})e^{j\omega t} = \underline{f}e^{j\omega t}, \tag{4.3}$$

where $\underline{f} = \widehat{f}e^{j\theta_0}$. The complex quantity $\underline{f}$ can be represented by a vector in the complex plane. If $\widehat{f}$ is a constant, the vector is analogous to the conventional phasor

---

[2]This function can represent a three-phase signal, or three time-varying parameters, for example, inductances.

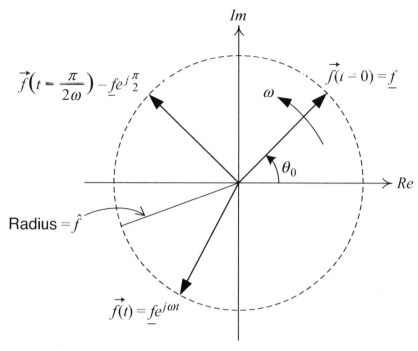

FIGURE 4.1   Space-phasor representation in the complex plane.

that is used to analyze linear circuits under steady-state sinusoidal conditions, and the tip of $\overrightarrow{f}(t)$ moves along the circumference of a circle centered at the complex plane origin (Fig. 4.1). Based on (4.3), the space phasor $\overrightarrow{f}(t)$ is the same phasor $\underline{f}$ that rotates counterclockwise with the angular speed ω. It should be noted that $\overrightarrow{f}(t)$ retains the form expressed by (4.3) even if $\widehat{f}$ is not a constant; if $\widehat{f}$ is a function of time, the corresponding phasor $\underline{f}$ is also a complex-valued function of time.

The definition of the space phasor can be extended to include a variable-frequency three-phase function [43]. Consider the three-phase function

$$f_a(t) = \widehat{f}(t) \cos \left[\theta(t) \right],$$

$$f_b(t) = \widehat{f}(t) \cos \left[\theta(t) - \frac{2\pi}{3} \right],$$

$$f_c(t) = \widehat{f}(t) \cos \left[\theta(t) - \frac{4\pi}{3} \right], \tag{4.4}$$

where

$$\theta(t) = \theta_0 + \int_0^t \omega(\tau) d\tau \tag{4.5}$$

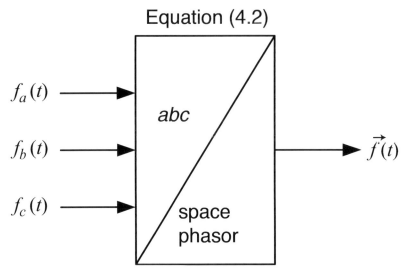

FIGURE 4.2   *abc*-frame to space-phasor signal transformer.

and $\omega(t)$ is a time-varying frequency. Based on (4.2), the space phasor corresponding to (4.4) is expressed as

$$\overrightarrow{f}(t) = \widehat{f}(t)e^{j\theta(t)}. \tag{4.6}$$

As (4.6) suggests, a space phasor in its most general form embeds information on the amplitude, phase angle, and frequency of the corresponding three-phase function. The space phasors represented by (4.3) and (4.6) are identical if $\omega(t)$ is a constant. Based on (4.2), we define the *abc-frame to space-phasor signal transformer* of Figure 4.2.

The real-valued components $f_a(t)$, $f_b(t)$, and $f_c(t)$ can be retrieved from the corresponding space phasor, based on the following equations:

$$f_a(t) = Re\left\{\overrightarrow{f}(t)e^{-j0}\right\},$$
$$f_b(t) = Re\left\{\overrightarrow{f}(t)e^{-j\frac{2\pi}{3}}\right\},$$
$$f_c(t) = Re\left\{\overrightarrow{f}(t)e^{-j\frac{4\pi}{3}}\right\}, \tag{4.7}$$

where $Re\{\cdot\}$ is the real-part operator. Based on (4.7), $f_a(t)$, $f_b(t)$, and $f_c(t)$ are projections of, respectively, $\overrightarrow{f}(t)$, $\overrightarrow{f}(t)e^{-j\frac{2\pi}{3}}$, and $\overrightarrow{f}(t)e^{-j\frac{4\pi}{3}}$ on the real axis of the complex plane. Figure 4.3 illustrates the block diagram of a *space phasor to abc-frame signal transformer*, based on (4.7).

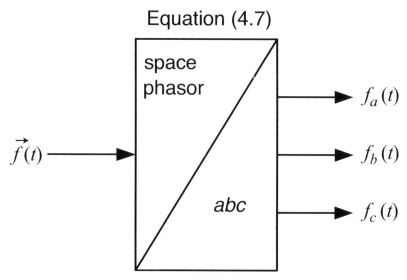

FIGURE 4.3   Space phasor to *abc*-frame signal transformer.

4.2.2   Changing the Amplitude and Phase Angle of a Three-phase Signal

Control of a VSC system typically involves tracking of sinusoidal commands. Since a sinusoidal signal is characterized by its amplitude and phase, it may be necessary in some applications to stipulate changes in the amplitude and/or the phase of the reference commands or control signals. This can be conveniently achieved through the space-phasor concept.

Consider a three-phase signal f_{abc} whose equivalent space phasor is $\vec{f}(t)$. Assume that the objective is to find a system that receives the space phasor $\vec{f}(t)$ as an input and generates the space phasor $\vec{f}'(t)$ corresponding to a new three-phase signal with the following properties:

- The phase angle of each component of $f'_{abc}(t)$ is shifted by $\phi(t)$ with respect to that of the corresponding component of f_{abc}, where $\phi(t)$ is an arbitrary function of time.
- The amplitude of each component of $f'_{abc}(t)$ is $A(t)$ times that of the corresponding component of f_{abc}, where $A(t)$ is an arbitrary function of time.

Let us call the aforementioned system the *space-phasor phase-shifter/scaler*, which enables the desired phase shift and amplitude scaling, based on

$$\vec{f}'(t) = \vec{f}(t)A(t)e^{j\phi(t)}. \tag{4.8}$$

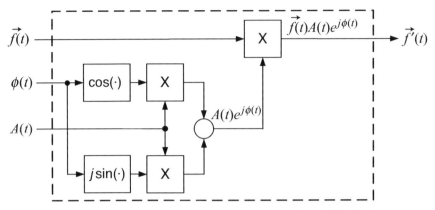

FIGURE 4.4   Block diagram of the space-phasor phase-shifter/scaler.

Figure 4.4 illustrates a block representation of the space-phasor phase-shifter/scaler.

To understand the mechanism of the space-phasor phase-shifter/scaler of Figure 4.4, consider the three-phase signal (4.1). Based on (4.3), the corresponding space phasor is $\overrightarrow{f}(t) = \widehat{f}e^{j\theta_0}e^{j\omega t}$. Figure 4.4 shows that $\overrightarrow{f}'(t) = \overrightarrow{f}(t)A(t)e^{j\phi(t)} = A(t)\widehat{f}e^{j(\omega t+\theta_0+\phi(t))}$. Then, based on (4.7), it can be shown that the three-phase signal corresponding to $\overrightarrow{f}'(t)$ is

$$f_a'(t) = A(t)\widehat{f}\cos\left[(\omega t + \theta_0) + \phi(t)\right],$$

$$f_b'(t) = A(t)\widehat{f}\cos\left[\left(\omega t + \theta_0 - \frac{2\pi}{3}\right) + \phi(t)\right],$$

$$f_c'(t) = A(t)\widehat{f}\cos\left[\left(\omega t + \theta_0 - \frac{4\pi}{3}\right) + \phi(t)\right].$$

The space-phasor phase-shifter/scaler of Figure 4.4 can be modified to receive/deliver *abc*-frame signals rather than space phasors. This is achieved by augmenting the space-phasor phase-shifter/scaler of Figure 4.4 with the abc-frame to space-phasor and space-phasor to abc-frame signal transformers. The modified space-phasor phase-shifter/scaler is illustrated in Figure 4.5.

One application of the space-phasor phase-shifter/scaler of Figure 4.5 is in *voltage-controlled* VSC systems that are usually encountered in high-power applications, for example, in flexible AC transmission systems (FACTS) controllers [1, 44, 45]. Figure 4.6 illustrates a simplified schematic diagram of a voltage-controlled VSC system in which a three-phase VSC is interfaced with an AC system via a three-phase inductor, L. If the resistance of the inductor is neglected, the real- and reactive-power components exchanged between the VSC system and the AC system, that is, P_s and Q_s, can be effectively controlled by the phase angle and the amplitude of the VSC

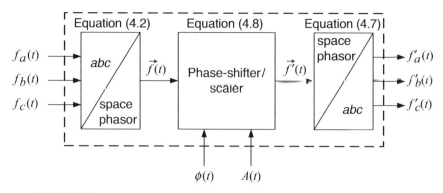

FIGURE 4.5   Block diagram of the modified space-phasor phase shifter/scaler.

AC-side terminal voltage, $V_{t\text{-}abc}$, relative to those of the AC system voltage, $V_{s\text{-}abc}$ [46, 47]. Thus, $\phi(t)$ ($A(t)$) is commanded by a (another) feedback loop that processes the error between the real power (reactive power) and its respective reference value, to regulate the real power (reactive power). The output of the space-phasor phase-shifter/scaler corresponds to the terminal voltage to be reproduced by the VSC and is delivered to the VSC pulse-width modulation (PWM) switching scheme.

Example 4.1 illustrates the operation of the space-phasor phase-shifter/scaler of Figure 4.4.

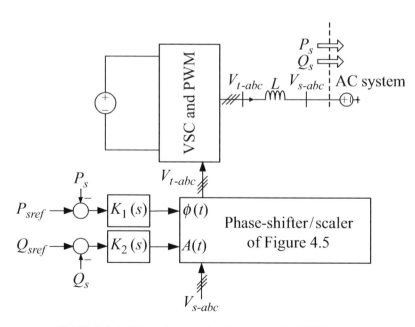

FIGURE 4.6   Block diagram of voltage-controlled VSC system.

EXAMPLE 4.1   Imposing Amplitude and Phase Changes on a Three-Phase Signal

Let the following three-phase signal be given to the space-phasor phase-shifter/scaler of Figure 4.5:

$$f_a(t) = \cos(377t),$$

$$f_b(t) = \cos\left(377t - \frac{2\pi}{3}\right),$$

$$f_c(t) = \cos\left(377t - \frac{4\pi}{3}\right).$$

The system of Figure 4.5 first converts f_{abc} to $\overrightarrow{f}(t)$, based on an *abc*-frame to space-phasor signal transformation. Then $\overrightarrow{f}(t)$ is converted to $\overrightarrow{f}'(t)$ by means of a space-phasor phase-shifter/scaler of Figure 4.4. Finally, $\overrightarrow{f}'(t)$ is transformed to f'_{abc} by means of a space-phasor to *abc*-frame transformation.

Assume that $A(t)$ is a step function changing from 1 to 1.5 at $t = 66$ ms, while $\phi(t) \equiv 0$. Figure 4.7 shows the output of the system of Figure 4.5, that is, $f'_{abc}(t)$. It is observed that at $t = 66$ ms the amplitude of $f_{abc}(t)$ suddenly changes from 1 to 1.5 without any shift in its phase angle. Now, consider the case where $\phi(t)$ is a step function changing from 0 to π, at $t = 18$ ms, while $A(t) \equiv 1$. Figure 4.8 shows that the polarity of $f_{abc}(t)$ becomes opposite at $t = 18$ ms, but the amplitude is preserved.

The space-phasor phase-shifter/scaler of Figure 4.5 can also be used to change the frequency of a three-phase signal. For instance, provide the three-phase signal (4.1)

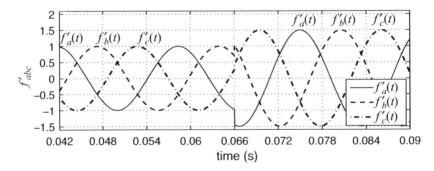

FIGURE 4.7   A step change in the amplitude of a three-phase signal

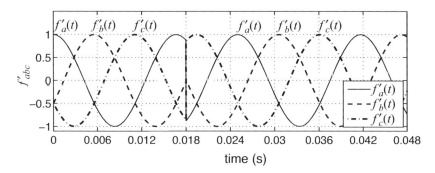

FIGURE 4.8   A step change in the phase angle of a three-phase signal.

to the space-phasor phase-shifter/scaler of Figure 4.5. Then, the output is

$$f'_a(t) = A(t)\widehat{f}\cos\left[(\omega t + \theta_0) + \phi(t)\right],$$

$$f'_b(t) = A(t)\widehat{f}\cos\left[\left(\omega t + \theta_0 - \frac{2\pi}{3}\right) + \phi(t)\right],$$

$$f'_c(t) = A(t)\widehat{f}\cos\left[\left(\omega t + \theta_0 - \frac{4\pi}{3}\right) + \phi(t)\right].$$

If $A(t) = 1$ and $\phi(t)$ is determined based on

$$\phi(t) = \int_0^t \Delta\omega(\tau)d\tau, \tag{4.9}$$

the output of the space-phasor phase-shifter/scaler becomes

$$f'_a(t) = \widehat{f}\cos\left(\omega t + \theta_0 + \int_0^t \Delta\omega(\tau)d\tau\right),$$

$$f'_b(t) = \widehat{f}\cos\left(\omega t + \theta_0 - 2\pi/3 + \int_0^t \Delta\omega(\tau)d\tau\right),$$

$$f'_c(t) = \widehat{f}\cos\left(\omega t + \theta_0 - 4\pi/3 + \int_0^t \Delta\omega(\tau)d\tau\right), \tag{4.10}$$

where $\Delta\omega(t)$ is an arbitrary function of time. The signal $f'_{abc}(t)$ described by (4.10) represents a three-phase signal of which the frequency is $\omega + \Delta\omega(t)$ and, thus, can be varied by $\Delta\omega(t)$. A special case corresponds to $\Delta\omega(t) = -\omega$, where the frequency of the output signal becomes zero and f'_{abc} dissolves into three DC signals. It should, however, be noted that the three DC signals still constitute a balanced three-phase signal. Example 4.2 illustrates this operation.

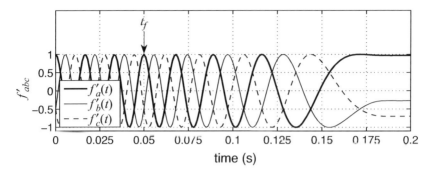

FIGURE 4.9   Changing the frequency of a three-phase signal; Example 4.2.

EXAMPLE 4.2   Changing the Frequency of a Three-Phase Signal

Assume that the following three-phase signal is the input to the space-phasor phase-shifter/scaler of Figure 4.5:

$$f_a(t) = \cos(377t),$$

$$f_b(t) = \cos\left(377t - \frac{2\pi}{3}\right),$$

$$f_c(t) = \cos\left(377t - \frac{4\pi}{3}\right).$$

Also, let $\Delta\omega(t)$ be

$$\Delta\omega(t) = \begin{cases} 0, & t < t_f, \\ -(2\pi)(445)(t - t_f), & t_f \leq t \leq 0.1848, \\ -377, & t \geq 0.1848, \end{cases}$$

where $t_f = 0.05$ s. Figure 4.9 shows the corresponding output, that is, $f'_{abc}(t)$, and indicates that the frequency is constant (at 60 Hz) until $t_f = 0.05$ s. However, the frequency decreases at the rate of 445 Hz/s until it reaches zero at $t = 0.1848$ s. Thereafter, the three-phase output signal is frozen and its components settle at constant values.

4.2.3   Generating a Controllable-Amplitude/Controllable-Frequency Three-Phase Signal

In some applications, both the amplitude and the frequency of a three-phase signal must be controlled. This objective can be achieved following the concept of space phasors, as shown in the block diagram of Figure 4.10. The signal generator of Figure 4.10 is conceptually the same space-phasor phase-shifter/scaler of Figure 4.5 except that the *abc*-frame to space-phasor transformation block is omitted, and

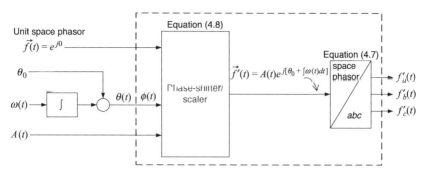

FIGURE 4.10 Block diagram of a controllable-frequency/amplitude three-phase signal generator.

$\vec{f}(t)$ in this case is the unit space phasor with phase angle zero, that is, $\vec{f} = 1e^{j0}$. The function of the signal generator of Figure 4.10 is to shift the unit space phasor by angle $\theta(t)$, which is composed of a time-varying component and a constant component. The time-varying component is the integral of a prespecified angular frequency, for example, $\omega(t)$, and causes the unit space-phasor to rotate with the same frequency. The constant component specifies the initial phase angle of the rotating space phasor. The length of the rotating space phasor and, thus, the amplitude of the resulting three-phase signal can be controlled by $A(t)$. The rotating space phasor is finally converted to the corresponding three-phase signal, based on (4.7).

An interesting special case takes place when $\omega(t)$ is a negative constant value, for example, $\omega(t) = -\omega$. For the sake of simplicity, let us assume that $A(t) \equiv 1$. Thus, the rotating space phasor will have the form of $\vec{f'}(t) = e^{j(-\omega t + \theta_0)}$. Then, based on (4.7) the corresponding three-phase signal is

$$f'_a(t) = \cos(-\omega t + \theta_0),$$

$$f'_b(t) = \cos\left(-\omega t + \theta_0 - \frac{2\pi}{3}\right),$$

$$f'_c(t) = \cos\left(-\omega t + \theta_0 - \frac{4\pi}{3}\right). \tag{4.11}$$

Equation (4.11) represents a three-phase signal that has a negative frequency. Since a negative frequency has no physical meaning, we use the identity $\cos(-\theta) = \cos\theta$ to have a positive-frequency three-phase signal. Thus, (4.11) can be rewritten as

$$f'_a(t) = \cos(\omega t - \theta_0),$$

$$f'_b(t) = \cos\left(\omega t - \theta_0 + \frac{2\pi}{3}\right),$$

$$f'_c(t) = \cos\left(\omega t - \theta_0 + \frac{4\pi}{3}\right),$$

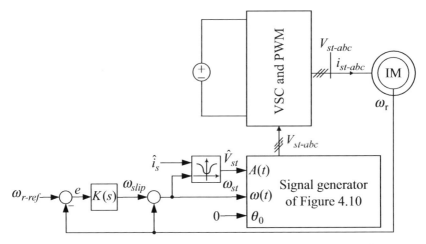

FIGURE 4.11   Block diagram of a variable-speed induction motor drive.

which in turn is equivalent to

$$f_a'(t) = \cos(\omega t - \theta_0),$$

$$f_b'(t) = \cos\left(\omega t - \theta_0 - \frac{4\pi}{3}\right),$$

$$f_c'(t) = \cos\left(\omega t - \theta_0 - \frac{2\pi}{3}\right). \qquad (4.12)$$

Equation (4.12) indicates that the phase sequence in the output signal is reversed from the abc to the acb. In other words, a space phasor with a negative frequency corresponds to a negative-sequence three-phase signal.

One known application of the signal generator of Figure 4.10 is in variable-speed asynchronous motor drives [48]. Figure 4.11 illustrates a simplified schematic diagram of such a motor drive in which a three-phase VSC controls the machine stator voltage V_{st-abc} [16, 43]. The machine speed, ω_r, is compared with the reference command, and the error is processed by a compensator. The output of the compensator, ω_{slip}, corresponds to the slip frequency and is added to ω_r to determine the required stator frequency, ω_{st}, as shown in Figure 4.11. On the other hand, the amplitude of the stator voltage is determined through a nonlinear static function, based on the stator frequency and the amplitude of the stator current. Thus, at a small torque (resulting in a small stator current), $\widehat{V}_{st}$ is almost proportional to ω_{st} (constant V/f operation). However, $\widehat{V}_{st}$ is further boosted at higher values of torque, to compensate for the stator resistive voltage drop. The required values of ω_{st} and $\widehat{V}_{st}$ are delivered to the three-phase signal generator of Figure 4.10 whose output corresponds to the stator voltage to be synthesized by the VSC through the PWM strategy.

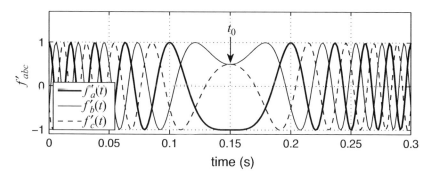

FIGURE 4.12   Reversing the sequence of the phases of a three-phase signal.

EXAMPLE 4.3   Reversing Phase Sequence of a Three-Phase Signal

Consider the three-phase signal generator of Figure 4.10 with $\theta_0 \equiv 0$, $A(t) \equiv 1$, and $\omega(t) = 377 - 2513t$. Thus, the frequency is initially 60 Hz but decreases as a ramp function with a slope of 400 Hz/s. The frequency crosses zero at $t = 0.15$ s and becomes negative thereafter.

Figure 4.12 shows the output waveform of the signal generator of Figure 4.10. Until $t = 0.15$ s, the positive frequency is decreased. Therefore, the period of each sinusoidal constituent increases, but the phase sequence does not change. At $t = 0.15$ s, the frequency crosses zero and becomes negative thereafter. Consequently, as Figure 4.12 illustrates, the phase sequence is reversed and the output signal period decreases as the absolute value of the negative frequency becomes larger.

4.2.4   Space-Phasor Representation of Harmonics

In an electrical energy system, voltage and current waveforms are often periodic functions of time and may include harmonic components. Nonlinearities and switching processes are the two main reasons for the existence of harmonics, although some control instabilities can also lead to the generation of harmonics. Harmonics cause distortion in the corresponding fundamental waveforms and usually have adverse impact on the efficiency and performance of systems. This section applies the space-phasor definition of Section 4.2.1 to harmonics.

Consider the following three-phase signal:

$$f_a(t) = \widehat{f}_1 \cos(\omega t) + \widehat{f}_n \cos(n\omega t),$$

$$f_b(t) = \widehat{f}_1 \cos\left(\omega t - \frac{2\pi}{3}\right) + \widehat{f}_n \cos\left(n\omega t - \frac{2n\pi}{3}\right),$$

$$f_c(t) = \widehat{f}_1 \cos\left(\omega t - \frac{4\pi}{3}\right) + \widehat{f}_n \cos\left(n\omega t - \frac{4n\pi}{3}\right), \qquad (4.13)$$

where $\widehat{f_1}$ is the amplitude of the fundamental, n is the order of the harmonic, and $\widehat{f_n}$ is the amplitude of the harmonic. Applying the definition of the space phasor, (4.4), one obtains

$$\overrightarrow{f}(t) = \widehat{f_1} e^{j\omega t} + \overrightarrow{f_n}(t), \tag{4.14}$$

where

$$\overrightarrow{f_n}(t) = \left(\frac{\widehat{f_n}}{3}\right) \left[1 + e^{-j(n-1)\frac{2\pi}{3}} + e^{-j(n-1)\frac{4\pi}{3}} \right] e^{jn\omega t}$$

$$+ \left(\frac{\widehat{f_n}}{3}\right) \left[1 + e^{j(n+1)\frac{2\pi}{3}} + e^{j(n+1)\frac{4\pi}{3}} \right] e^{-jn\omega t}. \tag{4.15}$$

Equation (4.14) shows that $\overrightarrow{f}(t)$ is composed of two space phasors. The first space phasor corresponds to the fundamental component of the three-phase signal and rotates counterclockwise with the angular frequency ω. The second space phasor, $\overrightarrow{f_n}(t)$, corresponds to the harmonic component. Based on (4.15), $\overrightarrow{f_n}(t) \equiv 0$ if n is an integer multiple of 3. Otherwise, $\overrightarrow{f_n}(t)$ is a space phasor with the length $\widehat{f_n}$ and rotates with the angular frequency $n\omega$. However, the direction of its rotation depends on the harmonic order. We call $\overrightarrow{f_n}(t)$ a positive-sequence space-phasor if it rotates counterclockwise. Thus, we refer to the three-phase waveform corresponding to a positive-sequence space phasor as a positive-sequence harmonic. Similarly, we call $\overrightarrow{f_n}(t)$ a negative-sequence space phasor if it rotates clockwise and refer to its corresponding three-phase waveform as a negative-sequence harmonic. Harmonics with orders of multiple of 3 are called zero-sequence harmonics as they correspond to $\overrightarrow{f_n}(t) \equiv 0$. Based on (4.15), a set of positive- and negative-sequence harmonics are listed in Table 4.1[3].

4.3   SPACE-PHASOR REPRESENTATION OF THREE-PHASE SYSTEMS

The three-phase to space-phasor transformation and its inverse transformation were discussed in Section 4.2. Using the space-phasor concept, the procedures to dynamically change the amplitude and frequency of a signal were also presented. The signals under consideration can be reference signals, feedback signals, control signals, and so on. Since we have identified the tools to process signals in the space-phasor domain, it is warranted to introduce a methodology to model a generic three-phase system in the space-phasor domain. In the subsequent developments, for simplicity and without loss of generality, we illustrate examples dealing with linear systems. In particular,

[3] The discussion of Section 4.2.4 is valid only for balanced periodic waveforms. The phase sequences of the harmonics of an unbalanced three-phase waveform do not necessarily conform to the result of Table 4.1.

TABLE 4.1 Positive- and Negative-Sequence Harmonics

Positive-Sequence Harmonic $\overrightarrow{f_n}(t) = \widehat{f_n} e^{jn\omega t}$	Negative-Sequence Harmonic $\overrightarrow{f_n}(t) = \widehat{f_n} e^{-jn\omega t}$
$n = 1$	$n = 2$
4	5
7	8
10	11
13	14
19	17
22	20
25	23
28	26
31	32

we identify three classes of three-phase systems: symmetrical and decoupled, symmetrical and coupled, and asymmetrical.

4.3.1 Decoupled Symmetrical Three-Phase Systems

Consider the three-phase system of Figure 4.13 in which each phase of the output y_{abc} is controlled by the corresponding phase of the input u_{abc}. The system of Figure 4.13 is composed of three decoupled identical subsystems. The system is symmetrical since the input/output relationships of the three phases retain their original expressions if

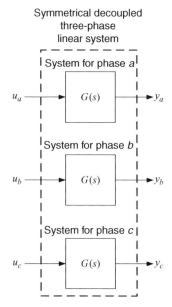

FIGURE 4.13 Block diagram of a symmetrical, decoupled, linear, three-phase system.

index a is replaced by b, index b is replaced by c, and index c is replaced by a in the phase equations. Let the input/output relationship of each subsystem be described by a transfer function, $G(s)$, as

$$Y_a(s) = G(s)U_a(s),$$

$$Y_b(s) = G(s)U_b(s),$$

$$Y_c(s) = G(s)U_c(s), \tag{4.16}$$

where $G(s) = (k_m s^m + k_{m-1}s^{m-1} + \cdots + k_0)/(s^n + l_{n-1}s^{n-1} + \cdots + l_0)$ is a rational transfer function; it can be verified based on (4.16) that the system of Figure 4.13 is symmetrical. The time-domain equations of the three-phase system are

$$\frac{d^n y_a}{dt^n} + l_{n-1}\frac{d^{n-1} y_a}{dt^{n-1}} + \cdots + l_0 y_a = k_m \frac{d^m u_a}{dt^m} + k_{m-1}\frac{d^{m-1} u_a}{dt^{m-1}} + \cdots + k_0 u_a, \tag{4.17}$$

$$\frac{d^n y_b}{dt^n} + l_{n-1}\frac{d^{n-1} y_b}{dt^{n-1}} + \cdots + l_0 y_b = k_m \frac{d^m u_b}{dt^m} + k_{m-1}\frac{d^{m-1} u_b}{dt^{m-1}} + \cdots + k_0 u_b, \tag{4.18}$$

$$\frac{d^n y_c}{dt^n} + l_{n-1}\frac{d^{n-1} y_c}{dt^{n-1}} + \cdots + l_0 y_c = k_m \frac{d^m u_c}{dt^m} + k_{m-1}\frac{d^{m-1} u_c}{dt^{m-1}} + \cdots + k_0 u_c. \tag{4.19}$$

Multiplying both sides of (4.17), (4.18), and (4.19), respectively, by $\frac{2}{3}e^{j0}$, $\frac{2}{3}e^{j\frac{2\pi}{3}}$, and $\frac{2}{3}e^{j\frac{4\pi}{3}}$, one obtains

$$\frac{d^n}{dt^n}\left(\frac{2}{3}e^{j0}y_a\right) + l_{n-1}\frac{d^{n-1}}{dt^{n-1}}\left(\frac{2}{3}e^{j0}y_a\right) + \cdots + l_0\left(\frac{2}{3}e^{j0}y_a\right)$$
$$= k_m\frac{d^m}{dt^m}\left(\frac{2}{3}e^{j0}y_a\right) + k_{m-1}\frac{d^{m-1}}{dt^{m-1}}\left(\frac{2}{3}e^{j0}y_a\right) + \cdots + k_0\left(\frac{2}{3}e^{j0}y_a\right), \tag{4.20}$$

$$\frac{d^n}{dt^n}\left(\frac{2}{3}e^{j\frac{2\pi}{3}}y_b\right) + l_{n-1}\frac{d^{n-1}}{dt^{n-1}}\left(\frac{2}{3}e^{j\frac{2\pi}{3}}y_b\right) + \cdots + l_0\left(\frac{2}{3}e^{j\frac{2\pi}{3}}y_b\right)$$
$$= k_m\frac{d^m}{dt^m}\left(\frac{2}{3}e^{j\frac{2\pi}{3}}y_b\right) + k_{m-1}\frac{d^{m-1}}{dt^{m-1}}\left(\frac{2}{3}e^{j\frac{2\pi}{3}}y_b\right) + \cdots + k_0\left(\frac{2}{3}e^{j\frac{2\pi}{3}}y_b\right), \tag{4.21}$$

$$\frac{d^n}{dt^n}\left(\frac{2}{3}e^{j\frac{4\pi}{3}}y_c\right) + l_{n-1}\frac{d^{n-1}}{dt^{n-1}}\left(\frac{2}{3}e^{j\frac{4\pi}{3}}y_c\right) + \cdots + l_0\left(\frac{2}{3}e^{j\frac{4\pi}{3}}y_c\right)$$

$$= k_m\frac{d^m}{dt^m}\left(\frac{2}{3}e^{j\frac{4\pi}{3}}y_c\right) + k_{m-1}\frac{d^{m-1}}{dt^{m-1}}\left(\frac{2}{3}e^{j\frac{4\pi}{3}}y_c\right) + \cdots + k_0\left(\frac{2}{3}e^{j\frac{4\pi}{3}}y_c\right).$$

(4.22)

Adding the corresponding sides of (4.20), (4.21), and (4.22), and using (4.2), one concludes that

$$\frac{d^n}{dt^n}\overrightarrow{y} + l_{n-1}\frac{d^{n-1}}{dt^{n-1}}\overrightarrow{y} + \cdots + l_0\overrightarrow{y} = k_m\frac{d^m}{dt^m}\overrightarrow{u} + k_{m-1}\frac{d^{m-1}}{dt^{m-1}}\overrightarrow{u} + \cdots + k_0\overrightarrow{u}.$$

(4.23)

Equation (4.23) represents the system of Figure 4.13 in the space-phasor domain. It is noted that the system input–output relationship in the space-phasor domain has the same form as that for each subsystem in the *abc*-frame. Equation (4.23) provides a compact representation of the original three-phase system. It can be observed that (4.23) possesses the same form as each of (4.17), (4.18), and (4.19). Therefore, the space-phasor equations of a symmetrical, decoupled, linear, three-phase system can be conveniently derived by replacing the time-domain variables with the corresponding space-phasor variables in any set of equations corresponding to the three phases.

The foregoing procedure to transform the differential equations of a three-phase system to space-phasor domain can also be readily applied to state-space equations, as illustrated in Example 4.4.

EXAMPLE 4.4   Space-Phasor State-Space Equations of a Three-Phase Circuit

Figure 4.14 illustrates a simplified circuit diagram of a current-controlled three-phase VSC system of which each phase is interfaced with the corresponding phase of an AC system. The AC system is represented by a voltage source v_{sabc} in series with three decoupled inductors, one per phase. The inductance of each inductor is L_s. In the circuit of Figure 4.14, v_{abc} signifies the voltage of the point of common coupling (PCC), and i_{sabc} and i_{abc} represent the AC system current and the VSC current, respectively. Since, in practice, i_{abc} contains harmonic components, the capacitors C are used to provide bypass paths for the harmonics and prevent them from penetrating into the AC system. It is assumed that the fundamental component of i_{abc} can be controlled by a PWM scheme. This, in turn, enables the control of real and reactive power that the VSC system delivers to the AC system.

Based on the aforementioned description of the circuit of Figure 4.14, i_{abc} is the control variable whereas v_{abc} and i_{sabc} are the state variables. Since there is no control over v_{sabc}, it is regarded as the disturbance input. Depending on

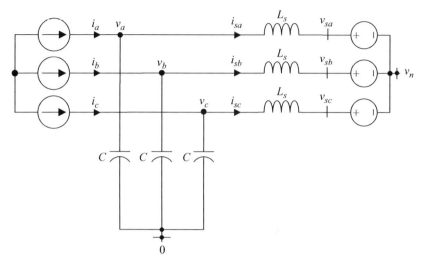

FIGURE 4.14   Three-phase circuit of Example 4.4.

the application, the output is a combination of state variables. Selecting the common point of the capacitors, that is, node 0, as the reference voltage node, the voltage of the AC system neutral point is assumed to be $v_n(t)$. Thus, the following state-space equations can be written:

$$C\frac{dv_a}{dt} = i_a - i_{sa},$$
(4.24)

$$C\frac{dv_b}{dt} = i_b - i_{sb},$$
(4.25)

$$C\frac{dv_c}{dt} = i_c - i_{sc},$$
(4.26)

$$L_s\frac{di_{sa}}{dt} = v_a - v_{sa} - v_n,$$
(4.27)

$$L_s\frac{di_{sb}}{dt} = v_b - v_{sb} - v_n,$$
(4.28)

$$L_s\frac{di_{sc}}{dt} = v_c - v_{sc} - v_n.$$
(4.29)

Equations (4.24)–(4.26) are expressed in the space-phasor domain as

$$C\frac{d\vec{v}}{dt} = \vec{i} - \vec{i_s}.$$
(4.30)

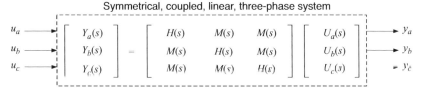

FIGURE 4.15   Block diagram of a symmetrical, coupled, linear, three-phase system.

The space-phasor equivalent of (4.27)–(4.29) assumes a form slightly different from the original equations. Since $(2/3)(e^{j0} + e^{j\frac{2\pi}{3}} + e^{j\frac{4\pi}{3}}) \equiv 0$, one concludes that $(2/3)(e^{j0}v_n + e^{j\frac{2\pi}{3}}v_n + e^{j\frac{4\pi}{3}}v_n) \equiv 0$. Therefore, v_n is eliminated in the process of conversion from the abc-frame to the space-phasor domain, and

$$L_s \frac{d\overrightarrow{i_s}}{dt} = \overrightarrow{v} - \overrightarrow{v_s}. \tag{4.31}$$

4.3.2   Coupled Symmetrical Three-Phase Systems

Figure 4.15 represents a block diagram of a coupled symmetrical three-phase system. In this section, we demonstrate that, despite the coupling, a space-phasor representation can be formulated for the system using a procedure similar to the one adopted for the decoupled system of Figure 4.13. The coupled equations representing the system of Figure 4.15 are

$$Y_a(s) = H(s)U_a(s) + M(s)U_b(s) + M(s)U_c(s),$$
$$Y_b(s) = M(s)U_a(s) + H(s)U_b(s) + M(s)U_c(s),$$
$$Y_c(s) = M(s)U_a(s) + M(s)U_b(s) + H(s)U_c(s), \tag{4.32}$$

where $H(s)$ and $M(s)$ are self- and mutual transfer functions, respectively. Since $u_a + u_b + u_c \equiv 0$, so is $U_a + U_b + U_c \equiv 0$. Therefore, (4.32) can be rewritten as

$$Y_a(s) = [H(s) - M(s)]U_a(s),$$
$$Y_b(s) = [H(s) - M(s)]U_b(s),$$
$$Y_c(s) = [H(s) - M(s)]U_c(s). \tag{4.33}$$

The equation set (4.33) corresponds to the symmetrical decoupled three-phase system of Figure 4.13 in which $G(s) = H(s) - M(s)$. Therefore, based on the discussions in Section 4.3.1, the transfer function from the input space phasor to the output space phasor is also $G(s) = H(s) - M(s)$.

EXAMPLE 4.5   Space-Phasor Equations of a Mutually Coupled Three-Phase Inductor

Consider the three-phase inductor of Example 4.4 (Fig. 4.14), where each inductor is mutually coupled with the other two inductors, and the mutual inductances are equal to M. Therefore, (4.31) is modified to

$$(L_s - M)\frac{d\overrightarrow{i_s}}{dt} = \overrightarrow{v} - \overrightarrow{v_s}. \tag{4.34}$$

4.3.3   Asymmetrical Three-Phase Systems

The space-phasor system representation introduced in Sections 4.3.1 and 4.3.2 cannot be developed for asymmetrical three-phase systems. The reason is that in an asymmetrical system no transfer function can be found to describe the output space phasor as a function of the input space phasor. In other words, the real and imaginary components of the output space phasor in an asymmetrical system are related to both the real and imaginary components of the input space phasor, through different transfer functions[4] . Consequently, the modeling must be carried out in either the $\alpha\beta$-frame or the dq-frame, as discussed in subsequent sections.

4.4   POWER IN THREE-WIRE THREE-PHASE SYSTEMS

In this section, based on the space-phasor concept we define real-, reactive-, and apparent-power components exchanged with a three-wire, three-phase port. In contrast to the conventional phasor concept, the definitions of power components based on the space-phasor theory are also applicable to dynamic and/or variable-frequency scenarios.

Consider the balanced three-phase network of Figure 4.16 whose terminal voltages and currents are v_{abc} and i_{abc}, respectively. v_{abc} and/or i_{abc} are not necessarily balanced, but $i_a + i_b + i_c = 0$. The instantaneous total (real) power in the time domain is expressed as

$$P(t) = v_a(t)i_a(t) + v_b(t)i_b(t) + v_c(t)i_c(t). \tag{4.35}$$

In (4.35), v_{abc} and i_{abc} can be expressed in terms of their corresponding space phasors based on (4.7). Thus,

$$P(t) = Re\left\{\overrightarrow{v}(t)e^{j0}\right\} Re\left\{\overrightarrow{i}(t)e^{j0}\right\}$$

$$+Re\left\{\overrightarrow{v}(t)e^{-j\frac{2\pi}{3}}\right\} Re\left\{\overrightarrow{i}(t)e^{-j\frac{2\pi}{3}}\right\}$$

$$+Re\left\{\overrightarrow{v}(t)e^{-j\frac{4\pi}{3}}\right\} Re\left\{\overrightarrow{i}(t)e^{-j\frac{4\pi}{3}}\right\}. \tag{4.36}$$

---

[4]We will demonstrate this in Section 4.5.5, Example 4.7.

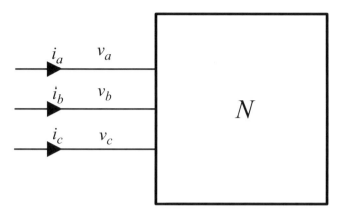

FIGURE 4.16   A three-wire three-phase network.

Based on the identity $Re\{\alpha\}Re\{\beta\} = (Re\{\alpha\beta\} + Re\{\alpha\beta^*\})/2$, (4.36) is expressed as

$$P(t) = \frac{Re\left\{\overrightarrow{v}(t)\,\overrightarrow{i}(t)\right\} + Re\left\{\overrightarrow{v}(t)\,\overrightarrow{i}\,^*(t)\right\}}{2}$$

$$+ \frac{Re\left\{\overrightarrow{v}(t)\,\overrightarrow{i}(t)e^{-j\frac{4\pi}{3}}\right\} + Re\left\{\overrightarrow{v}(t)\,\overrightarrow{i}\,^*(t)\right\}}{2}$$

$$+ \frac{Re\left\{\overrightarrow{v}(t)\,\overrightarrow{i}(t)e^{-j\frac{8\pi}{3}}\right\} + Re\left\{\overrightarrow{v}(t)\,\overrightarrow{i}\,^*(t)\right\}}{2}.$$

$$(4.37)$$

Since $e^{j0} + e^{-j\frac{4\pi}{3}} + e^{-j\frac{8\pi}{3}} \equiv 0$, (4.37) is simplified to

$$P(t) = Re\left\{\frac{3}{2}\overrightarrow{v}(t)\,\overrightarrow{i}\,^*(t)\right\}. \qquad (4.38)$$

It should be noted that (4.38) is developed with no assumption on the frequencies or the amplitudes of v_{abc} and i_{abc}. The frequencies and amplitudes can assume any arbitrary functions of time. Moreover, there is no requirement for the frequencies of v_{abc} and i_{abc} to be equal. Furthermore, (4.38) is obtained with no assumption on the harmonic contents of v_{abc} and i_{abc}, and it is valid under both transient and steady-state conditions. The only assumption is that the port corresponds to a three-wire circuit, that is, $i_a + i_b + i_c = 0$.

If v_{abc} and i_{abc} are distortion-free sinusoidal waveforms with constant amplitudes and equal constant frequencies, (4.38) is reduced to the expression for the real power in the conventional phasor analysis. This analogy is the main incentive behind the definitions of the instantaneous reactive power and the instantaneous complex power, as

$$Q(t) = Im \left\{ \frac{3}{2} \vec{v}(t) \vec{i}^*(t) \right\},$$ (4.39)

$$S(t) = P(t) + jQ(t) = \frac{3}{2} \vec{v}(t) \vec{i}^*(t).$$ (4.40)

Under a steady-state, balanced, sinusoidal condition, the instantaneous reactive power assumes the same expression as that for the conventional reactive power.

EXAMPLE 4.6   Instantaneous Power Absorbed by a Three-Phase Inductor Bank

Consider a set of three inductors with no mutual couplings, as shown in Figure 4.17. The inductors carry the balanced three-phase current i_{abc}. The instantaneous power absorbed by the inductors is formulated as follows.

Since the system is decoupled and symmetrical, the following space-phasor equation is valid for the inductor voltage:

$$\vec{v}(t) = L \frac{d\vec{i}}{dt}.$$

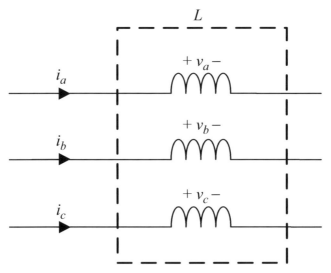

FIGURE 4.17   Balanced three-phase inductors of Example 4.6.

Based on (4.38), the instantaneous real power of the inductor set is

$$P_L(t) = \left(\frac{3L}{2}\right) Re\left\{\frac{d\vec{i}}{dt}\,\vec{i}^*\right\}. \tag{4.41}$$

Under a steady-state condition, $\vec{i}(t) = \widehat{i}e^{j\omega t}$, $d\vec{i}/dt = \widehat{i}(j\omega)e^{j\omega t}$, and $\vec{i}^* = \widehat{i}e^{-j\omega t}$. Thus,

$$P_L(t) = \left(\frac{3L}{2}\right) Re\left\{\left(j\omega\widehat{i}e^{j\omega t}\right)\left(\widehat{i}e^{-j\omega t}\right)\right\} = \left(\frac{3L}{2}\right) Re\left\{j\omega\widehat{i}^2\right\} \equiv 0,$$

which is consistent with the fact that an ideal inductor does not absorb or deliver real power in the steady state.

4.5 αβ-FRAME REPRESENTATION AND CONTROL OF THREE-PHASE SIGNALS AND SYSTEMS

Previous sections introduced the concept of space phasor for representation of three-phase signals, and also demonstrated that a symmetrical three-phase system can be described by a set of space-phasor equations. It was noted that a space phasor is a complex-valued function of time that can be conveniently expressed in the polar coordinate system. Such a representation is particularly useful when dynamics of amplitude and phase of the system variables are of interest. However, for control design and implementation purposes, it is preferred to map space phasors and space-phasor equations in the Cartesian coordinate system where one deals with real-valued functions of time. Moreover, as explained in Section 4.3.3, an asymmetrical three-phase system cannot be directly expressed in the space-phasor domain. Hence, in this section we introduce the mapping of a space phasor onto the Cartesian coordinate system, which is commonly referred to as αβ-*frame* in the technical literature.

4.5.1 αβ-Frame Representation of a Space Phasor

Consider the space phasor

$$\vec{f}(t) = \frac{2}{3}\left[e^{j0}f_a(t) + e^{j\frac{2\pi}{3}}f_b(t) + e^{j\frac{4\pi}{3}}f_c(t)\right], \tag{4.42}$$

where $f_a + f_b + f_c \equiv 0$. $\vec{f}(t)$ can be decomposed into its real and imaginary components as

$$\vec{f}(t) = f_\alpha(t) + jf_\beta(t), \tag{4.43}$$

where f_α and f_β are referred to as α- and β-axis components of $\overrightarrow{f}(t)$, respectively. Substituting for $\overrightarrow{f}(t)$ from (4.43) in (4.42) and equating the corresponding real and imaginary parts of both sides of the resultant, we deduce

$$\begin{bmatrix} f_\alpha(t) \\ f_\beta(t) \end{bmatrix} = \frac{2}{3} \mathbf{C} \begin{bmatrix} f_a(t) \\ f_b(t) \\ f_c(t) \end{bmatrix}, \tag{4.44}$$

where

$$\mathbf{C} = \begin{bmatrix} 1 & -\frac{1}{2} & -\frac{1}{2} \\ 0 & \frac{\sqrt{3}}{2} & -\frac{\sqrt{3}}{2} \end{bmatrix}. \tag{4.45}$$

Equation (4.44) can be graphically represented by the *abc-frame to αβ-frame signal transformer* of Figure 4.18. The signal transformer of Figure 4.18 is an equivalent of that of Figure 4.2. Based on (4.7), f_{abc} can also be expressed in terms of $f_{\alpha\beta}$ as

$$f_a(t) = Re\left\{ \left[f_\alpha(t) + jf_\beta(t) \right] e^{-j0} \right\} = f_\alpha(t),$$

$$f_b(t) = Re\left\{ \left[f_\alpha(t) + jf_\beta(t) \right] e^{-j\frac{2\pi}{3}} \right\} = -\frac{1}{2} f_\alpha(t) + \frac{\sqrt{3}}{2} f_\beta(t),$$

$$f_c(t) = Re\left\{ \left[f_\alpha(t) + jf_\beta(t) \right] e^{-j\frac{4\pi}{3}} \right\} = -\frac{1}{2} f_\alpha(t) - \frac{\sqrt{3}}{2} f_\beta(t). \tag{4.46}$$

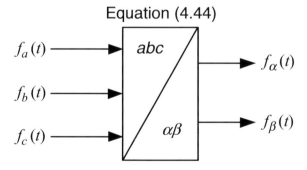

FIGURE 4.18   The *abc*-frame to *αβ*-frame signal transformer.

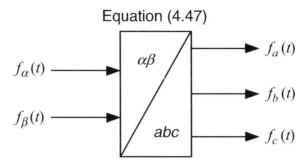

FIGURE 4.19   The $\alpha\beta$-frame to abc-frame signal transformer

Equation (4.46) can be written in the following matrix form:

$$
\begin{bmatrix} f_a(t) \\ f_b(t) \\ f_c(t) \end{bmatrix} = \begin{bmatrix} 1 & 0 \\ -\frac{1}{2} & \frac{\sqrt{3}}{2} \\ -\frac{1}{2} & -\frac{\sqrt{3}}{2} \end{bmatrix} \begin{bmatrix} f_\alpha(t) \\ f_\beta(t) \end{bmatrix} = \mathbf{C}^T \begin{bmatrix} f_\alpha(t) \\ f_\beta(t) \end{bmatrix}, \tag{4.47}
$$

where $\mathbf{C}$ is defined by (4.45) and superscript T denotes matrix transposition. Similarly, (4.47) can be represented by the *$\alpha\beta$-frame to abc-frame signal transformer* of Figure 4.19, which is equivalent to the block diagram of Figure 4.3.

Equations (4.44) and (4.47) introduce the matrix transformations from the abc-frame to the $\alpha\beta$-frame, and vice versa, respectively. Considering Figure 4.1, one can conclude that $f_\alpha(t)$ and $f_\beta(t)$ are the projections of $\overrightarrow{f}(t)$ on the real axis and the imaginary axis, respectively. Thus, we can rename the real and the imaginary axes in Figure 4.1 to α-axis and β-axis, respectively, as shown in Figure 4.20.

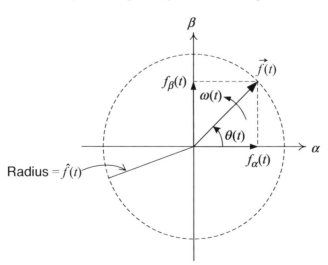

FIGURE 4.20   The $\alpha\beta$-frame components of a space phasor.

The following relationships are deduced from Figure 4.20:

$$\widehat{f}(t) = \sqrt{f_\alpha^2(t) + f_\beta^2(t)}, \tag{4.48}$$

$$\cos[\theta(t)] = \frac{f_\alpha(t)}{\widehat{f}(t)} = \frac{f_\alpha(t)}{\sqrt{f_\alpha^2(t) + f_\beta^2(t)}}, \tag{4.49}$$

$$\sin[\theta(t)] = \frac{f_\beta(t)}{\widehat{f}(t)} = \frac{f_\beta(t)}{\sqrt{f_\alpha^2(t) + f_\beta^2(t)}}. \tag{4.50}$$

Alternatively, (4.49) and (4.50) can be written as

$$f_\alpha(t) = \widehat{f}(t) \cos[\theta(t)], \tag{4.51}$$

$$f_\beta(t) = \widehat{f}(t) \sin[\theta(t)], \tag{4.52}$$

that is, f_α and f_β are sinusoidal functions of time with the amplitude $\widehat{f}$ and the frequency $\omega = d\theta/dt$.

It can be verified that

$$\frac{2}{3}\mathbf{CC}^T \begin{bmatrix} f_\alpha \\ f_\beta \end{bmatrix} = \begin{bmatrix} 1 & 0 \\ 0 & 1 \end{bmatrix} \begin{bmatrix} f_\alpha \\ f_\beta \end{bmatrix} = \begin{bmatrix} f_\alpha \\ f_\beta \end{bmatrix}. \tag{4.53}$$

Equation (4.53) can be used in conjunction with (4.47) to derive the $\alpha\beta$-frame dynamic equations of a three-phase system from the abc-frame equations; the procedure is illustrated in Section 4.5.5. It can also be shown that

$$\frac{2}{3}\mathbf{C}^T\mathbf{C} \begin{bmatrix} f_a \\ f_b \\ f_c \end{bmatrix} = \frac{2}{3} \underbrace{\begin{bmatrix} 1 & -\frac{1}{2} & -\frac{1}{2} \\ -\frac{1}{2} & 1 & -\frac{1}{2} \\ -\frac{1}{2} & -\frac{1}{2} & 1 \end{bmatrix} \begin{bmatrix} f_a \\ f_b \\ f_c \end{bmatrix}}_{f_a + f_b + f_c \equiv 0} = \begin{bmatrix} f_a \\ f_b \\ f_c \end{bmatrix}. \tag{4.54}$$

Equations (4.54) and (4.44) can be used to transform a set of equations in the $\alpha\beta$-frame into the corresponding sets of abc-frame equations.

4.5.2 Realization of Signal Generators/Conditioners in $\alpha\beta$-Frame

For implementation purposes, the systems and signals have to be expressed in terms of real-valued functions, that is, in $\alpha\beta$-frame. In this section, we seek equivalent block diagrams for (i) the space-phasor phase-shifter/scaler of Section 4.2.2 and (ii) the three-phase signal generator of Section 4.2.3.

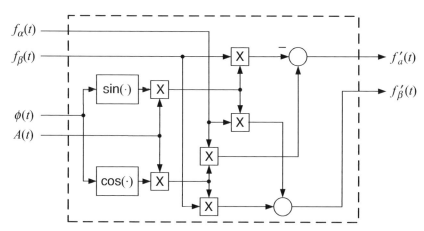

FIGURE 4.21   Block diagram of a space-phasor phase-shifter/scaler in αβ-frame, equivalent to Figure 4.4.

4.5.2.1 Space-Phasor Phase-Shifter/Scaler in αβ-Frame

Section 4.2.2 introduced the space-phasor phase-shifter/scalers of Figure 4.4 and 4.5 and discussed their applications. To find their corresponding equivalent blocks in αβ-frame, we use the Euler's identity, $e^{j(\cdot)} = \cos(\cdot) + j\sin(\cdot)$, and deduce

$$\begin{bmatrix} f'_\alpha(t) \\ f'_\beta(t) \end{bmatrix} = A(t) \begin{bmatrix} \cos\phi(t) & -\sin\phi(t) \\ \sin\phi(t) & \cos\phi(t) \end{bmatrix} \begin{bmatrix} f_\alpha(t) \\ f_\beta(t) \end{bmatrix}. \tag{4.55}$$

In the αβ-frame, a phase-shifter/scaler equivalent to that of Figure 4.4 can be realized based on (4.55), as illustrated in Figure 4.21. The phase-sihfter/scaler of Figure 4.21 can further be augmented with an *abc-* to αβ-frame signal transformer at the input, and an αβ- to *abc*-frame signal transformer at the output, as shown in Figure 4.22. Thus, the space-phasor phase-shifter/scaler of Figure 4.22 is equivalent to that of Figure 4.5.

4.5.2.2 Three-Phase Signal Generator in αβ-Frame

Section 4.2.3 introduced the controllable-frequency/amplitude three-phase signal generator of Figure 4.10 and discussed its application. To find an equivalent signal generator in the αβ-frame, the Euler's identity $e^{j(\cdot)} = \cos(\cdot) + j\sin(\cdot)$ is used again. The block diagram of the signal generator in the αβ-frame is shown in Figure 4.23.

4.5.3 Formulation of Power in αβ-Frame

Section 4.4 presented the space-phasor expressions for the instantaneous real- and reactive-power components. To obtain equivalent expressions in terms of αβ-frame variables, we substitute in (4.38) and (4.40) for $\overrightarrow{v}(t) = v_\alpha + jv_\beta$ and $\overrightarrow{i}^*(t) =$

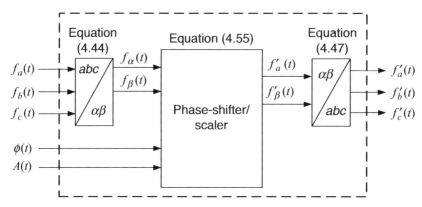

FIGURE 4.22   Block diagram of a space-phasor phase-shfter/scaler in $\alpha\beta$-frame, equivalent to Figure 4.5

$i_\alpha - ji_\beta$, and obtain

$$P(t) = \frac{3}{2}\left[v_\alpha(t)i_\alpha(t) + v_\beta(t)i_\beta(t)\right] \qquad (4.56)$$

and

$$Q(t) = \frac{3}{2}\left[-v_\alpha(t)i_\beta(t) + v_\beta(t)i_\alpha(t)\right]. \qquad (4.57)$$

4.5.4   Control in $\alpha\beta$-Frame

Figure 4.24 illustrates the generic control block diagram of a three-phase VSC system in $\alpha\beta$-frame. The control plant may consist of three-phase electric machines, VSCs,

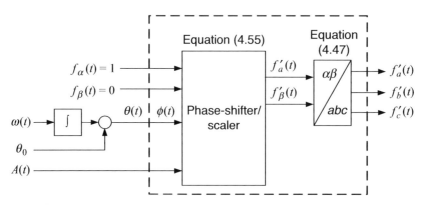

FIGURE 4.23   Block diagram of a controllable-frequency/amplitude three-phase signal generator in $\alpha\beta$-frame, equivalent to Figure 4.10.

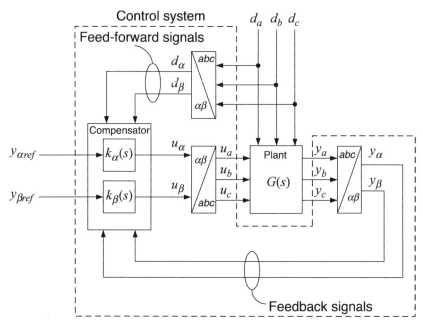

FIGURE 4.24   Block diagram of the $\alpha\beta$-frame control scheme of a typical three-phase system.

interface reactors, transformers, harmonic filters, loads, and sources. The control input u_{abc} and the output y_{abc} are identified on the basis of the application and control/operation requirements. Commonly, u_{abc} and y_{abc} are the three-phase PWM modulating signal and a three-phase voltage or current, respectively. The input–output relationship is assumed to be described by a matrix transfer function $G(s)$[5] and the objective is to control y_{abc} by u_{abc}, in the presence of the disturbance input d_{abc}.

To exercise the control in $\alpha\beta$-frame, y_{abc} is measured and transformed to equivalent $\alpha\beta$-frame signals. Dynamics of the transducers, if not negligible, can also be included in $G(s)$. Moreover, if the disturbance input is measurable, it can be transformed to the $\alpha\beta$-frame and taken into account as a feed-forward signal. Based on the reference, feedback, and feed-forward signals, a set of compensators generate the control signals u_α and u_β that are transformed back to u_{abc} and delivered to the actual three-phase control plant.

To design the compensators, the plant dynamics must be formulated in $\alpha\beta$-frame, as illustrated in Figure 4.25. The $\alpha\beta$-frame description of a symmetrical plant is provided by (4.58) and (4.59), and the equations represent two decoupled subsystems that can be independently controlled. Thus, as shown in Figure 4.24, two decoupled compensators, $k_\alpha(s)$ and $k_\beta(s)$, are used to control the α- and β-axis subsystems.

The $\alpha\beta$-frame description of an asymmetrical system represents a multi-input–multi-output (MIMO) plant in which the α- and β-axis subsystems are coupled.

---

[5]Without loss of generality, a linear control plant has been assumed here.

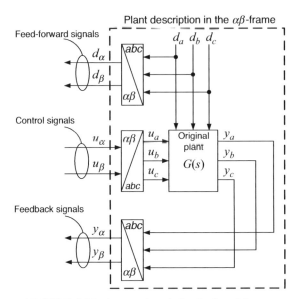

FIGURE 4.25   System description in the $\alpha\beta$-frame.

Consequently, in general, the compensators cannot be split, and a MIMO compensator is inevitable. However, under special conditions, the α- and β-axes dynamics may be decoupled by using appropriate feed-forward techniques. A number of these techniques are discussed in the subsequent chapters. Regardless of the plant properties, the compensator design task must take into consideration (i) the stability of each subsystem, (ii) disturbance rejection capability, and (iii) capability to track the reference commands $y_{\alpha ref}$ and $y_{\beta ref}$ with small errors.

It should be noted that the control system of Figure 4.24 is usually used in conjunction with an outer control loop, in a nested loop structure, to control other variables. For example, in a grid-connected VSC system where $y_{abc} = i_{abc}$ is the VSC current, $i_{\alpha ref}$ and $i_{\beta ref}$ are determined according to (4.56) and (4.57) to control the required real- and reactive-power components exchanged with the AC side.

In should be noted that $y_{\alpha ref}$ and $y_{\beta ref}$, in general, are sinusoidal functions of time. Therefore, to ensure satisfactory tracking performance, either the closed-loop control system must have an adequately large bandwidth or the compensators must include complex-conjugate poles at the frequency of the commands. Consequently, the control system of Figure 4.24 is not the best candidate for variable-frequency applications. For such applications, a dq-frame-based control is usually adopted. The dq-frame-based modeling and control are discussed in Section 4.6.

4.5.5   Representation of Systems in $\alpha\beta$-Frame

The $\alpha\beta$-frame control of a three-phase VSC system, based on the generic block diagram of Figure 4.24, requires that the control plant equations expressing the outputs in terms of the control and disturbance inputs, which is the model of the control plant, be developed in the $\alpha\beta$-frame to enable the design and optimization of the compensators. The following section describes the modeling procedures.

4.5.5.1 Symmetrical Three-Phase Systems As discussed in Section 4.3.1, the space-phasor equations of a decoupled symmetrical three-phase system can be directly derived from the *abc*-frame equations corresponding to only one phase of the system. It was further demonstrated in Section 4.3.2 that, even if the symmetrical three-phase system is coupled, an equivalent decoupled system representation can be deduced for it. Therefore, in $\alpha\beta$-frame terms, it can be asserted that in a symmetrical three-phase system the α- and β-axis components of the output space phasor are decoupled and controlled by α- and β-frame components of the input space phasor, respectively. Hence, the original three-phase system can be considered as two decoupled subsystems, that is, the α-axis and β-axis subsystems, while the transfer functions of both subsystems are the same. To verify this, let us consider (4.23) in which we substitute for $\overrightarrow{y} = y_\alpha + jy_\beta$ and $\overrightarrow{u} = u_\alpha + ju_\beta$, and split the real and imaginary parts as

$$\frac{d^n}{dt^n} y_\alpha + l_{n-1}\frac{d^{n-1}}{dt^{n-1}} y_\alpha + \cdots + l_0 y_\alpha = k_m \frac{d^m}{dt^m} u_\alpha + k_{m-1}\frac{d^{m-1}}{dt^{m-1}} u_\alpha + \cdots + k_0 u_\alpha,$$

(4.58)

$$\frac{d^n}{dt^n} y_\beta + l_{n-1}\frac{d^{n-1}}{dt^{n-1}} y_\beta + \cdots + l_0 y_\beta = k_m \frac{d^m}{dt^m} u_\beta + k_{m-1}\frac{d^{m-1}}{dt^{m-1}} u_\beta + \cdots + k_0 u_\beta.$$

(4.59)

Equations (4.58) and (4.59) indicate that a balanced three-phase linear system, which is described by three dynamic equations, can be equivalently described by two dynamic equations. Based on (4.58) and (4.59), one can control u_α and u_β to control the three-phase system. This indicates that in the $\alpha\beta$-frame we have to implement only two controllers instead of three, that is, one for the α-axis subsystem, (4.58), and the other for the β-axis subsystem, (4.59). The control requires transformation of the feedback signals from the *abc*-frame to the $\alpha\beta$-frame, and transformation of the control signals back from the $\alpha\beta$-frame to the *abc*-frame. However, these transformations are algebraic and thus of low computational burden.

4.5.5.2 Asymmetrical Three-Phase Systems As discussed in Section 4.3.3, no space-phasor representation can be developed for an asymmetrical three-phase system. In an asymmetrical system, the α- and β-axis components of the output space phasor are coupled to both components of the input space phasor. Consequently, the α- and β-axis subsystems are coupled, and the system represented in $\alpha\beta$-frame is a MIMO system. Moreover, the transfer functions of the two subsystems are different. Consequently, the control design task for an asymmetrical three-phase system is more involved than that for a symmetrical system.

EXAMPLE 4.7   Space-Phasor Equations of Three Mutually Coupled Inductors

Consider an asymmetrical three-phase system governed by the following matrix equation:

$$
\begin{bmatrix} v_a \\ v_b \\ v_c \end{bmatrix} = \begin{bmatrix} L_s & M_{ab} & M_{ac} \\ M_{ab} & L_s & M_{bc} \\ M_{ac} & M_{bc} & L_s \end{bmatrix} \frac{d}{dt} \begin{bmatrix} i_a \\ i_b \\ i_c \end{bmatrix}, \tag{4.60}
$$

which represents three mutually coupled inductors with unequal mutual inductances. To derive the $\alpha\beta$-frame equations, based on (4.47), the abc-frame voltage and current vectors are expressed in $\alpha\beta$-frame. Hence,

$$
\mathbf{C}^T \begin{bmatrix} v_\alpha \\ v_\beta \end{bmatrix} = \begin{bmatrix} L_s & M_{ab} & M_{ac} \\ M_{ab} & L_s & M_{bc} \\ M_{ac} & M_{bc} & L_s \end{bmatrix} \frac{d}{dt} \left(\mathbf{C}^T \begin{bmatrix} i_\alpha \\ i_\beta \end{bmatrix} \right)
$$

$$
= \begin{bmatrix} L_s & M_{ab} & M_{ac} \\ M_{ab} & L_s & M_{bc} \\ M_{ac} & M_{bc} & L_s \end{bmatrix} \mathbf{C}^T \frac{d}{dt} \begin{bmatrix} i_\alpha \\ i_\beta \end{bmatrix}, \tag{4.61}
$$

where $\mathbf{C}$ is given by (4.45). Premultiplying both sides of (4.61) by $(2/3)\mathbf{C}$ and using the identity (4.53) at the left-hand side of the resultant, we deduce

$$
\begin{bmatrix} v_\alpha \\ v_\beta \end{bmatrix} = \frac{2}{3}\mathbf{C} \begin{bmatrix} L_s & M_{ab} & M_{ac} \\ M_{ab} & L_s & M_{bc} \\ M_{ac} & M_{bc} & L_s \end{bmatrix} \mathbf{C}^T \frac{d}{dt} \begin{bmatrix} i_\alpha \\ i_\beta \end{bmatrix}. \tag{4.62}
$$

Substituting in (4.62) for C and C^T, based on (4.45), one deduces

$$
\begin{bmatrix} v_\alpha \\ v_\beta \end{bmatrix} = \frac{2}{3} \begin{bmatrix} \frac{3}{2}L_s - (M_{ab} + M_{ac}) + \frac{1}{2}M_{bc} & \frac{\sqrt{3}}{2}(M_{ab} - M_{ac}) \\ \frac{\sqrt{3}}{2}(M_{ab} - M_{ac}) & \frac{3}{2}(L_s - M_{bc}) \end{bmatrix} \frac{d}{dt} \begin{bmatrix} i_\alpha \\ i_\beta \end{bmatrix}. \tag{4.63}
$$

Equation (4.63) indicates that v_α and v_β the functions of both i_α and i_β, through different transfer functions. However, if the mutual inductances are identical and equal to M, the three inductors constitute a symmetrical (coupled) three-phase system, and (4.63) is simplified to

$$
\begin{bmatrix} v_\alpha \\ v_\beta \end{bmatrix} = \begin{bmatrix} (L_s - M) & 0 \\ 0 & (L_s - M) \end{bmatrix} \frac{d}{dt} \begin{bmatrix} i_\alpha \\ i_\beta \end{bmatrix}, \tag{4.64}
$$

which represents two identical decoupled subsystems in $\alpha\beta$-frame. Equation (4.64) is $\alpha\beta$-frame equivalent of the space-phasor equation (4.34), Example 4.5.

4.6 *dq*-FRAME REPRESENTATION AND CONTROL OF THREE-PHASE SYSTEMS

Section 4.5.4 presented the structure of a three-phase control system in $\alpha\beta$-frame (Fig. 4.24), for which the compensators are designed based on the $\alpha\beta$-frame description of the control plant, as shown in Figure 4.25. The control in $\alpha\beta$-frame has the feature of reducing the number of required control loops from three to two. However, the reference, feedback, and feed-forward signals are in general sinusoidal functions of time. Therefore, to achieve a satisfactory performance and small steady-state errors, the compensators may need to be of high orders, and the closed-loop bandwidths must be adequately larger than the frequency of the reference commands. Consequently, the compensator design is not a straightforward task, especially if the operating frequency is variable. The *dq*-frame-based control offers a solution to this problem.

In *dq*-frame, the signals assume DC waveforms under steady-state conditions. This, in turn, permits utilization of compensators with simpler structures and lower dynamic orders. Moreover, zero steady-state tracking error can be achieved by including integral terms in the compensators. A *dq*-frame representation of a three-phase system is also more suitable for analysis and control design tasks. For example, the *abc*-frame representation of a salient-pole synchronous machine includes time-varying self- and mutual inductances. However, if the machine equations are transformed to a proper *dq*-frame, the time-varying inductances manifest themselves as constant parameters. The *dq*-frame representation and control of signals and systems are discussed in the following subsections.

4.6.1 *dq*-Frame Representation of a Space Phasor

For the space phasor $\overrightarrow{f} = f_\alpha + jf_\beta$, the $\alpha\beta$- to *dq*-frame transformation is defined by

$$f_d + jf_q = (f_\alpha + jf_\beta)e^{-j\varepsilon(t)}, \tag{4.65}$$

which is equivalent to a phase shift in $\overrightarrow{f}(t)$ by the angle $-\varepsilon(t)$. The *dq*- to $\alpha\beta$-frame transformation can be obtained by multiplying both sides of (4.65) by $e^{j\varepsilon(t)}$. Thus,

$$f_\alpha + jf_\beta = (f_d + jf_q)e^{j\varepsilon(t)}. \tag{4.66}$$

To highlight the usefulness of the transformation given by (4.65), assume that $\overrightarrow{f}$ has the following general form:

$$\overrightarrow{f}(t) = f_\alpha + jf_\beta = \widehat{f}(t)e^{j\left[\theta_0 + \int \omega(\tau)d\tau\right]},$$

where $\omega(t)$ is the (time-varying) frequency and θ_0 is the initial phase angle of the three-phase signal corresponding to $\overrightarrow{f}(t)$. If $\varepsilon(t)$ is chosen as

$$\varepsilon(t) = \varepsilon_0 + \int \omega(\tau)d\tau,$$

then based on (4.65) the dq-frame representation of $\overrightarrow{f}(t)$ becomes

$$f_d + jf_q = \widehat{f}(t)e^{j(\theta_0 - \varepsilon_0)},$$

which is stationary and, therefore, the constituents of its corresponding three-phase signal are DC quantities. Note that $\theta(t)$ and $\varepsilon(t)$ are not necessarily equal, but $d\theta(t)/dt = d\varepsilon(t)/dt$ must be ensured.

To better describe the dq-frame transformation, let us rewrite (4.66) as

$$\overrightarrow{f} = f_d(1 + 0 \cdot j)e^{j\varepsilon(t)} + f_q(0 + 1 \cdot j)e^{j\varepsilon(t)}. \tag{4.67}$$

An interpretation of (4.67) is that the vector $\overrightarrow{f}$ is represented by its components, that is, f_d and f_q, in an orthogonal coordinate system whose axes are along the unit vectors $(1 + 0 \cdot j)e^{j\varepsilon(t)}$ and $(0 + 1 \cdot j)e^{j\varepsilon(t)}$. In turn, $(1 + 0 \cdot j)$ and $(0 + 1 \cdot j)$ are the unit vectors along the α-axis and the β-axis of the $\alpha\beta$-frame, respectively. Therefore, as illustrated in Figure 4.26, one can consider $\overrightarrow{f}$ as a vector represented by the components f_d and f_q in a coordinate system that is rotated by $\varepsilon(t)$ with respect to the $\alpha\beta$-frame. We refer to this rotated coordinate system as a dq-frame. For the reason given above, the dq-frame is also known as *rotating reference frame*, in the technical literature. Usually, the rotational speed of the dq-frame is selected to be equal to that of $\overrightarrow{f}$.

Based on the Euler's identity $e^{j(\cdot)} = \cos(\cdot) + j\sin(\cdot)$, (4.65) can be written as

$$\begin{bmatrix} f_d(t) \\ f_q(t) \end{bmatrix} = \mathbf{R}\left[\varepsilon(t)\right]\begin{bmatrix} f_\alpha(t) \\ f_\beta(t) \end{bmatrix}, \tag{4.68}$$

where

$$\mathbf{R}\left[\varepsilon(t)\right] = \begin{bmatrix} \cos\varepsilon(t) & \sin\varepsilon(t) \\ -\sin\varepsilon(t) & \cos\varepsilon(t) \end{bmatrix}. \tag{4.69}$$

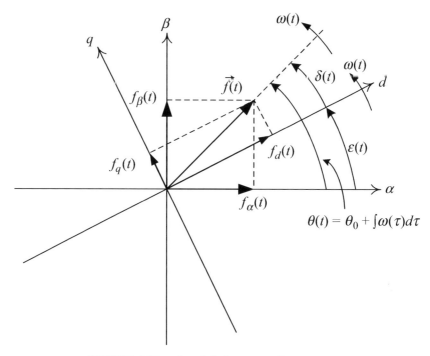

FIGURE 4.26 $\alpha\beta$- and *dq*-frame coordinate systems.

Similarly, the *dq*- to $\alpha\beta$-frame transformation (4.66) can be rewritten as

$$
\begin{bmatrix} f_\alpha(t) \\ f_\beta(t) \end{bmatrix} = \mathbf{R}^{-1}\left[\varepsilon(t)\right] \begin{bmatrix} f_d(t) \\ f_q(t) \end{bmatrix}
$$

$$
= \mathbf{R}\left[-\varepsilon(t)\right] \begin{bmatrix} f_d(t) \\ f_q(t) \end{bmatrix}, \tag{4.70}
$$

where

$$
\mathbf{R}^{-1}\left[\varepsilon(t)\right] = \mathbf{R}\left[-\varepsilon(t)\right] = \begin{bmatrix} \cos\varepsilon(t) & -\sin\varepsilon(t) \\ \sin\varepsilon(t) & \cos\varepsilon(t) \end{bmatrix}. \tag{4.71}
$$

It can also be verified that

$$
\mathbf{R}^{-1}\left[\varepsilon(t)\right] = \mathbf{R}^T\left[\varepsilon(t)\right]. \tag{4.72}
$$

A direct transformation from the abc-frame to the dq-frame can be obtained by substitution of $[f_\alpha \ f_\beta]^T$ from (4.44) in (4.68), as

$$\begin{bmatrix} f_d(t) \\ f_q(t) \end{bmatrix} = \frac{2}{3} \mathbf{T}[\varepsilon(t)] \begin{bmatrix} f_a(t) \\ f_b(t) \\ f_c(t) \end{bmatrix}, \tag{4.73}$$

where

$$\mathbf{T}[\varepsilon(t)] = \mathbf{R}[\varepsilon(t)]\mathbf{C} = \begin{bmatrix} \cos[\varepsilon(t)] & \cos\left[\varepsilon(t) - \frac{2\pi}{3}\right] & \cos\left[\varepsilon(t) - \frac{4\pi}{3}\right] \\ \sin[\varepsilon(t)] & \sin\left[\varepsilon(t) - \frac{2\pi}{3}\right] & \sin\left[\varepsilon(t) - \frac{4\pi}{3}\right] \end{bmatrix}. \tag{4.74}$$

Similarly, a direct transformation from the dq-frame to the abc-frame can be obtained by substituting for $[f_\alpha \ f_\beta]^T$ from (4.70) in (4.47) as

$$\begin{bmatrix} f_a(t) \\ f_b(t) \\ f_c(t) \end{bmatrix} = \mathbf{T}[\varepsilon(t)]^T \begin{bmatrix} f_d(t) \\ f_q(t) \end{bmatrix}, \tag{4.75}$$

where

$$\mathbf{T}[\varepsilon(t)]^T = \mathbf{C}^T \mathbf{R}[-\varepsilon(t)] = \begin{bmatrix} \cos[\varepsilon(t)] & \sin[\varepsilon(t)] \\ \cos\left[\varepsilon(t) - \frac{2\pi}{3}\right] & \sin\left[\varepsilon(t) - \frac{2\pi}{3}\right] \\ \cos\left[\varepsilon(t) - \frac{4\pi}{3}\right] & \sin\left[\varepsilon(t) - \frac{4\pi}{3}\right] \end{bmatrix}. \tag{4.76}$$

Based on Figure 4.26, one deduces

$$\widehat{f}(t) = \sqrt{f_d^2(t) + f_q^2(t)}, \tag{4.77}$$

$$\cos[\delta(t)] = \frac{f_d(t)}{\widehat{f}(t)} = \frac{f_d(t)}{\sqrt{f_d^2(t) + f_q^2(t)}}, \tag{4.78}$$

$$\sin[\delta(t)] = \frac{f_q(t)}{\widehat{f}(t)} = \frac{f_q(t)}{\sqrt{f_d^2(t) + f_q^2(t)}}, \tag{4.79}$$

$$\theta(t) = \varepsilon(t) + \delta(t). \tag{4.80}$$

It can be verified that

$$\frac{2}{3}\mathbf{T}\mathbf{T}^T \begin{bmatrix} f_d \\ f_q \end{bmatrix} = \begin{bmatrix} 1 & 0 \\ 0 & 1 \end{bmatrix} \begin{bmatrix} f_d \\ f_q \end{bmatrix} = \begin{bmatrix} f_d \\ f_q \end{bmatrix}. \tag{4.81}$$

Equation (4.81) can be used in conjunction with (4.75) to formulate *dq*-frame equations for a three-phase system, based on *abc*-frame equations. It can also be shown that

$$\frac{2}{3}\mathbf{T}^T\mathbf{T} \begin{bmatrix} f_a \\ f_b \\ f_c \end{bmatrix} = \underbrace{\frac{2}{3} \begin{bmatrix} 1 & -\frac{1}{2} & -\frac{1}{2} \\ -\frac{1}{2} & 1 & -\frac{1}{2} \\ -\frac{1}{2} & -\frac{1}{2} & 1 \end{bmatrix} \begin{bmatrix} f_a \\ f_b \\ f_c \end{bmatrix}}_{f_a + f_b + f_c \equiv 0} = \begin{bmatrix} f_a \\ f_b \\ f_c \end{bmatrix}. \tag{4.82}$$

Equations (4.82) and (4.73) can be used to transform a set of *dq*-frame equations to an equivalent set of equations in *abc*-frame.

4.6.2   Formulation of Power in *dq*-Frame

Section 4.4 presented the space-phasor expressions for the instantaneous real- and reactive-power components. In this section, we derive analogous expressions in terms of *dq*-frame variables. Based on (4.66), we substitute for $\overrightarrow{v}(t) = (v_d + jv_q)e^{j\varepsilon(t)}$ and $\overrightarrow{i}^*(t) = (i_d - ji_q)e^{-j\varepsilon(t)}$ in (4.38) and (4.40), and deduce

$$P(t) = \frac{3}{2} \left[v_d(t)i_d(t) + v_q(t)i_q(t) \right] \tag{4.83}$$

and

$$Q(t) = \frac{3}{2} \left[-v_d(t)i_q(t) + v_q(t)i_d(t) \right]. \tag{4.84}$$

Equations (4.83) and (4.84) suggest that if $v_q = 0$, the real- and reactive-power components are proportional to i_d and i_q, respectively. This property is widely employed in the control of grid-connected three-phase VSC systems, as discussed in Chapter 8.

4.6.3   Control in *dq*-Frame

Figure 4.27 illustrates the generic control block diagram of a three-phase VSC system in *dq*-frame. The control system of Figure 4.27 is an extension of its $\alpha\beta$-frame counterpart, that of Figure 4.24, in which the feedback, feed-forward, and control signals are, in general, sinusoidal functions of time. Therefore, to enable the compensators to process DC rather than sinusoidal signals, we alter the control system of Figure 4.24

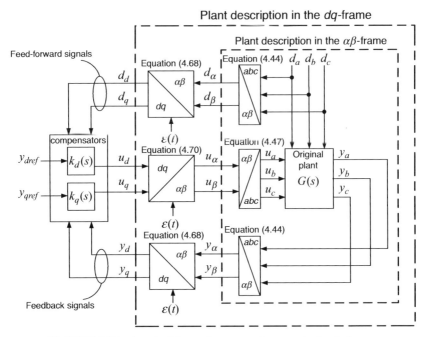

FIGURE 4.27   A typical three-phase control system in the *dq*-frame.

by cascading each *abc*- to *αβ*-frame signal transformer by an *αβ*- to *dq*-frame signal transformer. Thus, the compensators process (DC quantities) d_{dq} and y_{dq}, and provide the (DC) control signals u_{dq}. Finally, u_{dq} is transformed back to the *αβ*-frame by means of a *dq*- to *αβ*-frame signal transformer. Hence, the equivalent control plant in *dq*-frame has the inputs u_{dq}, the disturbances d_{dq}, and the outputs y_{dq}.

In the control system of Figure 4.27, $\varepsilon(t)$ represents the angle for the *αβ*/*dq*-frame transformations. In general, $\varepsilon(t) = \varepsilon_0 + \int \omega(\tau)d\tau$, where $\omega(t)$ is the frequency and ε_0 is a constant. In the special case of a constant-frequency VSC system, $\omega(t)$ is equal to the AC system operating frequency, for example, ω_0, and $\varepsilon(t) = \varepsilon_0 + \omega_0 t$. There are strict choices for $\varepsilon(t)$ in some specific applications. For example,

- In a grid-connected VSC system, the real- and reactive-power components exchanged between the VSC and the grid become proportional to the *d*- and *q*-axis components of the converter current, respectively, if $\varepsilon(t)$ is made equal to the angle of the grid voltage space phasor. As discussed in Chapter 8, the angle of the grid voltage is estimated and provided by a phase-locked loop (PLL) [49]. In this case, the reference commands i_{dref} and i_{qref} are the outputs of the real- and reactive-power compensators, respectively.

- In an asynchronous machine drive, the goal is to regulate the machine flux at a constant value and to control the torque dynamically. In *dq*-frame, the flux and torque are nonlinear functions of both the d- and q-axis components of the stator current, that is, i_{sd} and i_{sq}. However, as discussed in Chapter 10, if $\varepsilon(t)$ is equal to the angle of the rotor flux space phasor, the flux and torque become decoupled and proportional to i_{sd} and i_{sq}, respectively. The rotor flux is not measurable and, consequently, its angle is obtained based on estimation techniques [43]. The reference commands, that is, i_{sdref} and i_{sqref}, are the outputs of the flux and torque control loops, respectively. This control method is referred to as *vector control* or *field-oriented control*, in the technical literature.

- The vector-control strategy is also applicable to a synchronous machine [43]. However, in the case of a synchronous machine, to decouple the machine flux and torque, $\varepsilon(t)$ is chosen to be equal to the rotor electrical angle, which can be obtained from a shaft encoder or through an estimation process [50–52]. Similar to the case of an asynchronous machine drive, the machine flux and torque become proportional to i_{sd} and i_{sq}, respectively.

4.6.4   Representation of Systems in *dq*-Frame

The *dq*-frame control of a three-phase VSC system, based on the generic block diagram of Figure 4.27, requires that the control plant equations expressing the outputs in terms of control and disturbance inputs, that is, the model of the control plant, be developed in *dq*-frame, to enable the design and optimization of the compensators. The next section describes the modeling procedures.

4.6.4.1   *Derivation of dq-Frame Model from Space-Phasor Equations*   As discussed in Section 4.3, the space-phasor equations of symmetrical three-phase systems can be developed by inspection of *abc*-frame equations. Most VSC systems treated in this book belong to this class of symmetrical systems. Given the space-phasor equations, a *dq*-frame model can be developed based on the following steps:

- In the space-phasor equations, $\overrightarrow{d}$, $\overrightarrow{y}$, and $\overrightarrow{u}$ are, respectively, substituted by $\overrightarrow{d} = (d_d + jd_q)e^{j\varepsilon(t)}$, $\overrightarrow{y} = (y_d + jy_q)e^{j\varepsilon(t)}$, and $\overrightarrow{u} = (u_d + ju_q)e^{j\varepsilon(t)}$.
- If encountered, the derivative $d\varepsilon/dt$ in the equations is replaced by ω.
- The resultant set of equations is rearranged in terms of $d_d + jd_q$, $y_d + jy_q$, and $u_d + ju_q$.
- The equations are decomposed into the real and imaginary components.
- $\varepsilon(t)$ is defined as a new state variable for which the state-space equation is $d\varepsilon/dt = \omega(t)$, where $\omega(t)$ is a new control input. Note that this step is not necessary if $\omega(t)$ is a constant parameter.

Examples 4.8 and 4.9 illustrate the steps.

EXAMPLE 4.8   dq-Frame Model of the Three-Phase System of Example 4.4

Consider the circuit of Figure 4.14 for which the space-phasor equations are given by (4.30) and (4.31). Substituting for $\overrightarrow{v} = v_{dq}e^{j\varepsilon(t)}$, $\overrightarrow{i} = i_{dq}e^{j\varepsilon(t)}$, and $\overrightarrow{i_s} = i_{sdq}e^{j\varepsilon(t)}$ in (4.30), we obtain

$$C \frac{d}{dt}\left(v_{dq}e^{j\varepsilon(t)}\right) = \left(i_{dq}e^{j\varepsilon(t)}\right) - \left(i_{sdq}e^{j\varepsilon(t)}\right), \tag{4.85}$$

where $f_{dq} = f_d + jf_q$. Equation (4.85) can be rewritten as

$$\left(C\frac{dv_{dq}}{dt}\right)e^{j\varepsilon(t)} + \left(j\omega v_{dq}\right)e^{j\varepsilon(t)} = \left(i_{dq}\right)e^{j\varepsilon(t)} - \left(i_{sdq}\right)e^{j\varepsilon(t)}, \tag{4.86}$$

where

$$\frac{d\varepsilon}{dt} = \omega(t). \tag{4.87}$$

Eliminating $e^{j\varepsilon(t)}$ from both sides of (4.86) and decomposing the resultant into the real and imaginary parts, we conclude that

$$C\frac{dv_d}{dt} = C\omega(t)v_q + i_d - i_{sd}, \tag{4.88}$$

$$C\frac{dv_q}{dt} = -C\omega(t)v_d + i_q - i_{sq}. \tag{4.89}$$

Following a similar procedure for (4.31), we obtain

$$L_s\frac{di_{sd}}{dt} = L_s\omega(t)i_{sq} + v_d - v_{sd}, \tag{4.90}$$

$$L_s\frac{di_{sq}}{dt} = -L_s\omega(t)i_{sd} + v_q - v_{sq}. \tag{4.91}$$

Equations (4.87)–(4.91) constitute a dq-frame model for the circuit of Figure 4.14.

EXAMPLE 4.9   dq-Frame Model of a Second-Order Three-Phase System

Assume that the dynamics of a balanced three-phase system are described by the following space-phasor equation:

$$\frac{d^2}{dt^2}\overrightarrow{y} = \overrightarrow{u}. \tag{4.92}$$

Let $\vec{y} = y_{dq}e^{j\varepsilon(t)}$ and $\vec{u} = u_{dq}e^{j\varepsilon(t)}$. Thus, we obtain

$$\frac{d^2 y_{dq}}{dt^2} + j\left(2\omega\frac{dy_{dq}}{dt}\right) + \left(j\dot{\xi} - \omega^2\right)y_{dq} = u_{dq}, \tag{4.93}$$

where

$$\frac{d\varepsilon}{dt} = \omega(t), \tag{4.94}$$

$$\frac{d\omega}{dt} = \xi(t). \tag{4.95}$$

Decomposing (4.93) into real and imaginary parts, we obtain

$$\frac{d^2 y_d}{dt^2} - 2\omega\frac{dy_q}{dt} - \omega^2 y_d - \xi y_q = u_d, \tag{4.96}$$

$$\frac{d^2 y_q}{dt^2} + 2\omega\frac{dy_d}{dt} + \xi y_d - \omega^2 y_q = u_q. \tag{4.97}$$

Equations (4.94)–(4.97) represent the dynamics of the system of (4.92) in *dq*-frame. Based on (4.94)–(4.97), $\varepsilon(t)$ and $\omega(t)$ are two new state variables, and $\xi(t)$ is a new control variable. Note that although the original three-phase system is linear, its description in the *dq*-frame becomes nonlinear if the frequency, ω, is not constant.

4.6.4.2 *Derivation of dq-Frame Model from abc-Frame Equations*   As explained in Section 4.3, the model of an asymmetrical system cannot be expressed in the space-phasor form. Rather, an asymmetrical system is usually modeled in either the $\alpha\beta$-frame or the *dq*-frame. If the equations of the physical system are provided in *abc*-frame, they can be expressed in $\alpha\beta$-frame using the approach presented in Example 4.7 or Example 4.10. Then, to develop the *dq*-frame model, the α- and β-axis variables can be eliminated from the $\alpha\beta$-frame model using (4.70). If the derivative of the *dq*-frame angle ε appears in the equations, it is replaced by $\omega = d\varepsilon/dt$. Thus, the new state equation $d\varepsilon/dt = \omega$ is introduced to the system equations. The next example illustrates these steps.

EXAMPLE 4.10   *dq*-Frame Model of Permanent-Magnet Synchronous Machine

Based on the model of a conventional salient-pole synchronous machine (SM) [53], equations of a permanent-magnet synchronous machine (PMSM) can be derived by introducing the following changes to those of the SM: (i) the stator flux equations are preserved, but the equations describing the damper windings fluxes are eliminated, (ii) the damper winding currents are replaced by zero, and (iii) the field current is replaced by a constant parameter corresponding to the permanent magnet flux. This procedure yields[6]

$$
\begin{bmatrix} \lambda_a \\ \lambda_b \\ \lambda_c \end{bmatrix} = \mathbf{L} \begin{bmatrix} i_a \\ i_b \\ i_c \end{bmatrix} + \begin{bmatrix} \lambda_m \cos \theta_r \\ \lambda_m \cos \left(\theta_r - \frac{2\pi}{3} \right) \\ \lambda_m \cos \left(\theta_r - \frac{4\pi}{3} \right) \end{bmatrix},
\qquad (4.98)
$$

where λ_{abc}, i_{abc}, and θ_r are the stator flux, the stator current, and the rotor angle, respectively. λ_m represents the maximum flux generated by the rotor magnet, linked by a stator winding. $\mathbf{L}$, the inductance matrix, is defined as

$$
\mathbf{L} =
$$

$$
\frac{2}{3}
\begin{bmatrix}
a \cos 2\theta_r + b & a \cos 2 \left(\theta_r - \frac{\pi}{3} \right) - \frac{b}{2} & a \cos 2 \left(\theta_r - \frac{2\pi}{3} \right) - \frac{b}{2} \\
a \cos 2 \left(\theta_r - \frac{\pi}{3} \right) - \frac{b}{2} & a \cos 2 \left(\theta_r - \frac{2\pi}{3} \right) + b & a \cos 2\theta_r - \frac{b}{2} \\
a \cos 2 \left(\theta_r - \frac{2\pi}{3} \right) - \frac{b}{2} & a \cos 2\theta_r - \frac{b}{2} & a \cos 2 \left(\theta_r - \frac{4\pi}{3} \right) + b
\end{bmatrix},
$$

$$
(4.99)
$$

where $a = (L_d - L_q)/2$ and $b = (L_d + L_q)/2$. In turn, L_d and L_q are two constant inductance parameters that depend on the machine construction and geometry. Equations (4.98) and (4.99) indicate that each stator winding exhibits a variable self-inductance in addition to variable mutual inductances with the two other stator windings. The stator flux and terminal voltage are, in turn, related by the following state-space equations:

$$
\frac{d}{dt} \begin{bmatrix} \lambda_a \\ \lambda_b \\ \lambda_c \end{bmatrix} = \begin{bmatrix} -R_s & 0 & 0 \\ 0 & -R_s & 0 \\ 0 & 0 & -R_s \end{bmatrix} \begin{bmatrix} i_a \\ i_b \\ i_c \end{bmatrix} + \begin{bmatrix} v_{sa} \\ v_{sb} \\ v_{sc} \end{bmatrix} - \begin{bmatrix} v_n \\ v_n \\ v_n \end{bmatrix}, \quad (4.100)
$$

where R_s is the stator winding resistance and V_n is the voltage of the stator neutral point. Equations (4.98)–(4.100) describe the PMSM in the *abc*-frame. The model approximately represents a PMSM with interior (buried) magnets where the magnets are mounted inside a steel rotor core [54]. We express these equations first in $\alpha\beta$-frame and then in a selected *dq*-frame.

---

[6]Equations of a simplified, nonsalient-rotor PMSM are derived in Section A.5.

$\alpha\beta$-frame representation

To transform (4.98) to the $\alpha\beta$-frame, each *abc*-frame vector is expressed in the $\alpha\beta$-frame based on (4.47). Thus,

$$\mathbf{C}^T \begin{bmatrix} \lambda_\alpha \\ \lambda_\beta \end{bmatrix} = \mathbf{L}\mathbf{C}^T \begin{bmatrix} i_\alpha \\ i_\beta \end{bmatrix} + \begin{bmatrix} \lambda_m \cos\theta_r \\ \lambda_m \cos\left(\theta_r - \frac{2\pi}{3}\right) \\ \lambda_m \cos\left(\theta_r - \frac{4\pi}{3}\right) \end{bmatrix}. \tag{4.101}$$

Premultiplying both sides of (4.101) by $(2/3)\mathbf{C}$ and using the identity (4.53) at the left-hand side of the resultant, we deduce

$$\begin{bmatrix} \lambda_\alpha \\ \lambda_\beta \end{bmatrix} = \frac{2}{3}\mathbf{C}\mathbf{L}\mathbf{C}^T \begin{bmatrix} i_\alpha \\ i_\beta \end{bmatrix} + \frac{2}{3}\mathbf{C} \begin{bmatrix} \lambda_m \cos\theta_r \\ \lambda_m \cos\left(\theta_r - \frac{2\pi}{3}\right) \\ \lambda_m \cos\left(\theta_r - \frac{4\pi}{3}\right) \end{bmatrix}. \tag{4.102}$$

Substituting for $\mathbf{C}$ and $\mathbf{L}$ in (4.102), respectively from (4.45) and (4.99), we deduce

$$\begin{bmatrix} \lambda_\alpha \\ \lambda_\beta \end{bmatrix} = \begin{bmatrix} \left(\frac{L_d-L_q}{2}\right)\cos 2\theta_r + \left(\frac{L_d+L_q}{2}\right) & \left(\frac{L_d-L_q}{2}\right)\sin 2\theta_r \\ \left(\frac{L_d-L_q}{2}\right)\sin 2\theta_r & -\left(\frac{L_d-L_q}{2}\right)\cos 2\theta_r + \left(\frac{L_d+L_q}{2}\right) \end{bmatrix} \begin{bmatrix} i_\alpha \\ i_\beta \end{bmatrix}$$

$$+ \begin{bmatrix} \lambda_m \cos\theta_r \\ \lambda_m \sin\theta_r \end{bmatrix}. \tag{4.103}$$

Following a similar procedure, the $\alpha\beta$-frame representation of (4.100) is deduced as

$$\frac{d}{dt} \begin{bmatrix} \lambda_\alpha \\ \lambda_\beta \end{bmatrix} = \begin{bmatrix} -R_s & 0 \\ 0 & -R_s \end{bmatrix} \begin{bmatrix} i_\alpha \\ i_\beta \end{bmatrix} + \begin{bmatrix} v_{s\alpha} \\ v_{s\beta} \end{bmatrix}. \tag{4.104}$$

It should be noted that (4.100) represents a symmetrical system. Therefore, (4.104) could also be directly derived by inspection of the *abc*-frame equation of only one phase, as explained in Section 4.5.5.

dq-frame representation

Equation (4.103) in the $\alpha\beta$-frame model of the PMSM includes θ_r as a variable. Therefore, the analysis and control design based on the $\alpha\beta$-frame representation is not a straightforward task. However, transformation of the equations to dq-frame renders them time invariant and more suitable for analysis and control design if the dq-frame is synchronized to the rotor angle θ_r. This corresponds to $\varepsilon = \theta_r$ in (4.70) and (4.71). Thus,

$$
\begin{bmatrix} f_\alpha(t) \\ f_\beta(t) \end{bmatrix} = \mathbf{R}^{-1}[\theta_r] \begin{bmatrix} f_d(t) \\ f_q(t) \end{bmatrix}, \tag{4.105}
$$

where

$$
\mathbf{R}^{-1}[\theta_r] = \begin{bmatrix} \cos\theta_r & -\sin\theta_r \\ \sin\theta_r & \cos\theta_r \end{bmatrix}. \tag{4.106}
$$

Therefore, (4.103) can be rewritten as

$$
\mathbf{R}^{-1} \begin{bmatrix} \lambda_d \\ \lambda_q \end{bmatrix}
$$
$$
= \begin{bmatrix} \left(\frac{L_d-L_q}{2}\right)\cos 2\theta_r + \left(\frac{L_d+L_q}{2}\right) & \left(\frac{L_d-L_q}{2}\right)\sin 2\theta_r \\ \left(\frac{L_d-L_q}{2}\right)\sin 2\theta_r & -\left(\frac{L_d-L_q}{2}\right)\cos 2\theta_r + \left(\frac{L_d+L_q}{2}\right) \end{bmatrix} \mathbf{R}^{-1} \begin{bmatrix} i_d \\ i_q \end{bmatrix}
$$
$$
+ \begin{bmatrix} \lambda_m \cos\theta_r \\ \lambda_m \sin\theta_r \end{bmatrix}. \tag{4.107}
$$

Premultiplying both sides of (4.107) by $\mathbf{R}$, we deduce

$$
\begin{bmatrix} \lambda_d \\ \lambda_q \end{bmatrix}
$$
$$
= \mathbf{R} \begin{bmatrix} \left(\frac{L_d-L_q}{2}\right)\cos 2\theta_r + \left(\frac{L_d+L_q}{2}\right) & \left(\frac{L_d-L_q}{2}\right)\sin 2\theta_r \\ \left(\frac{L_d-L_q}{2}\right)\sin 2\theta_r & -\left(\frac{L_d-L_q}{2}\right)\cos 2\theta_r + \left(\frac{L_d+L_q}{2}\right) \end{bmatrix} \mathbf{R}^{-1} \begin{bmatrix} i_d \\ i_q \end{bmatrix}
$$
$$
+ \mathbf{R} \begin{bmatrix} \lambda_m \cos\theta_r \\ \lambda_m \sin\theta_r \end{bmatrix}, \tag{4.108}
$$

where

$$
\mathbf{R}[\theta_r] = \begin{bmatrix} \cos\theta_r & \sin\theta_r \\ -\sin\theta_r & \cos\theta_r \end{bmatrix}. \tag{4.109}
$$

Equation (4.108) can be further simplified to

$$
\begin{bmatrix} \lambda_d \\ \lambda_q \end{bmatrix} = \begin{bmatrix} L_d & 0 \\ 0 & L_q \end{bmatrix} \begin{bmatrix} i_d \\ i_q \end{bmatrix} + \begin{bmatrix} \lambda_m \\ 0 \end{bmatrix}.
\tag{4.110}
$$

Similarly (4.104) can be rewritten as

$$
\frac{d}{dt}\left\{ \mathbf{R}^{-1} \begin{bmatrix} \lambda_d \\ \lambda_q \end{bmatrix} \right\} = \begin{bmatrix} -R_s & 0 \\ 0 & -R_s \end{bmatrix} \mathbf{R}^{-1} \begin{bmatrix} i_d \\ i_q \end{bmatrix} + \mathbf{R}^{-1} \begin{bmatrix} v_{sd} \\ v_{sq} \end{bmatrix},
\tag{4.111}
$$

which can be rearranged as

$$
\mathbf{R}^{-1}\frac{d}{dt} \begin{bmatrix} \lambda_d \\ \lambda_q \end{bmatrix} + \frac{d\mathbf{R}^{-1}}{dt} \begin{bmatrix} \lambda_d \\ \lambda_q \end{bmatrix} = \begin{bmatrix} -R_s & 0 \\ 0 & -R_s \end{bmatrix} \mathbf{R}^{-1} \begin{bmatrix} i_d \\ i_q \end{bmatrix} + \mathbf{R}^{-1} \begin{bmatrix} v_{sd} \\ v_{sq} \end{bmatrix}.
\tag{4.112}
$$

Premultiplying both sides of (4.112) by **R** and rearranging the resultant, we deduce

$$
\begin{aligned}
\frac{d}{dt} \begin{bmatrix} \lambda_d \\ \lambda_q \end{bmatrix} &= -\mathbf{R}\frac{d\mathbf{R}^{-1}}{dt} \begin{bmatrix} \lambda_d \\ \lambda_q \end{bmatrix} + \mathbf{R} \begin{bmatrix} -R_s & 0 \\ 0 & -R_s \end{bmatrix} \mathbf{R}^{-1} \begin{bmatrix} i_d \\ i_q \end{bmatrix} + \begin{bmatrix} v_{sd} \\ v_{sq} \end{bmatrix} \\
&= -\left(\mathbf{R}\frac{d\mathbf{R}^{-1}}{d\theta_r} \right)\left(\frac{d\theta_r}{dt} \right) \begin{bmatrix} \lambda_d \\ \lambda_q \end{bmatrix} + \mathbf{R} \begin{bmatrix} -R_s & 0 \\ 0 & -R_s \end{bmatrix} \mathbf{R}^{-1} \begin{bmatrix} i_d \\ i_q \end{bmatrix} + \begin{bmatrix} v_{sd} \\ v_{sq} \end{bmatrix}.
\end{aligned}
\tag{4.113}
$$

Substituting for $\mathbf{R}^{-1}$ and $\mathbf{R}$ in (4.113), one concludes

$$
\frac{d}{dt} \begin{bmatrix} \lambda_d \\ \lambda_q \end{bmatrix} = \begin{bmatrix} 0 & \omega_r \\ -\omega_r & 0 \end{bmatrix} \begin{bmatrix} \lambda_d \\ \lambda_q \end{bmatrix} + \begin{bmatrix} -R_s & 0 \\ 0 & -R_s \end{bmatrix} \begin{bmatrix} i_d \\ i_q \end{bmatrix} + \begin{bmatrix} v_{sd} \\ v_{sq} \end{bmatrix},
\tag{4.114}
$$

where

$$
\frac{d\theta_r}{dt} = \omega_r.
\tag{4.115}
$$

To derive an expression for the machine torque, the principle of power balance can be used. Based on (4.83), the power delivered to the machine stator is expressed in *dq*-frame as

$$
P_e = \frac{3}{2} \begin{bmatrix} i_d \\ i_q \end{bmatrix}^T \begin{bmatrix} v_{sd} \\ V_{sq} \end{bmatrix}.
\tag{4.116}
$$

Substituting for $[v_{sd}\ v_{sq}]^T$ from (4.114) in (4.116), we have

$$P_e = \frac{3}{2}\begin{bmatrix} i_d \\ i_q \end{bmatrix}^T \left\{ \begin{bmatrix} R_s i_d \\ R_s i_q \end{bmatrix} + \frac{d}{dt}\begin{bmatrix} \lambda_d \\ \lambda_q \end{bmatrix} + \begin{bmatrix} -\omega_r \lambda_q \\ \omega_r \lambda_d \end{bmatrix} \right\}, \qquad (4.117)$$

which can be simplified as

$$P_e = \underbrace{\frac{3}{2}R_s\left(i_d^2 + i_q^2\right)}_{P_{loss}} + \underbrace{\frac{3}{2}\left(i_d\frac{d\lambda_d}{dt} + i_q\frac{d\lambda_q}{dt}\right)}_{P_{stored}} + \underbrace{\omega_r\frac{3}{2}\left(\lambda_d i_q - \lambda_q i_d\right)}_{P_{gap}}.$$

$$(4.118)$$

On the right-hand side of (4.118), the first term represents the ohmic power loss of the stator windings, whereas the second term corresponds to the power stored/released by the magnetic field. The third term, however, represents the power delivered to the machine air gap, P_{gap}, which is responsible for the machine torque. Thus,

$$T_e = \frac{P_{gap}}{\omega_r} = \frac{3}{2}\left(\lambda_d i_q - \lambda_q i_d\right). \qquad (4.119)$$

It is common to express the torque expression (4.119) in terms of space phasors as

$$T_e = \frac{3}{2}Im\left\{ \overrightarrow{i}\ \overrightarrow{\lambda}^* \right\}. \qquad (4.120)$$

Substituting for λ_d and λ_q from (4.110) in (4.119), we deduce

$$T_e = \frac{3}{2}\left(L_d - L_q\right)i_d i_q + \frac{3}{2}\lambda_m i_q. \qquad (4.121)$$

Equations (4.110), (4.114), (4.115), and (4.121) constitute a dq-frame model for the PMSM. Since the dq-frame is synchronized to θ_r, the model is also known as the model in rotor-field coordinates. As (4.110), (4.114), (4.115), and (4.121) indicate, the dq-frame model of the PMSM is time invariant since the parameters are constants, but it is nonlinear as the products of the state variables are present in the equations.

5 Two-Level, Three-Phase Voltage-Sourced Converter

5.1 INTRODUCTION

Chapter 2 introduced the principles of operation of the DC/AC half-bridge converter and based on the method of averaging presented an equivalent circuit and a dynamic model for the converter. In Chapter 3, the control of the half-bridge converter was studied. Chapter 4 introduced the *abc-* to $\alpha\beta$-frame transformation to reduce the complexity of the control problem for a three-phase converter system. Chapter 4 further presented the *abc-* to *dq*-frame transformation to enable the control scheme of a three-phase converter system to process DC rather than sinusoidal signals and thus to provide a simpler structure.

This chapter employs the half-bridge converter as the building block for the three-phase voltage-sourced converter (VSC). A three-phase VSC can be realized based on a variety of configurations. However, in this book we consider two main configurations: the two-level VSC and the three-level neutral-point clamped (NPC) VSC. This chapter introduces the two-level three-phase VSC as the composition of three identical half-bridge converters. In Chapter 6, we will introduce the three-level NPC three-phase VSC as the composition of six identical half-bridge converters. Hereinafter, throughout this book we refer to the two-level three-phase VSC as *two-level VSC* and to the three-level NPC three-phase VSC as *three-level NPC.*[1]

5.2 TWO-LEVEL VOLTAGE-SOURCED CONVERTER

5.2.1 Circuit Structure

Figure 5.1 shows a schematic diagram of the two-level VSC [55]. The two-level VSC is composed of three identical half-bridge converters of Figure 2.13. The converter

---

[1] We shall use the generic term *VSC* for both the two-level VSC and the three-level NPC, in the context of system studies where the converter configuration is not the main focus.

Voltage-Sourced Converters in Power Systems, by Amirnaser Yazdani and Reza Iravani
Copyright © 2010 John Wiley & Sons, Inc.

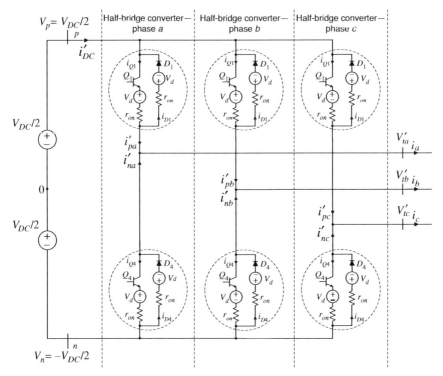

FIGURE 5.1   Schematic diagram of a nonideal two-level VSC.

of Figure 5.1 is called the two-level VSC since each of its AC-side terminals can assume either of the voltage levels $-V_{DC}$ and V_{DC}. The DC sides of the half-bridge converters are connected in parallel with a common DC-side voltage source. The AC-side terminal of each half-bridge converter is interfaced with one phase of a three-phase AC system (not shown in Fig. 5.1). The two-level VSC can provide a bidirectional power-flow path between the DC-side voltage source and the three-phase AC system. The AC system can be passive, for example, an RLC load, or active, for example, a synchronous machine. In the two-level VSC of Figure 5.1, we index the half-bridge converters by letters a, b, and c, to associate each with the corresponding phase of the AC system.

5.2.2   Principles of Operation

Chapter 2 introduced the pulse-width modulation (PWM) switching strategy and deduced that the AC-side terminal voltage of the nonideal half-bridge converter of

Figure 2.13 is given by (2.56), as[2]

$$V_t'(t) = m(t)\frac{V_{DC}}{2} - \frac{i(t)}{|i(t)|}V_e - r_e i(t), \tag{5.1}$$

where V_e and r_e are defined by (2.37) and (2.38), as

$$V_e = V_d - \left(\frac{Q_{rr} + Q_{tc}}{T_s}\right) r_{on} + V_{DC}\left(\frac{t_{rr}}{T_s}\right), \tag{5.2}$$

$$r_e = \left(1 - \frac{t_{rr}}{T_s}\right) r_{on} \approx r_{on}, \tag{5.3}$$

and T_s is the converter switching period.

The converter AC-side terminal voltage $V_t'(t)$ is controlled based on (5.1). The term $m(t)(V_{DC}/2)$ in (5.1) represents a dependent voltage source that can be controlled by the modulating signal, $m(t)$. The term $r_e i(t)$ in (5.1) can be regarded as an ohmic voltage drop. However, $(i(t)/|i(t)|)V_e$ represents a voltage offset whose polarity is dependent on the polarity of the AC-side current. If the current is negative, the offset is added to the terminal voltage, whereas it will be subtracted from the terminal voltage if the current is positive.

The applications studied in this book invariably concern sinusoidal waveforms. The (average of the) converter AC-side terminal voltage assumes a sinusoidal waveform if $m(t)$ is a sinusoidal function with the required amplitude and frequency. However, as (5.1) suggests, the offset term, that is, $(i(t)/|i(t)|)V_e$, introduces a dead zone in the control characteristic function from $m(t)$ to $V_t'(t)$, associated with the current zero crossing [16]. Consequently, $V_t'(t)$ is slightly distorted compared to a pure sinusoidal waveform. The distortion is negligible since V_e is typically only a few volts. This value is significantly smaller than the typical voltage levels for VSC systems. The impact of the distortion is further mitigated since $m(t)$ is often controlled by a closed-loop scheme that attempts to force the AC-side current to track a distortion-free sinusoidal command. For these reasons, we no longer tackle the voltage distortion issue in our subsequent formulations and approximate (5.1) by

$$V_t'(t) = m(t)\frac{V_{DC}}{2} - r_{on}i(t). \tag{5.4}$$

---

[2]Hereinafter, for the sake of compactness of notations, we drop the overbar signifying an averaged variable. However, it will be remembered that the variables are averaged over one switching cycle, unless otherwise noted.

In the two-level VSC of Figure 5.1, there are three identical half-bridge converters, one for each AC-side phase. Thus, the three AC-side terminal voltages are

$$V'_{ta}(t) = m_a(t)\frac{V_{DC}}{2} - r_{on}i_a(t), \tag{5.5}$$

$$V'_{tb}(t) = m_b(t)\frac{V_{DC}}{2} - r_{on}i_b(t), \tag{5.6}$$

$$V'_{tc}(t) = m_c(t)\frac{V_{DC}}{2} - r_{on}i_c(t). \tag{5.7}$$

Equations (5.5)–(5.7) indicate that, to obtain a balanced three-phase AC-side voltage and a balanced three-phase line current, $m_a(t)$, $m_b(t)$, and $m_c(t)$ must constitute a balanced three-phase signal; these are usually delivered by a closed-loop control scheme.

5.2.3   Power Loss of Nonideal Two-Level VSC

Based on (2.59), the power loss of the half-bridge DC/AC converter is given by

$$P_{loss} = V_{DC}\left(\frac{Q_{rr} + Q_{tc}}{T_s}\right) + V_e|i| + r_e i^2.$$

Thus, the power loss of the two-level VSC of Figure 5.1 is the summation of those of the three phases as

$$P_{loss} = \sum P_{loss(abc)} = 3V_{DC}\left(\frac{Q_{rr} + Q_{tc}}{T_s}\right)$$
$$+ V_e\left(|i_a| + |i_b| + |i_c|\right) + r_e\left(i_a^2 + i_b^2 + i_c^2\right). \tag{5.8}$$

If i_a, i_b, and i_c constitute a balanced three-phase sinusoidal waveform with a frequency of ω, due to the presence of the absolute value and square functions in (5.8), P_{loss} includes pulsating terms with the frequency 2ω. However, the average of (5.8) over one period of the sinusoid is

$$\langle P_{loss}\rangle_0 = V_{DC}\underbrace{\left[\frac{3(Q_{rr} + Q_{tc})}{T_s}\right]}_{i_{loss}} + \frac{6}{\pi}V_e\hat{i} + \frac{3}{2}r_e\hat{i}^2, \tag{5.9}$$

where the notation $\langle\ \rangle_0$ signifies the averaging operator over $T = 2\pi/\omega$, and $\hat{i}$ is the amplitude of the three-phase current.

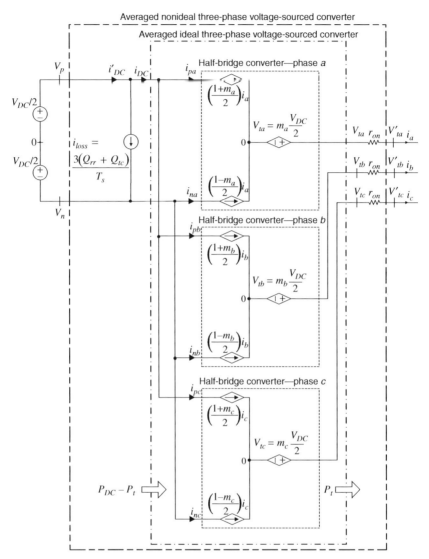

FIGURE 5.2   Averaged equivalent circuit of the nonideal two-level VSC of Figure 5.1.

5.3   MODELS AND CONTROL OF TWO-LEVEL VSC

5.3.1   Averaged Model of Two-Level VSC

Figure 5.2 illustrates the averaged equivalent circuit of the two-level VSC of Figure 5.1, as an extension of the averaged equivalent circuit of the half-bridge converter, that is, Figure 2.17. As discussed in Chapter 2, the averaged model of a nonideal half-bridge converter can be constructed by augmenting the averaged model of an ideal counterpart with two parasitic components; these are (i) the on-state resistance of one switch cell, connected in series with each AC-side terminal, and (ii) a current source connected in parallel with the converter DC side. While the former predominantly represents the converter conduction loss, the latter mainly represents the converter switching loss. Similarly, the averaged model for the two-level VSC of Figure 5.1 can also be regarded as the averaged model of an ideal two-level VSC (identified by dashed lines in Figure 5.2) supplemented by one on-state switch resistance in each phase, denoted by r_{on}, and an equivalent current source at the DC side, $i_{loss} = 3(Q_{rr} + Q_{tc})/T_s$. Since r_{on} and i_{loss} are approximately constant and independent of VSC voltages and currents, they can be lumped with AC and DC systems, respectively, interfaced with nonideal VSC. For example, in a drive system, r_{on} can be added to the motor stator resistance. Thus, we leave r_{on} and i_{loss} out from our subsequent developments and focus on the ideal two-level VSC.

Figure 5.3 illustrates the averaged equivalent circuit of an ideal two-level VSC, and the AC-side terminal voltages are

$$V_{ta}(t) = \frac{V_{DC}}{2} m_a(t), \tag{5.10}$$

$$V_{tb}(t) = \frac{V_{DC}}{2} m_b(t), \tag{5.11}$$

$$V_{tc}(t) = \frac{V_{DC}}{2} m_c(t), \tag{5.12}$$

where $m_{abc}(t)$ constitute a balanced three-phase signal, expressed as

$$m_a(t) = \widehat{m}(t) \cos\left[\varepsilon(t)\right], \tag{5.13}$$

$$m_b(t) = \widehat{m}(t) \cos\left[\varepsilon(t) - \frac{2\pi}{3}\right], \tag{5.14}$$

$$m_c(t) = \widehat{m}(t) \cos\left[\varepsilon(t) - \frac{4\pi}{3}\right], \tag{5.15}$$

where $\varepsilon(t)$ embeds the frequency and phase-angle information. The DC-side and AC-side terminal quantities of the two-level VSC are related based on the power balance principle, that is, $P_{DC}(t) = P_t(t)$. Therefore,

$$V_{DC}(t)i_{DC}(t) = V_{ta}(t)i_a(t) + V_{tb}(t)i_b(t) + V_{tc}(t)i_c(t). \tag{5.16}$$

For the reasons outlined in Chapter 4, a VSC system is usually controlled in $\alpha\beta$-frame or dq-frame. In these reference frames, the control is reduced to the control of two subsystems and, further, rapid control of the amplitude and/or frequency of the VSC AC-side voltage is straightforward. The following sections present the $\alpha\beta$- and dq-frame representations of the two-level VSC.

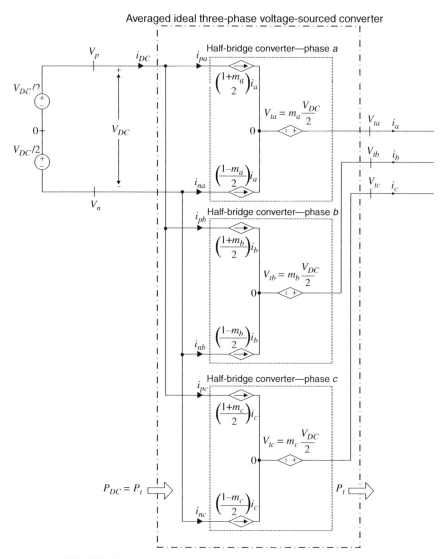

FIGURE 5.3   Averaged equivalent circuit of the ideal two-level VSC.

5.3.2   Model of Two-Level VSC in $\alpha\beta$-Frame

Equations (5.10)–(5.12) describe the relationships between a modulating signal and the corresponding AC-side terminal voltage of the two-level VSC. Equations (5.10)–(5.12) correspond to the space-phasor equation

$$\overrightarrow{V}_t(t) = \frac{V_{DC}}{2}\overrightarrow{m}(t), \tag{5.17}$$

which can be decomposed into real and imaginary parts, as

$$V_{t\alpha}(t) = \frac{V_{DC}}{2} m_{\alpha}(t), \tag{5.18}$$

$$V_{t\beta}(t) = \frac{V_{DC}}{2} m_{\beta}(t). \tag{5.19}$$

Equations (5.18) and (5.19) suggest that the α- and β-axis components of the converter AC-side terminal voltage are linearly proportional to the corresponding components of the modulating signal, while the proportionality constant is $V_{DC}/2$. Equations (5.18) and (5.19) also imply that the two-level VSC can be described by two subsystems in $\alpha\beta$-frame. The transfer function of each subsystem is the (time-varying) gain $V_{DC}/2$.

Based on (4.56), the AC-side terminal real power of the two-level VSC can be expressed in terms of the $\alpha\beta$-frame quantities as

$$P_t(t) = \frac{3}{2} \left[V_{t\alpha}(t) i_{\alpha}(t) + V_{t\beta}(t) i_{\beta}(t) \right]. \tag{5.20}$$

It then follows from the power balance principle that

$$V_{DC}(t) i_{DC}(t) = \frac{3}{2} \left[V_{t\alpha}(t) i_{\alpha}(t) + V_{t\beta}(t) i_{\beta}(t) \right]. \tag{5.21}$$

Equations (5.18), (5.19), and (5.21) constitute an $\alpha\beta$-frame model of the two-level VSC, as illustrated in Figure 5.4. In subsequent chapters, this model will be used for the analysis and control design.

Equations (5.18) and (5.19) correspond to two subsystems, the α- and β-axis subsystems, as illustrated in Figure 5.4(a). The two subsystems receive $m_{\alpha}(t)$ and $m_{\beta}(t)$ as the corresponding inputs and provide $V_{t\alpha}(t)$ and $V_{t\beta}(t)$ as the respective outputs. The two subsystems are decoupled and linear with a time-varying gain of

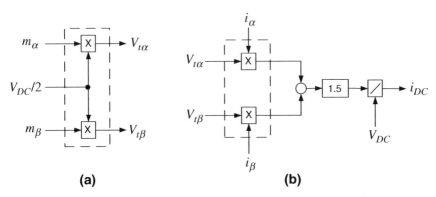

FIGURE 5.4   Control model of the ideal two-level VSC in the $\alpha\beta$-frame.

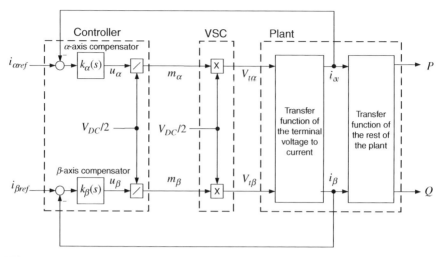

FIGURE 5.5   Generic control block diagram of a VSC system based on the $\alpha\beta$-frame control.

$V_{DC}(t)/2$. The DC-side current of the two-level VSC is determined according to (5.21), as shown in Figure 5.4(b). As discussed in Chapter 4, the α- and β-frame components of a three-phase signal embed the amplitude and frequency information. Thus, the amplitude and frequency of converter AC-side terminal voltage can be controlled by m_α and m_β. Therefore, the two-level VSC can be employed in applications where rapid and flexible control of amplitude and/or frequency (phase) is required.

Figure 5.5 illustrates a control block diagram of a generic *current-controlled* VSC system in $\alpha\beta$-frame. Figure 5.5 illustrates that a closed-loop scheme regulates i_α and i_β at their corresponding reference commands. This is achieved by controlling $V_{t\alpha}$ and $V_{t\beta}$ that, in turn, are controlled by $m_\alpha(t)$ and $m_\beta(t)$, respectively. Figure 5.5 also shows that the outputs of the compensators are scaled down by a factor of $V_{DC}/2$, to compensate for the gains of the α- and β-axis subsystems and, therefore, to make the loop gain independent of the DC-side voltage level.

In most applications, the current control strategy is an intermediate step toward the control of other variables. For example, in a grid-connected VSC system the objective is often to control the real and reactive power that the VSC system exchanges with the grid. This is fulfilled by controlling the α- and β-axis components of the VSC AC-side current. Hence, the overall control plant in the system of Figure 5.5 can be considered as the combination of two cascaded, two-input/two-output subplants. The first subplant has inputs $V_{t\alpha}$ and $V_{t\beta}$, while it provides outputs i_α and i_β; the second subplant receives i_α and i_β as its inputs and provides P and Q as final outputs. In case faster responses and/or superior disturbance rejection are required, P and Q themselves can be fed back and regulated by an outer control layer (not shown in Fig. 5.5) that issues the reference commands $i_{\alpha ref}$ and $i_{\beta ref}$.

5.3.3   Model and Control of Two-Level VSC in *dq*-Frame

In a current-controlled VSC system, optimization of the compensators is difficult since the variables are sinusoidal functions of time. Thus, the closed-loop control system must have an adequately large bandwidth to ensure command following with small steady-state errors, in addition to satisfactory disturbance rejection. Hence, the control design task is not straightforward, especially in variable-frequency applications. By contrast, in *dq*-frame the signals and variables are transformed to equivalent DC quantities. Hence, irrespective of the operating frequency, conventional PI compensators can be utilized for the control. Moreover, the $\alpha\beta$-frame representation of some asymmetrical three-phase systems includes time-varying parameters, for example, time-varying inductances in the $\alpha\beta$-frame model of a salient-pole synchronous machine, which make the control design task more involved. However, an appropriate *dq*-frame representation of such systems results in models with constant parameters. Therefore, a VSC system is preferably modeled and controlled in *dq*-frame.

A *dq*-frame representation of the two-level VSC can be developed following the procedures presented in Section 4.6.4. Substituting for $m(t) = (m_d + jm_q)e^{j\varepsilon(t)}$ and $V_t(t) = (V_{td} + jV_{tq})e^{j\varepsilon(t)}$ in (5.17), one deduces

$$(V_{td} + jV_{tq})e^{j\varepsilon(t)} = \frac{V_{DC}}{2}(m_d + jm_q)e^{j\varepsilon(t)}, \tag{5.22}$$

and concludes

$$V_{td}(t) = \frac{V_{DC}}{2}m_d(t), \tag{5.23}$$

$$V_{tq}(t) = \frac{V_{DC}}{2}m_q(t). \tag{5.24}$$

Equations (5.22) and (5.23) suggest that the *d*- and *q*-axis components of the VSC AC-side terminal voltage are linearly proportional to the corresponding components of the modulating signal, and the proportionality constant is $V_{DC}/2$. Equations (5.22) and (5.23) also imply that the two-level VSC can be described by two linear, time-varying subsystems in *dq*-frame. The transfer function of each subsystem is a time-varying gain, $V_{DC}/2$.

The AC-side terminal real power of the two-level VSC is formulated based on (4.83) as

$$P_t(t) = \frac{3}{2}\left[V_{td}(t)i_d(t) + V_{tq}(t)i_q(t)\right]. \tag{5.25}$$

Then, the principle of power balance requires that

$$V_{DC}(t)i_{DC}(t) = \frac{3}{2}\left[V_{td}(t)i_d(t) + V_{tq}(t)i_q(t)\right]. \tag{5.26}$$

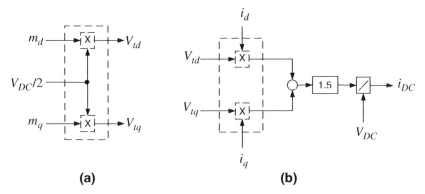

FIGURE 5.6   Control model of the ideal two-level VSC in dq-frame.

Equations (5.22), (5.23), and (5.26) constitute a model for the two-level VSC, expressed in dq-frame (Fig. 5.6). Comparing the block diagram of Figure 5.6 with that of Figure 5.4, one concludes that the models of the two-level VSC in dq-frame and $\alpha\beta$-frame are conceptually the same. Hence, the control block diagram of a current-controlled VSC system in dq-frame, illustrated in Figure 5.7, is similar to its $\alpha\beta$-frame counterpart, that is, Figure 5.5.

Figure 5.7 illustrates that a closed-loop scheme regulates i_d and i_q at their corresponding reference commands, by controlling V_{td} and V_{tq} that, in turn, are controlled by $m_d(t)$ and $m_q(t)$, respectively. Similar to the case of the $\alpha\beta$-frame control, the outputs of the compensators are multiplied by $2/V_{DC}$ to make the loop gain independent of the DC-side voltage level. If, for example, the current control is adopted as an intermediate step for the control of the real and reactive power exchanged between the VSC and an AC system, the commands i_{dref} and i_{qref} can be obtained from an outer layer of control (not shown in Figure 5.7). Another example of a nested control scheme is the one used to control the flux and torque of an asynchronous machine. In this case, i_d and i_q control the machine flux and torque, respectively. The composite control plant therefore consists of a subsystem that describes the dynamics of i_d and i_q as functions of V_{td} and V_{tq}, and a second subsystem that expresses the machine flux and torque in terms of i_d and i_q.

5.4   CLASSIFICATION OF VSC SYSTEMS

In this book, the generic VSC systems of Figures 5.5 and 5.7 are frequently encountered. Although the details are highly application specific, three main groups of VSC systems can be identified as follows:

- *Grid-Imposed Frequency VSC System*: In this group the VSC system is interfaced with a relatively large AC system, for example, a stiff utility grid. Therefore, the operating frequency is dictated by the AC system and is fairly constant. The grid-imposed frequency VSC systems are the subject of Chapters 7 and 8.

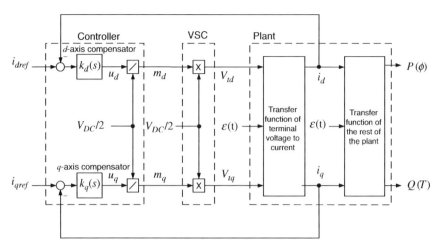

FIGURE 5.7 Generic control block diagram of a VSC system based on the *dq*-frame control.

- *Controlled-Frequency VSC System*: In this group, the AC system frequency is regulated by the control scheme of the VSC system, where the reference for the frequency may be obtained from a supervisory control system. The controlled-frequency VSC systems are treated in Chapter 9.
- *Variable-Frequency VSC System*: In a variable-frequency VSC system, the VSC is interfaced with an electric machine, and the operating frequency is a state variable of the overall VSC system, depends on the system operating point, and is not directly regulated. The variable-frequency VSC systems are studied in Chapter 10.

It should be noted that based on the foregoing classification, a VSC system can also be a composite system, that is, it includes a combination of the above-mentioned three classes. For example, a variable-speed wind-power unit employs a variable-frequency VSC system in conjunction with a grid-imposed frequency VSC system. The former VSC system controls a generator driven by a wind turbine, whereas the latter VSC system transfers the power to a (constant-frequency) utility grid. Another example of a composite VSC system is a back-to-back HVDC converter system that transfers energy from a utility grid to an off-grid island. Such an HVDC converter system is composed of a grid-imposed frequency VSC system and a controlled-frequency VSC system.

6 Three-Level, Three-Phase, Neutral-Point Clamped, Voltage-Sourced Converter

6.1 INTRODUCTION

Chapter 5 introduced the two-level, three-phase voltage-sourced converter (VSC) (Fig. 5.1) as the composition of three half-bridge converters. The two-level VSC is the dominant building block for a wide range of apparatus in medium/high-power applications. Figure 5.1 (and Fig. 2.13 for the half-bridge converter) illustrates that each switch cell in the two-level VSC must withstand the whole DC-side voltage in off state. Therefore, if the two-level VSC is employed for a high-power/high-voltage application, the switch cells must be rated for a high DC voltage level. If we consider a high voltage level for a specific application, then to implement the two-level VSC, probably the highest voltage switches that fulfill the voltage rating requirement are selected. Such switches are often the state of the art and typically expensive. Other than the cost and availability issues, beyond a voltage rating, even the highest voltage available switches may not fulfill the voltage requirements. The voltage ratings of the state-of-the-art switches, currently and most probably in the future, would not meet the voltage requirements of most utility applications.

One approach to achieving high-voltage switch cells is to connect a number of lower voltage switches in series. This approach has been extensively practiced in high-power/high-voltage applications such as HVDC converter systems. The main issue in this approach is the need for simultaneous gating and the snubber circuits for the switches that constitute a switch cell, to guarantee equal voltage sharing among the switches. Simultaneous gating calls for precise timings and carefully laid-out wiring to minimize the impact of parasitic effects. Snubber circuits are lossy and undesirable in terms of compact integration. In view of the foregoing issues, multilevel converters provide an alternative approach for high-power industrial and utility applications [56–58].

Voltage-Sourced Converters in Power Systems, by Amirnaser Yazdani and Reza Iravani
Copyright © 2010 John Wiley & Sons, Inc.

The three-level neutral-point clamped (NPC) VSC [1] [59] is a multilevel converter that offers an alternative to reduce (or even avoid) the number of series-connected switches. In the three-level NPC, each switch cell has to withstand half of the DC-side voltage. Thus, the number of switches to be connected in series can be reduced. Moreover, the three-level NPC can provide a three-phase AC voltage with a lower harmonic distortion compared to an equivalent two-level VSC.

In this chapter, the three-level half-bridge NPC is introduced as the building block for the three-level NPC. It is demonstrated that the three-level half-bridge NPC can be considered as the combination of two two-level half-bridge converters of Figure 2.13; one half-bridge converter generates a controlled positive AC voltage, whereas the other one generates a controlled negative AC voltage. Consequently, based on the developments of Chapter 2 for the two-level half-bridge converter, one can deduce the principles of operation, modeling, and control of the three-level half-bridge NPC. In the rest of this chapter, we combine three three-level half-bridge NPCs to construct a three-phase three-level NPC. We also present the averaged models for the three-level NPC in $\alpha\beta$ and dq-frames.

6.2   THREE-LEVEL HALF-BRIDGE NPC

Figure 6.1 shows a schematic diagram of a three-level half-bridge NPC that can be regarded as a combination of two two-level half-bridge converters of Figure 2.1 and two additional diodes, that is, D_2 and D_3. The first half-bridge converter is composed of the switch cells Q_{1-1}/D_{1-1} and Q_{4-1}/D_{4-1}, and the second one consists of the switch cells Q_{1-2}/D_{1-2} and Q_{4-2}/D_{4-2}. The net DC-side voltage is split into two equal halves and supplied by two identical voltage sources. The DC-side midpoint, 0, is connected to the three-level half-bridge NPC via the clamp diodes D_2 and D_3 (Fig. 6.1). All voltages in Figure 6.1 are expressed with reference to the DC-side midpoint. As for the two-level half-bridge converter, the switching function for a switch cell is defined as

$$s(t) = \begin{cases} 1, & \text{if the switch is commanded to conduct,} \\ 0, & \text{if the switch is blocked.} \end{cases}$$

We also note that the gating commands of Q_{1-1} and Q_{4-1} must be complementary, that is, $s_{1-1} + s_{4-1} \equiv 1$. Similarly, the gating commands of Q_{1-2} and Q_{4-2} are complementary, that is, $s_{1-2} + s_{4-2} \equiv 1$.

6.2.1   Generating Positive AC-Side Voltages

Assume that a positive voltage at the converter AC-side terminal t must be generated. Then, let $s_{1-2} \equiv 0$ (or equivalently $s_{4-2} \equiv 1$), corresponding to the circuit

---

[1]Hereinafter, we refer to the three-level half-bridge NPC converter and the three-level NPC VSC as *three-level half-bridge NPC* and *three-level NPC*, respectively.

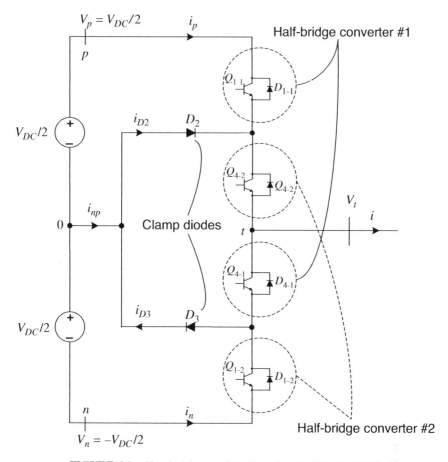

FIGURE 6.1   Circuit diagram of the three-level half-bridge NPC.

configuration shown in Figure 6.2(a). Thus, when $s_{1-1} = 1$ and $s_{4-1} = 0$, Q_{1-1} conducts if i is positive, whereas if i is negative D_{1-1} conducts. Consequently, for $s_{1-1} = 1$ and $s_{4-1} = 0$, $V_t = V_{DC}/2$, regardless of the polarity of i. On the other hand, for $s_{1-1} = 0$ and $s_{4-1} = 1$, if i is positive, D_2 conducts, whereas if i is negative, Q_{4-1} and D_3 conduct. Hence, for $s_{1-1} = 0$ and $s_{4-1} = 1$, $V_t = 0$, irrespective of the polarity of i. This analysis indicates that when $s_{1-2} \equiv 0$ and $s_{4-2} \equiv 1$, depending on switching states of Q_{1-1} and Q_{4-1}, the instantaneous AC-side terminal voltage is either $V_{DC}/2$ or zero. However, one can control the (positive) average of V_t by controlling the duty ratios of s_{1-1} and s_{4-1} based on a pulse-width modulation (PWM) switching strategy.

6.2.2   Generating Negative AC-Side Voltages

To generate a negative voltage at the converter AC-side terminal t, let $s_{1-1} \equiv 0$ (or equivalently $s_{4-1} \equiv 1$). Figure 6.2(b) shows the corresponding circuit configuration.

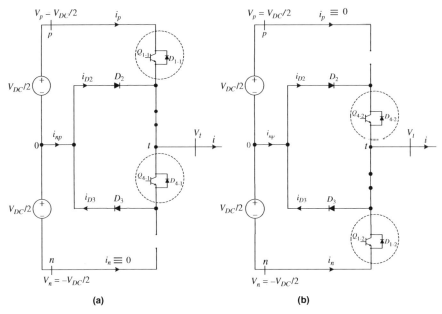

FIGURE 6.2 Subcircuits of the three-level half-bridge NPC corresponding to (a) positive AC-side voltage generation and (b) negative AC-side voltage generation.

Figure 6.2(b) illustrates that for $s_{1-2} = 1$ and $s_{4-2} = 0$, D_{1-2} conducts if i is positive, whereas if i is negative, Q_{1-2} conducts. Consequently, for $s_{1-2} = 1$ and $s_{4-2} = 0$, $V_t = -V_{DC}/2$, regardless of the polarity of i. On the other hand, for $s_{1-2} = 0$ and $s_{4-2} = 1$, if i is positive, Q_{4-2} and D_2 conduct, whereas if i is negative, D_3 conducts. Hence, for $s_{1-2} = 0$ and $s_{4-2} = 1$, $V_t = 0$, regardless of the polarity of i. Therefore, when $s_{1-1} \equiv 0$ and $s_{4-1} \equiv 1$, the instantaneous AC-side terminal voltage is either $-V_{DC}/2$ or zero, depending on the switching states of Q_{1-2} and Q_{4-2}. However, we can control the (negative) average of V_t by controlling the duty ratios of s_{1-2} and s_{4-2}, based on a PWM switching strategy.

The operation logic described here and in Section 6.2.1 justifies the term *three-level* for the half-bridge NPC of Figure 6.1; the AC-side terminal can assume any of the three voltage levels $-V_{DC}/2$, 0, or $V_{DC}/2$.

6.3   PWM SCHEME FOR THREE-LEVEL HALF-BRIDGE NPC

Sections 6.2.1 and 6.2.2 presented the principles of operation of the three-level half-bridge NPC. It was demonstrated that the converter is composed of two coordinated two-level half-bridge converters, one responsible for the positive and the other one for the negative AC-side voltage generation. In this section, we present a PWM scheme for the three-level half-bridge NPC to (i) coordinate the two half-bridge converters

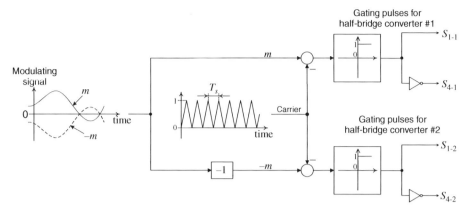

FIGURE 6.3   Schematic diagram of the PWM scheme for the three-level half-bridge NPC.

of Figure 6.2 and (ii) control the average value of the AC-side terminal voltage. The pulse-width modulator of Figure 6.3 fulfills the requirements.

The pulse-width modulator of Figure 6.3 is conceptually identical to that of Figure 2.2 presented for a two-level half-bridge converter. However, it utilizes a unipolar triangular carrier signal unlike the pulse-width modulator for the two-level half-bridge converter. The reason is that the two half-bridge converters that constitute a three-level half-bridge NPC, shown in Figure 6.2(a) and (b), must generate either a positive or a negative voltage (and not both) at their common AC side.

Figure 6.3 shows that the switching functions of Q_{1-1} and Q_{4-1} (Fig. 6.1) are obtained from the comparison of the modulating waveform m with a high-frequency (unipolar) carrier. Whenever m is larger than the carrier, a turn-on command is issued for Q_{1-1} and the turn-on command for Q_{4-1} is removed. However, to obtain switching functions for Q_{1-2} and Q_{4-2}, the negative of the modulating waveform, that is, $-m$, is compared with the carrier. Whenever $-m$ is bigger than the carrier, a turn-on command is issued for Q_{1-2} and the turn-on command for Q_{4-2} is removed.

When m is positive, $-m$ is negative and therefore smaller than the carrier signal. Consequently, $s_{1-2} \equiv 0$, $s_{4-2} \equiv 1$, and the three-level half-bridge NPC of Figure 6.1 is equivalent to the circuit of Figure 6.2(a). As explained in Section 6.2.1, the AC-side terminal voltage of the circuit of Figure 6.2(a) is either $V_{DC}/2$ or zero, depending on the states of Q_{1-1} and Q_{4-1}. The proportion of one carrier period over which Q_{1-1} is on (and Q_{4-1} is off) is proportional to the value of m. For a larger m, Q_{1-1} is on for a longer time period, and thus the average of V_t is more positive. If $m = m(t)$ is a positive function of time and its rate of change is considerably smaller than the carrier frequency, the average of V_t can be approximated as the product of $m(t)$ and the scale factor $V_{DC}/2$. For instance, if $m(t)$ represents the positive half-cycle of a sinusoidal waveform, the average of V_t is also the positive half-cycle of a sinusoid, $V_{DC}/2$ times $m(t)$. Figure 6.4(a) illustrates the PWM signals, the switching functions, and the AC-side terminal voltage of the converter, when m is positive.

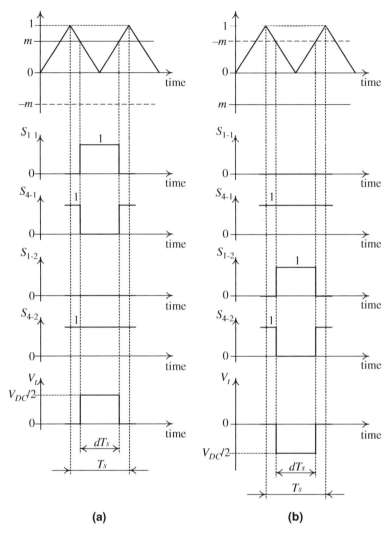

FIGURE 6.4 Switching functions of the three-level half-bridge NPC for (a) positive modulating signal and (b) negative modulating signal.

An analysis similar to the one described for a positive m can be conducted for a negative m. When m is negative and thus less than the carrier signal, then $s_{1-1} \equiv 0$, $s_{4-1} \equiv 1$, and the three-level half-bridge NPC of Figure 6.1 is equivalent to the circuit of Figure 6.2(b). As explained in Section 6.2.2, the AC-side terminal voltage of the circuit of Figure 6.2(a) is either $-V_{DC}/2$ or zero, depending on the states of Q_{1-2} and Q_{4-2}. The proportion of one carrier period over which Q_{1-2} is on (and Q_{4-2} is off) is determined by the value of $-m$. For a larger absolute value of m, $-m$ is more positive and thus Q_{1-2} conducts for a longer duration over one carrier period. This, in turn, corresponds to a larger magnitude of (the negative) average of V_t. If $m = m(t)$ is a negative function of time with a rate of change considerably smaller than the carrier frequency, the average of V_t is also a function of time proportional to $m(t)$, with the proportionality constant $V_{DC}/2$. For instance, if $m(t)$ is a negative half-cycle of a sinusoidal waveform, the average of V_t is also the negative half-cycle of a sinusoid, $V_{DC}/2$ times $m(t)$. Figure 6.4(b) illustrates the PWM signals, the switching functions, and the AC-side terminal voltage of the converter, when m is negative.

6.4 SWITCHED MODEL OF THREE-LEVEL HALF-BRIDGE NPC

6.4.1 Switched AC-Side Terminal Voltage

The previous sections demonstrated that the operational modes of a three-level half-bridge NPC are determined based on the polarity of the modulating signal m. Based on the discussions of Section 6.2.1, for a positive m the converter AC-side voltage can be expressed as

$$V_t(t) = V_p \cdot s_{1-1}(t) + 0 \cdot s_{4-1}(t), \quad m \geq 0, \tag{6.1}$$

where $s_{1-1}(t) + s_{4-1}(t) \equiv 1$. Similarly, based on the results of Section 6.2.2, for a negative m the converter AC-side voltage is

$$V_t(t) = V_n \cdot s_{1-2}(t) + 0 \cdot s_{4-2}(t), \quad m \leq 0, \tag{6.2}$$

where $s_{1-2}(t) + s_{4-2}(t) \equiv 1$. Equations (6.1) and (6.2) can be unified as

$$V_t(t) = V_p s_{1-1}(t)\mathrm{sgn}(m) + V_n s_{1-2}(t)\mathrm{sgn}(-m), \tag{6.3}$$

where the function $\mathrm{sgn}(\cdot)$ is defined as

$$\mathrm{sgn}(x) = \begin{cases} 1, & x \geq 0, \\ 0, & x < 0. \end{cases} \tag{6.4}$$

Figure 6.5 illustrates the switching functions and the instantaneous AC-side terminal voltage of the three-level half-bridge NPC when m is a sinusoidal function of time.

6.4.2 Switched DC-Side Terminal Currents

Based on Section 6.2.1, for a positive m we have $i_n \equiv 0$ (Fig. 6.2(a)). To determine i_p, we employ the principle of power balance as

$$V_p \cdot i_p = V_t(t) \cdot i(t), \quad m \geq 0. \tag{6.5}$$

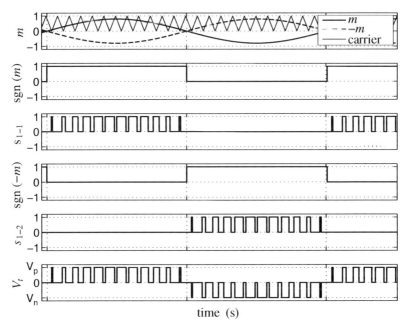

FIGURE 6.5   Switching functions and AC-side terminal voltage of the three-level half-bridge NPC for a sinusoidal modulating signal.

Substituting for V_t from (6.1) in (6.5), and eliminating V_p, we obtain

$$i_p = s_{1-1}(t) \cdot i(t), \quad m \geq 0. \tag{6.6}$$

Based on the results of Section 6.2.2, for a negative m we have $i_p \equiv 0$ (Fig. 6.2(b)). Analogous to (6.5), we determine i_n from

$$V_n \cdot i_n = V_t(t) \cdot i(t), \quad m \leq 0. \tag{6.7}$$

Substituting for V_t from (6.2) in (6.7), and eliminating V_n, we deduce

$$i_n = s_{1-2}(t) \cdot i(t), \quad m \leq 0. \tag{6.8}$$

The midpoint current i_{np} is determined through the application of KCL to the DC source midpoint, as

$$i_{np} = i - (i_p + i_n) = i(t) - \left[s_{1-1}\text{sgn}(m) + s_{1-2}\text{sgn}(-m)\right] i(t). \tag{6.9}$$

Equations (6.3), (6.4), (6.6), (6.8), and (6.9) constitute a switched model for the three-level half-bridge NPC of Figure 6.1. In the next section, these equations are averaged to develop an averaged model for the converter.

6.5 AVERAGED MODEL OF THREE-LEVEL HALF-BRIDGE NPC

6.5.1 Averaged AC-Side Terminal Voltage

Applying the averaging operator to both sides of (6.1), over one switching cycle, we deduce

$$\overline{V}_t(t) = \overline{V}_p d(t), \tag{6.10}$$

where $d(t)$ is the duty cycle of Q_{1-1}. As Figure 6.4(a) illustrates, d is equal to m. Thus,

$$\overline{V}_t(t) = \overline{V}_p m(t), \quad m \geq 0. \tag{6.11}$$

Similarly, application of the averaging process to both sides of (6.2), over one switching cycle, yields

$$\overline{V}_t(t) = \overline{V}_n d(t), \tag{6.12}$$

where $d(t)$ is the duty cycle of Q_{1-2}. As Figure 6.4(b) illustrates, d is equal to $-m$. Hence,

$$\overline{V}_t(t) = -\overline{V}_n m(t), \quad m \leq 0. \tag{6.13}$$

Equations (6.11) and (6.13) can be unified as

$$\overline{V}_t(t) = \left[\overline{V}_p \operatorname{sgn}(m) - \overline{V}_n \operatorname{sgn}(-m) \right] m(t), \tag{6.14}$$

where function $\operatorname{sgn}(\cdot)$ is defined by (6.4). For the special case where $\overline{V}_p = V_{DC}/2$ and $\overline{V}_n = -V_{DC}/2$, since $\operatorname{sgn}(m) + \operatorname{sgn}(-m) \equiv 1$, (6.14) reduces to

$$\overline{V}_t(t) = (V_{DC}/2)m(t). \tag{6.15}$$

Equation (6.15) is identical to (2.26) that was developed for the two-level half-bridge converter.

6.5.2 Averaged DC-Side Terminal Currents

Applying the averaging operator to both sides of (6.6), we deduce

$$\overline{i}_p = d(t) \cdot i(t) = m(t) \cdot i(t), \quad m \geq 0, \tag{6.16}$$

which can also be expressed as

$$\overline{i}_p = m(t)\operatorname{sgn}(m) \cdot i(t). \tag{6.17}$$

Similarly, the average of (6.8) is

$$\bar{i}_n = d(t) \cdot i(t) = -m(t) \cdot i(t), \quad m \leq 0, \tag{6.18}$$

which can be rewritten as

$$\bar{i}_n = -m(t)\mathrm{sgn}(-m) \cdot i(t). \tag{6.19}$$

The expression for the averaged midpoint current is derived by calculating the average of (6.9) as

$$\bar{i}_{np}(t) = i(t) - m(t)i(t) \left[\mathrm{sgn}(m) - \mathrm{sgn}(-m)\right]. \tag{6.20}$$

Equations (6.4), (6.15), (6.17), (6.19), and (6.20) constitute an averaged model for the three-level half-bridge NPC. These equations are used in the following sections for modeling the three-level NPC.

6.6   THREE-LEVEL NPC

6.6.1   Circuit Structure

Figure 6.6 shows a schematic diagram of the three-level NPC. The three-level NPC is composed of three identical three-level half-bridge NPCs of Figure 6.1. The DC sides of the half-bridge NPCs are connected in parallel with, and supplied by, a split voltage source. All the voltages are expressed with reference to the DC-side midpoint, that is, node 0. The AC-side terminal of each half-bridge NPC is connected to one phase of a three-phase AC system (not shown in Fig. 6.6), and the three-level NPC permits a bidirectional flow of power from the DC-side source to the three-phase AC system. The AC system can be a passive load, an electric machine, a utility power grid, and so on. In the three-level NPC of Figure 6.6, we index each half-bridge NPC by a, b, or c, where each letter corresponds to one phase of the three-phase system interfaced with the three-level NPC.

6.6.2   Principles of Operation

The AC-side terminal voltage of the three-level half-bridge NPC is governed by (6.15). The three-level NPC of Figure 6.6 is composed of three identical three-level half-bridge NPCs, one per phase. Thus, the voltages of the three phases are

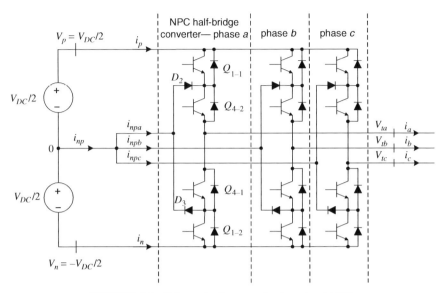

FIGURE 6.6   Schematic diagram of a three-level NPC.

expressed as

$$V_{ta}(t) = m_a(t)\frac{V_{DC}}{2}, \tag{6.21}$$

$$V_{tb}(t) = m_b(t)\frac{V_{DC}}{2}, \tag{6.22}$$

$$V_{tc}(t) = m_c(t)\frac{V_{DC}}{2}, \tag{6.23}$$

where for the sake of compactness we have dropped the averaging overbar from variables. Based on (6.21)–(6.23), one can control the converter AC-side terminal voltage $V_{tabc}(t)$ through $m_{abc}(t)$. In a VSC system, we often need to generate and control sinusoidal voltages and/or currents. Thus, to generate a balanced three-phase sinusoidal AC-side terminal voltage, m_{abc} must also be a balanced three-phase sinusoidal function of time with the required amplitude, phase angle, and frequency. $m_{abc}(t)$ is usually the output of a closed-loop control system that regulates i_{abc}. In general, m_{abc} assumes the following form:

$$m_a(t) = \widehat{m}(t)\cos\left[\varepsilon(t)\right], \tag{6.24}$$

$$m_b(t) = \widehat{m}(t)\cos\left[\varepsilon(t) - \frac{2\pi}{3}\right], \tag{6.25}$$

$$m_c(t) = \widehat{m}(t)\cos\left[\varepsilon(t) - \frac{4\pi}{3}\right], \tag{6.26}$$

where $\varepsilon(t)$ contains the phase-angle and frequency information.

In VSC systems, we also need to exercise fast changes in the amplitude and/or the phase angle of the AC-side terminal voltage. This, based on (6.21)–(6.23), is possible by stipulating rapid changes in amplitude and phase angle of m_{abc}, using the techniques presented in Chapter 4, based on the concepts of space phasor, $\alpha\beta$-frame, or dq-frame. Therefore, in this chapter the $\alpha\beta$- and dq-frame representations of the three-level NPC will also be introduced.

6.6.3 Midpoint Current

While the DC-side midpoint in the two-level VSC (i.e., node 0 in Fig. 5.1) is necessarily neither available nor accessible, but is considered merely as the potential reference node for circuit analysis, the DC-side midpoint of the three-level NPC of Figure 6.6 is a physical node and connected to the converter clamp diodes. In the three-level NPC, the total DC-side voltage is split into two equal halves, and the two halves are connected to the midpoint. In this section, we further investigate the characteristics of the midpoint current i_{np}.

The midpoint current of a three-level half-bridge NPC is given by (6.20). Thus, we deduce

$$i_{npa}(t) = i_a(t) - m_a(t)i_a(t)\left[\text{sgn}(m_a) - \text{sgn}(-m_a)\right], \tag{6.27}$$

$$i_{npb}(t) = i_b(t) - m_b(t)i_b(t)\left[\text{sgn}(m_b) - \text{sgn}(-m_b)\right], \tag{6.28}$$

$$i_{npc}(t) = i_c(t) - m_c(t)i_c(t)\left[\text{sgn}(m_c) - \text{sgn}(-m_c)\right], \tag{6.29}$$

where the averaging overbar has been dropped from the formulation, for the sake of compactness. The midpoint current of the three-level NPC of Figure 6.6 is given by

$$i_{np}(t) = [i_a(t) + i_b(t) + i_c(t)] - \left[f_a(t) + f_b(t) + f_c(t)\right], \tag{6.30}$$

where

$$f_a(t) = m_a(t)i_a(t)\left[\text{sgn}(m_a) - \text{sgn}(-m_a)\right], \tag{6.31}$$

$$f_b(t) = m_b(t)i_b(t)\left[\text{sgn}(m_b) - \text{sgn}(-m_b)\right], \tag{6.32}$$

$$f_c(t) = m_c(t)i_c(t)\left[\text{sgn}(m_c) - \text{sgn}(-m_c)\right]. \tag{6.33}$$

If a three-wire connection is assumed at the AC-side of the three-level NPC, then $i_a(t) + i_b(t) + i_c(t) \equiv 0$ and (6.30) reduces to

$$i_{np}(t) = -\left[f_a(t) + f_b(t) + f_c(t)\right]. \tag{6.34}$$

Thus, to evaluate i_{np}, we must calculate $f_a(t) + f_b(t) + f_c(t)$. Based on (6.31)–(6.33), f_a, f_b, and f_c are periodic functions of time and can be expressed by their corresponding Fourier series. If the three-level NPC operates under a balanced condition, $f_{abc}(t)$ forms a balanced set of waveforms, in which $f_b(t)$ and $f_c(t)$ are phase shifted, respectively, by $-2\pi/3$ and $-4\pi/3$ with respect to $f_a(t)$. Consequently, $\left[f_a(t) + f_b(t) + f_c(t)\right]$ is zero for any harmonic other than DC and multiples of 3. Under a steady state, $\widehat{m}(t) = \widehat{m}$ in (6.24)–(6.26), and the AC-side current of the three-level NPC is

$$i_a(t) = \widehat{i} \cos \left[\varepsilon(t) - \gamma\right], \tag{6.35}$$

$$i_b(t) = \widehat{i} \cos \left[\varepsilon(t) - \gamma - \frac{2\pi}{3}\right], \tag{6.36}$$

$$i_c(t) = \widehat{i} \cos \left[\varepsilon(t) - \gamma - \frac{4\pi}{3}\right], \tag{6.37}$$

where γ is the phase delay of the AC-side current components with respect to their corresponding AC-side terminal voltages, that is, γ is the power-factor angle of the three-level NPC. Substituting for m_a and i_a, respectively, from (6.24) and (6.35), in (6.31), we obtain

$$f_a(t) = \left(\frac{\widehat{m}\widehat{i}}{2}\right) \left[\cos \gamma + \cos(2\varepsilon - \gamma)\right] \left[\text{sgn}(m_a) - \text{sgn}(-m_a)\right]. \tag{6.38}$$

In (6.38), $\text{sgn}(m_a) - \text{sgn}(-m_a)$ is a periodic function, as shown in Figure 6.7, and thus can be expanded by a Fourier series as

$$\text{sgn}(m_a) - \text{sgn}(-m_a) = \left(\frac{4}{\pi}\right) \sum_{h=1,3,5,\ldots}^{+\infty} \frac{1}{h} \sin\left(\frac{h\pi}{2}\right) \cos(h\varepsilon). \tag{6.39}$$

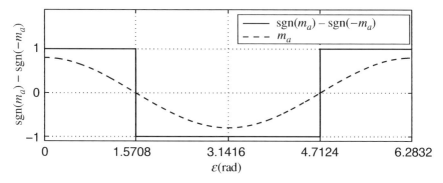

FIGURE 6.7 Function $(\text{sgn}(m_a) - \text{sgn}(-m_a))$ sketched for one period, based on m_a given by (6.24).

Substituting for $\text{sgn}(m_a) - \text{sgn}(-m_a)$ from (6.39) into (6.38), we obtain

$$
f_a(t) = \left(\frac{2\widehat{mi}}{\pi}\right) \cos\gamma \sum_{h=1,3,5,\ldots}^{+\infty} \frac{1}{h} \sin\left(\frac{h\pi}{2}\right) \cos(h\varepsilon)
$$

$$
+ \left(\frac{2\widehat{mi}}{\pi}\right) \sum_{h=1,3,5,\ldots}^{+\infty} \frac{1}{h} \sin\left(\frac{h\pi}{2}\right) \cos(h\varepsilon)\cos(2\varepsilon - \gamma). \tag{6.40}
$$

The term $\cos(h\varepsilon)\cos(2\varepsilon - \gamma)$ can be decomposed into $(1/2)\cos\left[(h-2)\varepsilon + \gamma\right] + (1/2)\cos\left[(h+2) - \gamma\right]$. Therefore, (6.40) can be rewritten as

$$
f_a(t) = \left(\frac{2\widehat{mi}}{\pi}\right) \cos\gamma \sum_{h=1,3,5,\ldots}^{+\infty} \frac{1}{h} \sin\left(\frac{h\pi}{2}\right) \cos(h\varepsilon)
$$

$$
+ \left(\frac{\widehat{mi}}{\pi}\right) \sum_{h=1,3,5,\ldots}^{+\infty} \frac{1}{h} \sin\left(\frac{h\pi}{2}\right) \cos\left[(h-2)\varepsilon + \gamma\right]
$$

$$
+ \left(\frac{\widehat{mi}}{\pi}\right) \sum_{h=1,3,5,\ldots}^{+\infty} \frac{1}{h} \sin\left(\frac{h\pi}{2}\right) \cos\left[(h+2)\varepsilon - \gamma\right]. \tag{6.41}
$$

$f_b(t)$ can be derived by replacing of ε with $(\varepsilon - 2\pi/3)$ in (6.41), and similarly, $f_c(t)$ is derived by replacing of ε with $(\varepsilon - 4\pi/3)$ in (6.41). Hence, $f_a(t) + f_b(t) + f_c(t)$ is evaluated as three times $f_a(t)$ for the DC component and the triple-n harmonics, but it is zero otherwise. Thus,

$$
f_a(t) + f_b(t) + f_c(t) = \left(\frac{6\widehat{mi}}{\pi}\right) \cos\gamma \sum_{h=3,9,15,\ldots}^{+\infty} \frac{1}{h} \sin\left(\frac{h\pi}{2}\right) \cos(h\varepsilon)
$$

$$
+ \left(\frac{3\widehat{mi}}{\pi}\right) \sum_{h=5,11,17,\ldots}^{+\infty} \frac{1}{h} \sin\left(\frac{h\pi}{2}\right) \cos\left[(h-2)\varepsilon + \gamma\right]
$$

$$
+ \left(\frac{3\widehat{mi}}{\pi}\right) \sum_{h=1,7,13,\ldots}^{+\infty} \frac{1}{h} \sin\left(\frac{h\pi}{2}\right) \cos\left[(h+2)\varepsilon - \gamma\right]. \tag{6.42}
$$

Substituting for $f_a(t) + f_b(t) + f_c(t)$ from (6.42) into (6.34), and introducing appropriate changes of variables for the second and third sets of terms, we deduce

$$
i_{np}(t) - \left(\frac{6\widehat{mi}}{\pi}\right) \cos \gamma \sum_{h=3,9,15,\ldots}^{+\infty} \frac{1}{h} \sin\left(\frac{h\pi}{2}\right) \cos(h\varepsilon)
$$

$$
+ \left(\frac{3\widehat{mi}}{\pi}\right) \sum_{h=3,9,15,\ldots}^{+\infty} \frac{1}{h+2} \sin\left(\frac{h\pi}{2}\right) \cos(h\varepsilon + \gamma)
$$

$$
+ \left(\frac{3\widehat{mi}}{\pi}\right) \sum_{h=3,9,15,\ldots}^{+\infty} \frac{1}{h-2} \sin\left(\frac{h\pi}{2}\right) \cos(h\varepsilon - \gamma). \qquad (6.43)
$$

Equation (6.43) indicates that the midpoint current has no DC component. Equation (6.43) also indicates that i_{np} includes odd triple-n harmonics, where the third harmonic is the dominant component. Therefore, the midpoint current can be approximated as

$$
i_{np}(t) \approx \left(\frac{2\widehat{mi}}{\pi}\right) \cos \gamma \cos(3\varepsilon) - \left(\frac{3\widehat{mi}}{5\pi}\right) \cos(3\varepsilon + \gamma) - \left(\frac{3\widehat{mi}}{\pi}\right) \cos(3\varepsilon - \gamma)
$$

$$
= \left(\frac{4\widehat{mi}}{5\pi}\right) \left[-2\cos \gamma \, \cos(3\varepsilon) - 3\sin \gamma \sin(3\varepsilon)\right]. \qquad (6.44)
$$

Equation (6.44) can be rewritten in the following compact form:

$$
i_{np}(t) \approx \left(\frac{4\widehat{mi}}{5\pi}\right) \sqrt{4 + 5\sin^2\gamma} \cos(3\varepsilon + \zeta)
$$

$$
= \left(\frac{4\widehat{mi}}{5\pi}\right) \sqrt{9 - 5\cos^2\gamma} \cos(3\varepsilon + \zeta), \qquad (6.45)
$$

where $\zeta = \pi - \tan^{-1}(1.5\tan\gamma)$. Equation (6.45) indicates that the amplitude of the midpoint current is a function of the converter operating point. Equation (6.45) also implies that the amplitude of i_{np} is linearly proportional to the converter AC-side terminal voltage and has a nonlinear relationship with the converter power factor. Thus, the amplitude has its largest value when the three-level NPC operates at zero power factor, that is, $\cos \gamma = 0$ and assumes its smallest value when it exchanges power at unity power factor. Based on (6.45), one can deduce the following estimate for the amplitude of the midpoint current:

$$
0.51\widehat{mi} \leq \widehat{i}_{np} \leq 0.76\widehat{mi}.
$$

This inequality indicates that the amplitude of the midpoint current can be as high as 76% of the amplitude of the converter AC-side current, corresponding to the scenario where the three-level NPC primarily exchanges reactive power with the AC-side system, for example, when utilized as a static compensator (STATCOM). However, even if the converter operates at (nearly) unity power factor, the amplitude of the midpoint current is not less than 0.51% of that of the converter AC-side current. Therefore, the DC-side voltage sources must accommodate a relatively large third-harmonic midpoint current, and this can be regarded as a disadvantage of the three-level NPC, as compared to its two-level VSC counterpart. The following example further highlights these conclusions.

EXAMPLE 6.1   Midpoint Current of Three-Level NPC

Consider the three-level NPC of Figure 6.6 that is interfaced with a three-phase voltage source via three series RL branches. Figure 6.8(a)–(c), respectively, illustrate the waveforms of the converter AC-side current and fundamental harmonic of the AC-side voltage, the switched waveform of the midpoint current, and the filtered waveform of the midpoint current.

Until $t = 1.5$ s, the converter fundamental voltage and AC-side current are in phase (unity power factor) (Fig. 6.8(a)) and 710 kW flows out of the converter AC-side terminals. At $t = 1.5$ s, the converter fundamental voltage is phase

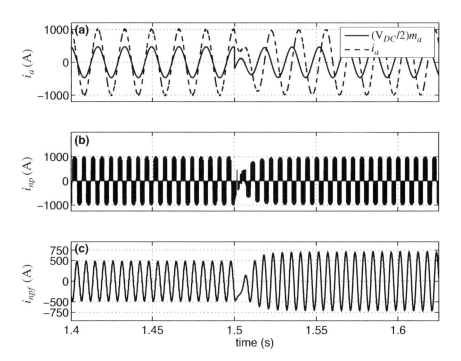

FIGURE 6.8 Midpoint current of the three-level NPC under unity and zero power-factor conditions; Example 6.1.

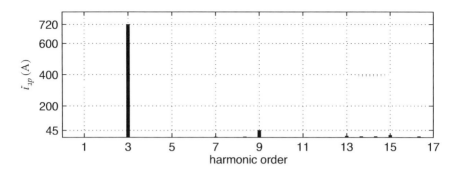

FIGURE 6.9   Spectrum of i_{np} when the three-level NPC operates at zero power factor; Example 6.1.

shifted by $\pi/2$ while the converter AC-side current is kept unchanged (Fig. 6.8(a)). Consequently, after $t = 1.5$ s, the power factor of the converter becomes zero while its apparent power remains unchanged. Since the amplitudes of the modulating signal and AC-side current remain unchanged, the amplitude of $i_{np}(t)$ does not change after $t = 1.5$ s (Fig. 6.8(b)). However, as Figure 6.8(c) illustrates, the amplitude of the third harmonic of i_{np} changes from about 487 A, corresponding to unity power-factor operation, to about 720 A, corresponding to zero power-factor operation.

After $t = 1.5$ s, the steady-state midpoint current exhibits the Fourier spectrum shown in Figure 6.9. It is noted that i_{np} possesses a dominant third harmonic. The next significant harmonic is the ninth, which is less than 7% of the third harmonic. Higher order harmonics, for example, 15th, 21st, and so on, are relatively insignificant. Therefore, i_{np} can be approximated by its third-harmonic component.

6.6.4   Three-Level NPC with Impressed DC-Side Voltages

To provide two identical DC-side voltage sources for the three-level NPC of Figure 6.6, the configuration of Figure 6.10 can be employed. This configuration is commonly used in industrial applications, for example, in variable-speed motor drives [60], due to its simplicity and ruggedness. In the converter system of Figure 6.10, power is supplied by the grid to two six-pulse diode-bridge rectifiers, and the three-level NPC controls the load, for example, an asynchronous machine. The DC side of each diode bridge is connected through a DC choke in parallel with one DC-side capacitor of the three-level NPC (Fig. 6.10). The diode bridges are supplied from the secondary and tertiary windings of a transformer. The bridge rectifiers in conjunction with the transformer constitute a 12-pulse rectifier system to ensure low current harmonic distortion at the grid side [55]. Since the resistances of DC reactors are low, the capacitor voltages are naturally maintained by bridge rectifiers at essentially half the net DC-bus voltage.

Salient demerits of the system of Figure 6.10 are the transformer size and the fact that the system cannot operate as a regenerative energy conversion unit since the

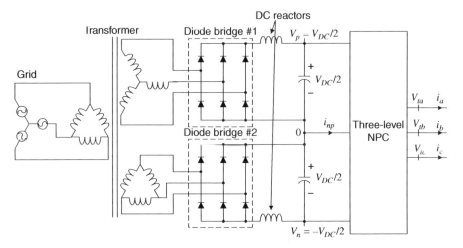

FIGURE 6.10   A three-level NPC supplied by a 12-pulse diode rectifier.

power flow is unidirectional, that is, only from the grid to the load. In the next section, we introduce an alternative configuration that can be utilized for converter systems that must operate in a bidirectional power-flow environment.

6.7   THREE-LEVEL NPC WITH CAPACITIVE DC-SIDE VOLTAGE DIVIDER

An alternative configuration for providing two equal partial DC-side voltage sources for the three-level NPC converter is shown in Figure 6.11. The three-level NPC of Figure 6.11 employs a capacitive voltage divider at its DC side. If the capacitors are (nominally) identical, the total DC-side voltage is equally divided between the capacitors. The configuration of Figure 6.11 can be augmented to constitute a back-to-back converter configuration, as shown in Figure 6.12, and thus to permit bidirectional power exchange between the two AC systems (not shown) that are interfaced with the three-level NPCs. Note that bidirectional power exchange is not possible with the system shown in Figure 6.10.

If no corrective control is employed, voltages of DC-side capacitors of VSC systems of Figures 6.11 and 6.12 drift from their nominal values, that is, half the total DC-bus voltage. The voltage drift occurs under both steady-state and transient conditions, even if the net DC-side voltage is maintained at a constant level. This problem arises due to tolerances of the converter components and asymmetries of gating commands of the switches. The second limitation of the converter systems of Figures 6.11 and 6.12 is that, due to the large third-harmonic component of the midpoint current (see Section 6.6.3 for the analysis), each partial DC-side voltage includes a third-harmonic ripple that in turn results in low-order voltage harmonics at the converter AC side. To limit the harmonic voltage, DC capacitors must be adequately large. The former issue and its solution are investigated in the next two sections while the latter is the subject of Section 6.7.5.

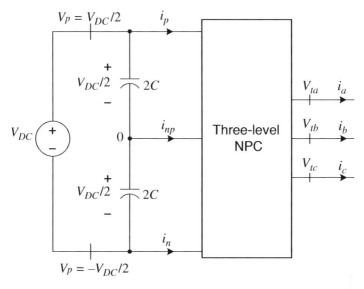

FIGURE 6.11   Three-level NPC with a capacitive DC-side voltage divider.

6.7.1   Partial DC-Side Voltage Drift Phenomenon

Equations (6.43) and (6.45) indicate that the midpoint current, $i_{np}(t)$, has no DC component. Equations (6.43) and (6.45) are developed based on the assumptions that (i) the three-level half-bridge NPCs that constitute the three-level NPC of Figure 6.6 are identical and (ii) their AC-side currents constitute balanced three-phase waveforms. Furthermore, it is implicitly assumed that the switching functions of each three-level half-bridge NPC are perfectly symmetrical in each period of the modulating

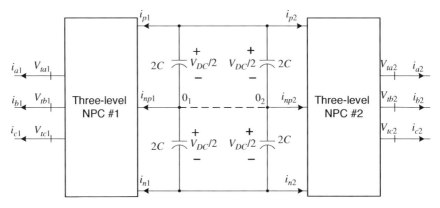

FIGURE 6.12   Three-level NPCs with corresponding capacitive DC-side voltage dividers, in a back-to-back converter system configuration.

waveform (see Fig. 6.5). In practice, tolerances of the circuit components in conjunction with implementation limitations, that is, digital word truncation errors, inherent sensor offsets, and so on, result in deviations from the ideal conditions. Therefore, inevitably, the midpoint current of each three-level half-bridge NPC and thus the midpoint current of the three-level NPC include DC components. Although the DC component is relatively small, it is integrated by the capacitors, and the partial DC-side voltages drift from the nominal values [61, 65].

A method to prevent the voltage drift phenomenon is to compensate for the DC component of i_{np} by means of an external converter [62]–[64]. In this method, the (DC) voltage difference between the two capacitors is processed by a compensator that specifies the amount of current the external converter should inject into the midpoint. The main shortcoming of this approach is the need for additional power hardware that adds to the system cost and complexity. An alternative approach is to modify the switching functions of each three-level half-bridge NPC, to nullify the DC component of i_{np} [61, 65–67]. This approach requires a more elaborate control strategy than the previous method, but it provides an economically viable and a technologically elegant solution to the problem. The next section presents the details of this method.

6.7.2   DC-Side Voltage Equalization

To introduce an intentional asymmetry to the switching functions to counteract the DC-side voltage imbalance, we modify the modulating waveforms of (6.24)–(6.26) as

$$m_a(t) = m_0 + \widehat{m}(t)\cos\left[\varepsilon(t)\right], \tag{6.46}$$

$$m_b(t) = m_0 + \widehat{m}(t)\cos\left[\varepsilon(t) - \frac{2\pi}{3}\right], \tag{6.47}$$

$$m_c(t) = m_0 + \widehat{m}(t)\cos\left[\varepsilon(t) - \frac{4\pi}{3}\right], \tag{6.48}$$

where $m_0(t)$ is a small offset added to the sinusoidal-varying components of $m_{abc}(t)$. Figure 6.13 illustrates the impact of the DC offset m_0 on the switching functions and operation of the three-level half-bridge NPC of Figure 6.1, when any of the modulating signals (6.46)–(6.48) is given to the PWM signal generator of Figure 6.3. Figure 6.13 shows that when a DC offset is added to the modulating signal, the time lapse of a period during which the pair (Q_{1-1}, Q_{4-1}) is pulse-width modulated will be different from the time lapse during which the pair (Q_{1-2}, Q_{4-2}) is pulse-width modulated. Consequently, the time lapse during which i_{np} is positive will be different from the one during which i_{np} is negative, and i_{np} will assume a nonzero mean value, that is, a DC component.

As presented in Section 6.6.3, the midpoint current is expressed as

$$i_{np}(t) = -\left[f_a(t) + f_b(t) + f_c(t)\right], \tag{6.49}$$

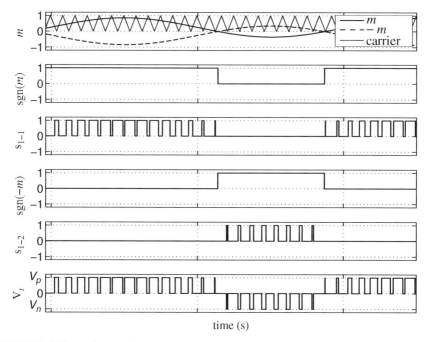

FIGURE 6.13 Switching functions and AC-side terminal voltage of the three-level half-bridge NPC when the sinusoidal modulating signal is supplemented by a DC offset.

where

$$f_a(t) = m_a(t)i_a(t) \left[\text{sgn}(m_a) - \text{sgn}(-m_a) \right],\tag{6.50}$$

$$f_b(t) = m_b(t)i_b(t) \left[\text{sgn}(m_b) - \text{sgn}(-m_b) \right],\tag{6.51}$$

$$f_c(t) = m_c(t)i_c(t) \left[\text{sgn}(m_c) - \text{sgn}(-m_c) \right].\tag{6.52}$$

To calculate $f_a(t) + f_b(t) + f_c(t)$, we note that based on (6.46)–(6.48), $m_a(t)$, $m_b(t)$, and $m_b(t)$ are identical in form but phase shifted with respect to each other by $-2\pi/3$. Therefore, since $i_{abc}(t)$ is a balanced three-phase waveform, $f_a(t)$, $f_b(t)$, and $f_c(t)$ are also identical in form but phase shifted by $-2\pi/3$. Consequently, $f_a(t) + f_b(t) + f_c(t)$ is three times $f_a(t)$ with respect to DC and triple-n harmonic components, and zero otherwise. Thus,

$$i_{np0}(t) = -3 f_{a0}(t),\tag{6.53}$$

where the subscript 0 denotes the DC component of the corresponding variable. Substituting for $i_a(t)$ and $m_a(t)$ in (6.50), respectively, from (6.35) and (6.46), we obtain

$$f_a(t) = m_0 \widehat{i} \cos(\varepsilon - \gamma) \left[\mathrm{sgn}(m_a) - \mathrm{sgn}(-m_a) \right]$$
$$+ \widehat{mi} \cos(\varepsilon - \gamma) \cos(\varepsilon) \left[\mathrm{sgn}(m_a) - \mathrm{sgn}(-m_a) \right]. \tag{6.54}$$

Using the identity $\cos(\alpha)\cos(\beta) = (1/2)\cos(\alpha - \beta) + (1/2)\cos(\alpha + \beta)$, (6.54) can be expanded as

$$f_a(t) = m_0 \widehat{i} \cos(\varepsilon - \gamma) \left[\mathrm{sgn}(m_a) - \mathrm{sgn}(-m_a) \right]$$
$$+ \frac{1}{2} \widehat{mi} \cos(\gamma) \left[\mathrm{sgn}(m_a) - \mathrm{sgn}(-m_a) \right]$$
$$+ \frac{1}{2} \widehat{mi} \cos(2\varepsilon - \gamma) \left[\mathrm{sgn}(m_a) - \mathrm{sgn}(-m_a) \right]. \tag{6.55}$$

The DC component of $f_a(t)$ can be calculated from

$$f_{a0} = \frac{1}{2\pi} \int_0^{2\pi} f_a(\theta) d\theta. \tag{6.56}$$

In (6.55), the function $\mathrm{sgn}(m_a) - \mathrm{sgn}(-m_a)$ is a two-level waveform that assumes either 1 or -1 over its period. Let us refer to the angles at which the levels change as θ_1 and θ_2, as Figure 6.14 illustrates. The dashed lines in Figure 6.14 illustrate the waveforms corresponding to a nonzero, small, m_0, whereas the solid lines correspond to $m_0 = 0$. As Figure 6.14 indicates, if m_0 is small, we can approximate θ_1 and θ_2 by $\pi/2$ and $3\pi/2$, respectively. Therefore, (6.56) can be rewritten as

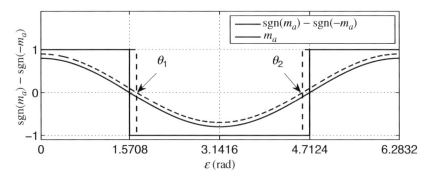

FIGURE 6.14 Function $(\mathrm{sgn}(m_a) - \mathrm{sgn}(-m_a))$ sketched for one period when m_a is supplemented with a DC offset based on (6.46).

$$f_{a0} = \frac{m_0\widehat{i}}{2\pi}\left(\int_0^{\pi/2}\cos(\theta-\gamma)d\theta - \int_{\pi/2}^{3\pi/2}\cos(\theta-\gamma)d\theta + \int_{3\pi/2}^{2\pi}\cos(\theta-\gamma)d\theta\right)$$

$$+\frac{m\widehat{i}}{4\pi}\cos\gamma\left(\int_0^{\pi/2}d\theta - \int_{\pi/2}^{3\pi/2}d\theta + \int_{3\pi/2}^{2\pi}d\theta\right)$$

$$+\frac{m\widehat{i}}{4\pi}\left(\int_0^{\pi/2}\cos(2\theta-\gamma)d\theta - \int_{\pi/2}^{3\pi/2}\cos(2\theta-\gamma)d\theta + \int_{3\pi/2}^{2\pi}\cos(2\theta-\gamma)d\theta\right)$$

$$= \frac{2\widehat{i}\cos\gamma}{\pi}m_0. \tag{6.57}$$

Substituting for f_{a0} from (6.57) in (6.53), one concludes

$$i_{np0}(t) = -\frac{6\widehat{i}\cos\gamma}{\pi}m_0(t). \tag{6.58}$$

Equation (6.58) suggests that the DC component of i_{np} can be controlled by m_0. Based on (6.58), the control transfer function is linear, but with a variable gain. The gain is zero if either the three-level NPC AC-side current is small or the converter operates at zero power factor. However, the gain is the largest if the three-level NPC operates at its rated capacity and unity power factor.

To equalize the DC components of the two DC-side voltages, a closed-loop scheme compares them and controls m_0 [61, 65]. Figure 6.15 shows a circuit model of the DC-side voltages and the midpoint current in which partial DC-side voltages are denoted by V_1 and V_2. Figure 6.15 indicates that i_{np} has two components: a third-harmonic component, i_{np3}, and a DC component, i_{np0}, which are formulated by (6.45) and (6.58), respectively. As an approximation, we assume that the capacitors are identical, each with a capacitance of $2C$.

If V_{DC}, that is, the net DC-side voltage, has no third-harmonic component, it can be replaced by a short circuit. Thus, the circuit of Figure 6.15 is simplified to the equivalent circuit of Figure 6.16(a), for the third-harmonic component. Based on Figure 6.16(a) and since the capacitors are identical, i_{np3} is equally divided between the two capacitors, and $i_1 = -i_2 = i_{np3}/2$, where i_{np3} is expressed by (6.45). Hence, in a steady state, the third-order harmonic components of V_1 and V_2 are given by

$$\langle V_1\rangle_3 = \widehat{V}_{r3}\sin(3\omega t + \zeta) \tag{6.59}$$

$$\langle V_2\rangle_3 = -\widehat{V}_{r3}\sin(3\omega t + \zeta), \tag{6.60}$$

where $\omega = d\varepsilon/dt$ and $\zeta = \pi - \tan^{-1}(1.5\tan\gamma)$, and the peak voltage ripple $\widehat{V}_{r3}$ is

$$\widehat{V}_{r3} = \frac{m\widehat{i}}{5\pi C\omega}\sqrt{9 - 5\cos^2\gamma}. \tag{6.61}$$

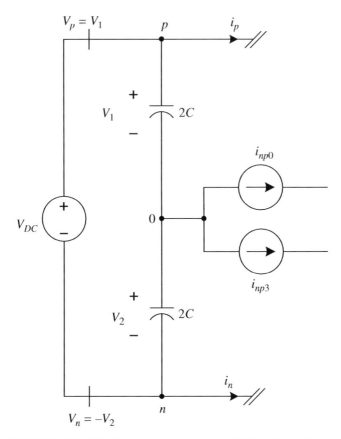

FIGURE 6.15   Circuit model representing partial DC-side voltages.

Equation (6.61) indicates that the amplitude of the third-order harmonic component of the partial DC-side voltages is a function of the converter operating point and inversely proportional to the capacitance of the DC-side capacitors.

Figure 6.16(b) illustrates a simplified circuit, equivalent to the circuit of Figure 6.15, for the DC analysis of the partial DC-side voltages and the midpoint current. Based on Figure 6.16(b), one can deduce that $i_1 - i_2 = i_{np0}$, and therefore

$$(2C)\frac{d}{dt}\langle V_1\rangle_0 - (2C)\frac{d}{dt}\langle V_2\rangle_0 = i_{np0}, \tag{6.62}$$

where $V_{DC} = \langle V_1\rangle_0 + \langle V_2\rangle_0$. Equation (6.62) can be expressed in the state-space form as

$$\frac{d}{dt}(\langle V_1 - V_2\rangle_0) = \left(\frac{1}{2C}\right)i_{np0}. \tag{6.63}$$

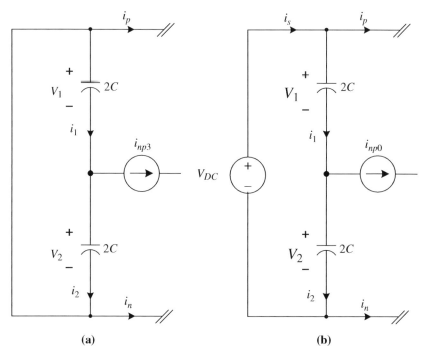

(a)   **(b)**

FIGURE 6.16 Circuit equivalents to the circuit of Figure 6.15: (a) for third-harmonic analysis; (b) for DC analysis.

Based on (6.63), the DC component of $V_1 - V_2$, that is, $\langle V_1 - V_2 \rangle_0$, is proportional to the integral of i_{np0}. Thus, the DC component of $V_1 - V_2$ is a constant value (preferably zero) if i_{np0} is zero. For an ideal condition, $i_{np} \equiv 0$ corresponds to $m_0 \equiv 0$. However, in practice, due to imperfections and intrinsic asymmetries of the system, m_0 must assume a nonzero value to ensure that $i_{np} \equiv 0$. Consequently, m_0 must be inevitably determined in a closed-loop system that attempts to regulate the DC component of $V_1 - V_2$ at zero.

Based on (6.63), the closed-loop voltage equalizing process requires $\langle V_1 - V_2 \rangle_0$ as the feedback signal. However, practically, only $V_1 - V_2$, and not its DC component, is available for measurement. To relate $\langle V_1 - V_2 \rangle_0$ to $V_1 - V_2$, one can express V_1 and V_2 as

$$V_1 = \langle V_1 \rangle_0 + \langle V_1 \rangle_3 = \langle V_1 \rangle_0 + \widehat{V}_{r3} \sin(3\omega t + \zeta), \tag{6.64}$$

$$V_2 = \langle V_2 \rangle_0 + \langle V_2 \rangle_3 = \langle V_2 \rangle_0 - \widehat{V}_{r3} \sin(3\omega t + \zeta). \tag{6.65}$$

Subtracting both sides of (6.64) from those of (6.65), we obtain

$$\langle V_1 - V_2 \rangle_0 = (V_1 - V_2) - 2\widehat{V}_{r3} \sin(3\omega t + \zeta). \tag{6.66}$$

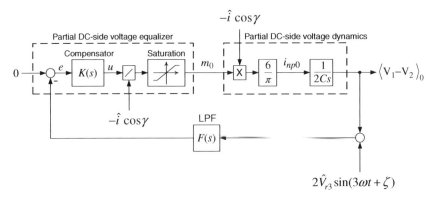

FIGURE 6.17   Control block diagram of the partial DC-side voltage equalizing scheme.

Equation (6.66) indicates that $\langle V_1 - V_2 \rangle_0$ is equal to $V_1 - V_2$ plus a triple-frequency component. Thus, $\langle V_1 - V_2 \rangle_0$ can be extracted through low-pass filtering of $V_1 - V_2$. The low-pass filter can be an independent component of the closed-loop system, although it can also be realized by a compensator that starts a steep gain roll-off at low frequencies.

Figure 6.17 illustrates a control block diagram of the closed-loop DC-side voltage equalizing scheme. It is noted that a saturation block limits m_0 to a small value such that $|m_0| \ll \hat{m}$. The division of the compensator output by $-\hat{i} \cos \gamma$ corresponds to a feed-forward compensation that makes the loop gain independent of the operating point. If, for simplicity, the feed-forward signal is replaced by a constant gain, its sign must still be included in the control loop as a multiplicative term. Otherwise, the loop gain assumes opposite signs in inverting and rectifying modes of operation, resulting in instability when the converter system moves from one mode of operation to another. In Section 7.4, we further examine the block diagram of Figure 6.17 and demonstrate the design process through Example 7.3.

6.7.3   Derivation of DC-Side Currents

Consider the three-level NPC of Figure 6.15 with node n as the circuit potential reference node. The instantaneous power, exchanged with the DC side of the three-level NPC, is

$$P_{DC}(t) = V_{DC}(t)i_p(t) + V_2(t)i_{np}(t) \approx V_{DC}(t)i_p(t) + \frac{V_{DC}(t)}{2}i_{np}(t). \quad (6.67)$$

If the converter losses are ignored, P_{DC} is equal to the power that flows out of the converter AC-side terminals. Thus,

$$V_{DC}(t)i_p(t) + \frac{V_{DC}(t)}{2}i_{np}(t) = P_t(t), \quad (6.68)$$

where

$$P_t(t) = V_{ta}(t)i_a(t) + V_{tb}(t)i_b(t) + V_{tc}(t)i_c(t). \tag{6.69}$$

Since $V_{tabc}(t)$ and $i_{abc}(t)$ are balanced waveforms, $P_t(t)$ is a DC variable. Therefore, based on (6.68), $V_{DC}i_p + (V_{DC}/2)i_{np}$ must also be a DC variable. Since $i_{np}(t)$ is a periodic function with no DC component, $i_p(t)$ must inevitably have a periodic component to cancel the periodic component of $i_{np}(t)$. Therefore, $i_p(t)$ can be expressed as

$$i_p(t) = i_{DC}(t) + \langle i_p \rangle_3(t). \tag{6.70}$$

Substituting for $P_t(t)$ and $i_p(t)$ in (6.68), respectively, from (6.69) and (6.70), we deduce

$$V_{DC}(t)i_{DC}(t) = V_{ta}(t)i_a(t) + V_{tb}(t)i_b(t) + V_{tc}(t)i_c(t) \tag{6.71}$$

and

$$\langle i_p \rangle_3(t) = -i_{np}(t)/2. \tag{6.72}$$

Therefore,

$$i_p(t) = i_{DC}(t) - i_{np}(t)/2. \tag{6.73}$$

Similarly, it can be shown that

$$i_n(t) = -i_{DC}(t) - i_{np}(t)/2. \tag{6.74}$$

6.7.4   Unified Models of Three-Level NPC and Two-Level VSC

A comparison between (6.71) and (5.16) indicates that the expression for the DC-side current of the two-level VSC is identical to the expression for (DC component of) the DC-side current of the three-level NPC. The reasons are that (i) in either converter configuration only the DC component of the DC-side current contributes to the power exchange and (ii) in the three-level NPC, the DC component of i_{np} is zero if the capacitor voltages are equal and stable. Furthermore, comparing (6.21)–(6.23) with their counterparts for the two-level VSC, that is, (5.10)–(5.12), one concludes that the AC-side terminal voltage of the three-level NPC assumes the same form as those of the two-level VSC. Hence, as long as the terminal voltage/current relationships are concerned, there exists an equivalent two-level VSC for a three-level NPC [61]. This *equivalence property* is illustrated in Figures 6.18 and 6.19.

Figure 6.18 shows a three-level NPC for which the equivalent two-level VSC is illustrated in Figure 6.19. Although the internal circuit structures and switching strategies of the two converters are different, they exhibit identical dynamic behavior if m_{abc} and V_{DC} are the same for both. It should be noted that the effective DC-bus capacitor of the equivalent two-level VSC, that is, Figure 6.19, is nominally half of each of the DC-side capacitors of the three-level NPC converter, that is, Figure 6.18. A VSC system can employ either the two-level VSC or the three-level NPC as its AC/DC power processor. Based on the equivalence property, hereinafter, we refer to a two-level VSC or a three-level NPC as *VSC* if the converter configuration is not the main focus.

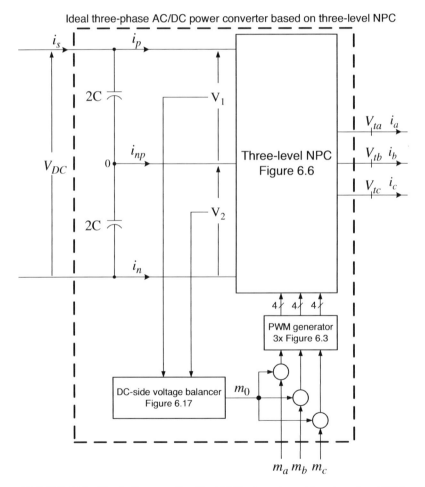

FIGURE 6.18 Block diagram of an ideal VSC whose kernel is a three-level NPC.

Ideal three-phase AC/DC power converter based on two-level VSC

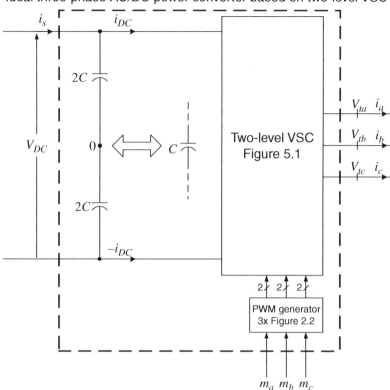

FIGURE 6.19   Block diagram of an ideal VSC whose kernel is a two-level VSC.

Based on the equivalence property, first the loops to control the terminal currents/voltages of a three-level NPC can be designed using the models and methods developed for the two-level VSC, as if a two-level VSC is the subject of control. Then, based on the block diagram of Figure 6.17, a DC-side voltage equalizing scheme, discussed in Section 6.7.2, is designed for the three-level NPC. The DC-side voltage balancing scheme can be designed independently of the other controllers, since the dynamics of partial DC-side voltages are decoupled from those of terminal currents/voltages.

6.7.5   Impact of DC Capacitors Voltage Ripple on AC-Side Harmonics

We conclude this chapter by showing that the third-harmonic components of DC capacitor voltages of the three-level NPC generate low-order voltage harmonics at the AC-side terminals of the converter. Let us first assume that the DC-side voltage balancer maintains the DC components of V_1 and V_2 at $V_{DC}/2$. Thus, (6.64) and (6.65) reduce to

$$V_1 = \frac{V_{DC}}{2} + \widehat{V}_{r3} \sin(3\omega t + \zeta), \tag{6.75}$$

$$V_2 = \frac{V_{DC}}{2} - \widehat{V}_{r3} \sin(3\omega t + \zeta). \tag{6.76}$$

Since $V_p = V_1$ and $V_n = -V_2$, we deduce

$$V_p = \frac{V_{DC}}{2} + \widehat{V}_{r3} \sin(3\omega t + \zeta), \tag{6.77}$$

$$V_n = -\frac{V_{DC}}{2} + \widehat{V}_{r3} \sin(3\omega t + \zeta). \tag{6.78}$$

Substituting for V_p and V_n in (6.14), from (6.77) and (6.78), and since $m(t) = \widehat{m} \cos(\omega t)$, we obtain

$$V_t(t) = \widehat{m} \frac{V_{DC}}{2} \cos(\omega t) \left[\mathrm{sgn}(m) + \mathrm{sgn}(-m) \right]$$
$$+ \widehat{V}_{r3} \widehat{m} \sin(3\omega t + \zeta) \cos(\omega t) \left[\mathrm{sgn}(m) - \mathrm{sgn}(-m) \right]. \tag{6.79}$$

Since $\mathrm{sgn}(m) + \mathrm{sgn}(-m) \equiv 1$, replacing $\mathrm{sgn}(m) - \mathrm{sgn}(-m)$ by its Fourier series expansion as provided by (6.39) and using the identity $\sin \alpha \cos \beta = (1/2) \sin(\alpha - \beta) + (1/2) \sin(\alpha + \beta)$, we obtain

$$V_t(t) = \widehat{m} \frac{V_{DC}}{2} \cos(\omega t)$$

$$- \left(\frac{\widehat{V}_{r3}\widehat{m}}{\pi} \right) \sum_{h=1,3,5,\ldots}^{+\infty} \frac{1}{h} \sin\left(\frac{h\pi}{2} \right) \sin\left[(h-4)\omega t - \zeta \right]$$

$$+ \left(\frac{\widehat{V}_{r3}\widehat{m}}{\pi} \right) \sum_{h=1,3,5,\ldots}^{+\infty} \frac{1}{h} \sin\left(\frac{h\pi}{2} \right) \sin\left[(h+2)\omega t + \zeta \right]$$

$$- \left(\frac{\widehat{V}_{r3}\widehat{m}}{\pi} \right) \sum_{h=1,3,5,\ldots}^{+\infty} \frac{1}{h} \sin\left(\frac{h\pi}{2} \right) \sin\left[(h-2)\omega t - \zeta \right]$$

$$+ \left(\frac{\widehat{V}_{r3}\widehat{m}}{\pi} \right) \sum_{h=1,3,5,\ldots}^{+\infty} \frac{1}{h} \sin\left(\frac{h\pi}{2} \right) \sin\left[(h+4)\omega t + \zeta \right]. \tag{6.80}$$

The term $\widehat{m}(V_{DC}/2)\cos(\omega t)$ in (6.80) corresponds to the fundamental voltage component of V_t if there is no ripple on the partial DC-side voltages. The rest of the terms in (6.80) represent the harmonics generated due to the third-harmonic ripples of the DC capacitor voltages. Equation (6.80) indicates that the amplitude of the fundamental component of V_t is slightly different from the one under the condition where no ripple exists. Moreover, the fundamental component of the voltage assumes a small phase shift as compared to the condition where no ripple exists. Furthermore, when the DC capacitor voltages have third-harmonic ripple components, odd voltage harmonics are generated at the AC-side terminals. The dominant voltage harmonic is the third harmonic followed by a smaller fifth harmonic. The magnitudes of higher order voltage harmonics are insignificant.

The amplitude change and the phase shift of the fundamental component of V_t is practically not an issue since the closed-loop scheme controlling the converter system corrects the modulating signals to establish the required fundamental voltage. As for the voltage harmonics, the third-order voltage harmonic is a zero-sequence harmonic and does not drive any current in a three-wire three-phase system. However, the presence of fifth- and possibly seventh-order harmonics can be problematic since these harmonics can deeply penetrate into the AC system. Since based on (6.80) the amplitudes of the harmonics are proportional to the peak value of the voltage ripple, to keep the ripple small, the capacitors must be adequately large as (6.61) suggests.

EXAMPLE 6.2   Low-Order AC-Side Voltage Harmonics

Consider the three-level NPC of Figure 6.6 that is interfaced with a three-phase voltage source via three series RL branches. The partial DC-side voltages are imposed and kept constant at 750 V, and the converter switching frequency is 34×60 Hz. Figure 6.20(a)–(c) illustrates the waveforms of, respectively, the converter AC-side fundamental voltage and sinusoidal current, a filtered sample of the midpoint current, and the partial DC-side voltages. Figure 6.21 illustrates the harmonic spectrum of the AC-side terminal voltage, confirming that there is no low-order voltage harmonic between the fundamental frequency and the switching side-band frequencies.

Figure 6.22 illustrates the same variables as those shown in Figure 6.20, but under the condition that a capacitive DC-side voltage divider is employed, while the net DC voltage is imposed and constant at 1500 V. Figure 6.22(c) shows that the partial DC-side voltages include third-harmonic ripple components each with an amplitude of about 50 V. Figure 6.23 illustrates the corresponding harmonic spectrum of the AC-side terminal voltage and indicates that the amplitude of the fundamental component is slightly different from that shown in Figure 6.21. Figure 6.23 further indicates that the third- and fifth-order harmonic components are also present in the AC-side terminal voltage spectrum.

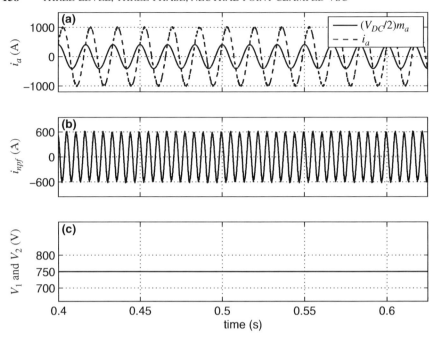

FIGURE 6.20   AC-side current and fundamental voltage, midpoint current, and partial DC-side voltages of the three-level NPC when partial DC-side voltages are impressed by constant-voltage sources; Example 6.2.

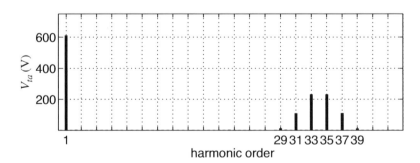

FIGURE 6.21   Harmonic spectrum of the AC-side terminal voltage of the three-level NPC of Example 6.2 when the partial DC-side voltages are impressed by constant-voltage sources.

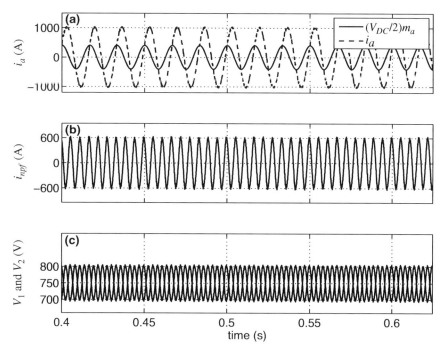

FIGURE 6.22   AC-side current and fundamental voltage, midpoint current, and partial DC-side voltages of the three-level NPC when a capacitive voltage divider is employed; Example 6.2.

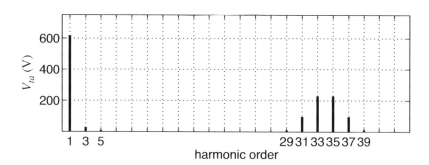

FIGURE 6.23   Harmonic spectrum of the AC-side terminal voltage of the three-level NPC of Example 6.2 when the capacitor voltages include third-harmonic ripple components.

7 Grid-Imposed Frequency VSC System: Control in $\alpha\beta$-Frame

7.1 INTRODUCTION

Chapter 5 presented dynamic models of the two-level VSC in $\alpha\beta$-frame and dq-frame and briefly introduced the control of a generic three-phase converter system in those frames. Chapter 6 introduced the three-level NPC as a generalization of the two-level VSC and established that the dynamic model of the three-level NPC is conceptually the same as that of the two-level VSC, except that the three-level NPC requires a DC-side voltage equalizing scheme to maintain the DC capacitor voltages, each at half the net DC-side voltage. Chapter 6 also presented a unified model for the three-level NPC and the two-level VSC which are generically called *VSC*. In this chapter, we introduce and investigate the control of a *grid-imposed frequency VSC system* in which the three-phase variables are sinusoidal functions of time with a frequency that is imposed by the AC system, for example, a power utility grid. The methodology is for the VSC and therefore covers the control of both the two-level VSC and the three-level NPC. This class of VSC systems is often adopted for real- and reactive-power control, or for DC voltage control. These functions constitute the main operational functions of electronically-coupled distributed generation (DG) units, VSC-based HVDC systems, and most FACTS controllers.

7.2 STRUCTURE OF GRID-IMPOSED FREQUENCY VSC SYSTEM

Figure 7.1 shows the schematic diagram of a grid-imposed frequency VSC system. The VSC can be a three-level NPC, with a DC-side voltage equalizer, or a two-level VSC. As Figure 7.1 shows, in either case, the VSC can be modeled as a lossless power processor with an equivalent DC-bus capacitor. A current source in parallel with the DC side represents the VSC switching power loss, and series resistances at the AC side, that is, switch on-state resistances, represent the conduction power loss. The DC side of the VSC may be interfaced with either a DC voltage source or the DC side of another power-electronic system. Each phase of the VSC is interfaced

Voltage-Sourced Converters in Power Systems, by Amirnaser Yazdani and Reza Iravani
Copyright © 2010 John Wiley & Sons, Inc.

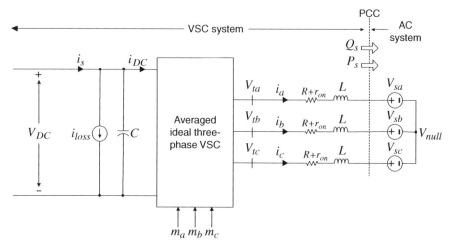

FIGURE 7.1   Schematic diagram of a grid-imposed frequency VSC system.

with the corresponding phase of an AC system, via a series RL branch. Initially, we assume an infinitely stiff AC system and, thus, represent it by an ideal, three-phase, voltage source, that is, V_{sabc}.[1] It is also assumed that V_{sabc} is balanced, sinusoidal, and with a constant frequency. The nodes where the VSC system and the AC system are interfaced are referred to as the point of common coupling (PCC) (Fig. 7.1). The VSC system of Figure 7.1 exchanges real- and reactive-power components $P_s(t)$ and $Q_s(t)$ with the AC system, at the PCC.

Depending on the control philosophy, the VSC system of Figure 7.1 can be used as either a *real-/reactive-power controller* or a *controlled DC-voltage power port*. In Chapter 12, we employ the real-/reactive-power controller as part of a back-to-back HVDC converter system. In Chapters 11, 12, and 13, we utilize the controlled DC-voltage power port as part of the static compensator (STATCOM), the back-to-back HVDC converter system, and variable-speed wind-power units, respectively.

7.3   REAL-/REACTIVE-POWER CONTROLLER

The grid-imposed frequency VSC system of Figure 7.1 can be employed as a real-/reactive-power controller. As such, the VSC DC side is connected in parallel to a DC voltage source (not shown in Fig. 7.1). The objective is to control the instantaneous real and reactive power delivered by the VSC system at the PCC, that is, $P_s(t)$ and $Q_s(t)$.

---

[1] In Chapter 11, we investigate the dynamics of the VSC system under nonstiff AC system conditions.

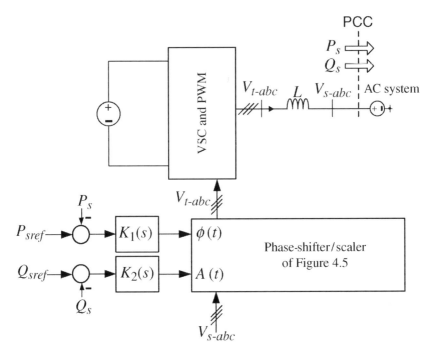

FIGURE 7.2   Schematic diagram of a voltage-controlled real-/reactive-power controller.

7.3.1 Current-Mode Versus Voltage-Mode Control

In the VSC system of Figure 7.1, P_s and Q_s can be controlled based on two distinct methods. The first approach is schematically illustrated in Figure 7.2 and is commonly referred to as *voltage-mode control*. The voltage-mode control has been mainly utilized in high-voltage/-power applications such as in FACTS controllers [44, 45], although industrial applications have also been reported [47].

In a voltage-controlled real-/reactive-power controller, the real and reactive power are controlled, respectively, by the phase angle and amplitude of the VSC AC-side terminal voltage relative to those of the PCC voltage [46]. If the amplitude and phase angle of V_{tabc} are close to those of V_{sabc}, the real and reactive power are almost decoupled and two independent compensators can be employed for their control (Fig. 7.2). Thus, the voltage-mode control has the merit of being simple and having a low number of control loops. However, since there is no control loop dedicated to the VSC line current, the VSC is not protected against overcurrents, and the current can undergo large excursions if the commands are changed rapidly or the AC system is subjected to a fault.

The second approach to control the real and reactive power in the VSC system of Figure 7.1 is referred to as *current-mode control*. In this approach, first the VSC AC-side current is controlled by a dedicated control scheme, through the VSC terminal voltage. Then, both real and reactive power are controlled by the phase angle and the amplitude of the VSC line current with respect to the PCC voltage. Thus, due to

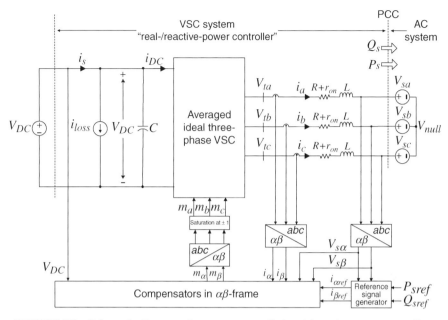

FIGURE 7.3   Schematic diagram of a current-controlled real-/reactive-power controller.

current regulation scheme, the VSC is protected against overload conditions. Other advantages of the current-mode control include the robustness against variations in parameters of the VSC and AC system, superior dynamic performance, and higher control precision [68]. We demonstrated the basics of the current-mode control strategy in Chapter 3 and will exclusively focus on this method throughout the rest of this book.

Figure 7.3 illustrates a schematic diagram of a current-controlled real-/reactive-power controller. Figure 7.3 illustrates that the control is performed in $\alpha\beta$-frame. Thus, P_s and Q_s are controlled by the line current components i_α and i_β. Feedback and feed-forward signals are first transformed to the $\alpha\beta$-frame and then processed by the compensators to produce control signals in $\alpha\beta$-frame. Finally, the control signals are transformed to the *abc*-frame and fed to the VSC (Fig. 7.3). To protect the VSC, the reference commands $i_{\alpha ref}$ and $i_{\beta ref}$ are limited by saturation blocks. It should be noted that the block diagram of Figure 7.3 is a special case of the general block diagram of Figure 4.24.

7.3.2   Dynamic Model of Real-/Reactive-Power Controller

Assume that the three-phase voltage of the AC system (Fig. 7.3) is

$$V_{sa}(t) = \widehat{V}_s \cos(\omega_0 t + \theta_0), \tag{7.1}$$

$$V_{sb}(t) = \widehat{V}_s \cos\left(\omega_0 t + \theta_0 - \frac{2\pi}{3}\right), \tag{7.2}$$

$$V_{sc}(t) = \widehat{V}_s \cos\left(\omega_0 t + \theta_0 - \frac{4\pi}{3}\right), \tag{7.3}$$

where $\widehat{V}_s$ is the peak value of the line-to-neutral voltage, ω_0 is the source (constant) frequency, and θ_0 is the initial phase angle. Based on (4.56) and (4.57), the real and reactive power delivered to the AC system are

$$P_s(t) = \frac{3}{2}\left[V_{s\alpha}(t)i_\alpha(t) + V_{s\beta}(t)i_\beta(t)\right], \tag{7.4}$$

$$Q_s(t) = \frac{3}{2}\left[-V_{s\alpha}(t)i_\beta(t) + V_{s\beta}(t)i_\alpha(t)\right], \tag{7.5}$$

where $V_{s\alpha}$ and $V_{s\beta}$ are the source $\alpha\beta$-frame components and cannot be controlled. Thus, based on (7.4) and (7.5), i_α and i_β must be controlled to exercise control over $P_s(s)$ and $Q_s(s)$. This calls for the following reference commands for i_α and i_β:

$$i_{\alpha ref}(t) = \frac{2}{3}\frac{V_{s\alpha}}{V_{s\alpha}^2 + V_{s\beta}^2}P_{sref}(t) + \frac{2}{3}\frac{V_{s\beta}}{V_{s\alpha}^2 + V_{s\beta}^2}Q_{sref}(t), \tag{7.6}$$

$$i_{\beta ref}(t) = \frac{2}{3}\frac{V_{s\beta}}{V_{s\alpha}^2 + V_{s\beta}^2}P_{sref}(t) - \frac{2}{3}\frac{V_{s\alpha}}{V_{s\alpha}^2 + V_{s\beta}^2}Q_{sref}(t). \tag{7.7}$$

If the control system ensures a fast command tracking performance, that is, $i_\alpha \approx i_{\alpha ref}$ and $i_\beta \approx i_{\beta ref}$, then $P_s \approx P_{sref}$ and $Q_s \approx Q_{sref}$. Equations (7.6) and (7.7) suggest that $P_s(t)$ and $Q_s(t)$ can be independently controlled. This is the salient feature of the $\alpha\beta$-frame control strategy.

Dynamics of the AC side of the VSC system of Figure 7.1 are described by

$$L\frac{di_a}{dt} = -(R + r_{on})i_a + V_{ta} - V_{sa} - V_{null}, \tag{7.8}$$

$$L\frac{di_b}{dt} = -(R + r_{on})i_b + V_{tb} - V_{sb} - V_{null}, \tag{7.9}$$

$$L\frac{di_c}{dt} = -(R + r_{on})i_c + V_{tc} - V_{sc} - V_{null}. \tag{7.10}$$

Then, based on (4.2) and since $e^{j0} + e^{j\frac{2\pi}{3}} + e^{j\frac{4\pi}{3}} \equiv 0$, (7.8)–(7.10) can be represented by the space-phasor equation

$$L\frac{d\overrightarrow{i}}{dt} = -(R + r_{on})\overrightarrow{i} + \overrightarrow{V_t} - \overrightarrow{V_s}, \tag{7.11}$$

in which V_{null} does not appear. Substituting for $\vec{f} = f_\alpha + jf_\beta$ in (7.11), and splitting the resultant into real and imaginary parts, we deduce

$$L\frac{di_\alpha}{dt} = -(R + r_{on})i_\alpha + V_{t\alpha} - V_{s\alpha}, \tag{7.12}$$

$$L\frac{di_\beta}{dt} = -(R + r_{on})i_\beta + V_{t\beta} - V_{s\beta}. \tag{7.13}$$

Expressing $V_{t\alpha}$ and $V_{t\beta}$ in terms of m_α and m_β, respectively, in (7.12) and (7.13), we deduce

$$L\frac{di_\alpha}{dt} = -(R + r_{on})i_\alpha + \frac{V_{DC}}{2}m_\alpha - V_{s\alpha}, \tag{7.14}$$

$$L\frac{di_\beta}{dt} = -(R + r_{on})i_\beta + \frac{V_{DC}}{2}m_\beta - V_{s\beta}. \tag{7.15}$$

Equations (7.12) and (7.13), or their equivalents (7.14) and (7.15), form a basis for the control of the VSC system of Figure 7.3.

7.3.3   Current-Mode Control of Real-/Reactive-Power Controller

Based on (7.14) and (7.15), Figure 7.4 illustrates a control block diagram of the real-/reactive-power controller of Figure 7.3. Figure 7.4 shows that the control plant

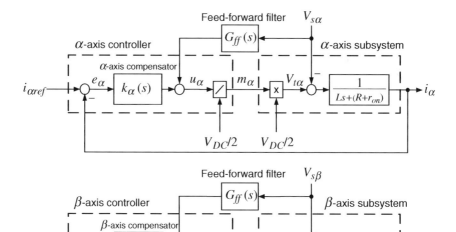

FIGURE 7.4   Control block diagram of the current-control loops of the VSC system of Figure 7.3, in $\alpha\beta$-frame.

consists of two subsystems: One is α-axis subsystem described by (7.14) and the other is β-axis subsystem described by (7.15). It is noted that α- and β-axis subsystems are decoupled and controlled independently. Thus, the reference commands $i_{\alpha ref}$ and $i_{\beta ref}$ are determined using (7.6) and (7.7), based on the required real- and reactive-power components. Since V_{sabc} is sinusoidal, so are $V_{s\alpha}$, $V_{s\beta}$, $i_{\alpha ref}$, and $i_{\beta ref}$. Therefore, to ensure command tracking with small steady-state errors, the bandwidth of the closed-loop control system must be selected to be adequately larger than the AC system frequency. Alternatively, the loop gain can be made very large at the AC system frequency. The latter approach corresponds to the inclusion of one pair of complex-conjugate poles in each of the compensators $k_\alpha(s)$ and $k_\beta(s)$; the compensators require additional poles and zeros to ensure desired phase and gain margins (see Example 3.6), and typically are of high dynamic orders. Therefore, to simplify the design process, a class of compensators with one pair of complex-conjugate poles and only two gain parameters has been proposed in the literature [69]. Thus, this class of compensators, referred to as *stationary-frame generalized integrator* [39] or *proportional+resonant compensator* (P+Resonant) [40], is analogous to the conventional PI compensator as it has only two parameters to tune.

It should be emphasized that each control loop in Figure 7.4, that is, the α- or β-axis control loop, is equivalent to the control loop presented in Figure 3.6 for the half-bridge converter. Therefore, the compensator design for α- and β-axis loops should follow the same steps as outlined for the system of Figure 3.6, in Example 3.6.

EXAMPLE 7.1   Dynamic Performance of Real-/Reactive-Power Controller

Consider the real-/reactive-power controller of Figure 7.3 with the parameters $L = 100$ μH, $R = 0.75$ mΩ, $r_{on} = 0.88$ mΩ, $V_d = 1.0$ V, $V_{DC} = 1450$ V, and $f_s = 3420$ Hz. Assume that the VSC system is connected at the PCC to a stiff utility grid with a line-to-line rms voltage of 480 V, corresponding to $\widehat{V}_s = 391$ V, and with $\omega_0 = 377$ rad/s. Following the guidelines of Example 3.6, one obtains the following compensator for each control loop in Figure 7.4:

$$k_\alpha(s) = k_\beta(s)$$

$$= 1258 \left(\frac{s + 16.34}{s^2 + (377)^2} \right) \left(\frac{s + 966}{s + 5633} \right) \left(\frac{s + 2}{s + 0.05} \right) \quad [\Omega]. \quad (7.16)$$

Note that the compensator includes one pair of complex-conjugate poles at $\omega_0 = 377$ rad/s, to ensure zero steady-state error, and it provides a closed-loop bandwidth of about $\omega_b = 3820$ rad/s (see Example 3.6 for details). The transfer function of the feed-forward filter is $G_{ff}(s) = 1/(8 \times 10^{-6} s + 1)$. Figure 7.5 illustrates the time response of the closed-loop system to the start-up process and step changes in P_{sref} and Q_{sref}.

For the real-/reactive-power controller of Figure 7.3, initially the gating pulses are blocked and the controllers are inactive. At $t = 0.1$ s, the gating

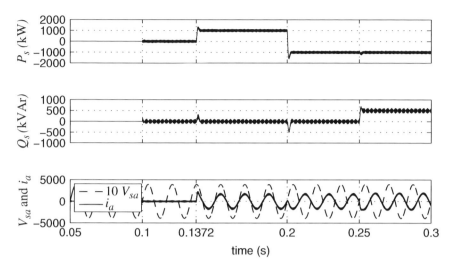

FIGURE 7.5   Dynamic responses of real and reactive power; Example 7.1.

pulses are unblocked and the controllers are activated while $P_{sref} = Q_{sref} \equiv 0$. Therefore, the line current remains at zero and no (average) real or reactive power is exchanged; small ripples in instantaneous values of i_a, P_s and Q_s (Fig. 7.5) are due to the pulse-width modulation (PWM) switching harmonics. At $t = 0.1372$ s and $t = 0.20$ s, P_{sref} is changed from 0 to 1.0 MW and from 1.0 to -1.0 MW, respectively. At $t = 0.25$ s, Q_{sref} is changed from 0 to 500 kVAr.

Figure 7.5 illustrates that P_s and Q_s rapidly follow their respective commands. However, the responses of P_s and Q_s are not perfectly decoupled from each other. The reason is that a perfect decoupled control of P_s and Q_s requires i_α and i_β to instantly follow their corresponding reference commands issued based on (7.6) and (7.7). However, due to the limited bandwidth of α- and β-axis closed-loop systems, i_α and i_β are limited in terms of the speed of response and thus P_s and Q_s are somewhat coupled. Figure 7.5 also shows the grid phase-a voltage waveform, that is, V_{sa}, and the converter phase-a current waveform, that is, i_a. Figure 7.5 shows that i_a is (i) in phase with V_{sa} when $(P_s, Q_s) = (1.0\,\text{MW}, 0)$, (ii) $180°$ lagging V_{sa} when $(P_s, Q_s) = (-1.0\,\text{MW}, 0)$, and (iii) $153°$ lagging V_{sa} when $(P_s, Q_s) = (-1.0\,\text{MW}, 0.5\,\text{MVAr})$.

To provide more details into the $\alpha\beta$-frame control, we show in Figure 7.6 a number of important variables of the control system of Figure 7.4. Figure 7.6 shows that $i_{\alpha ref}$ (and $i_{\beta ref}$) is zero until $t = 0.1372$ s. Therefore, the α-axis (and β-axis) controller imposes u_α (and u_β) to be equal to $V_{s\alpha}$ (and $V_{s\beta}$). This ensures that there is no voltage drop across the interface inductors and, thus, i_α (and i_β) remains zero. Figure 7.6 also shows that $i_{\alpha ref}$ (and $i_{\beta ref}$) is changed to a nonzero value at $t = 0.1372$ s. Consequently, the α-axis (and β-axis) controller accordingly changes u_α (and u_β) with respect to $V_{s\alpha}$ (and $V_{s\beta}$),

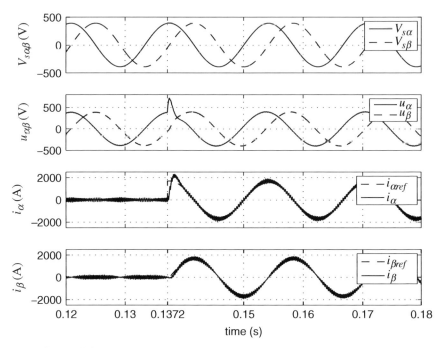

FIGURE 7.6   Dynamic responses of α-axis and β-axis controllers; Example 7.1.

to generate the required voltage drop across the interface inductors and regulate i_α (and i_β) at its reference command. It should be noted that the outputs of α- and β-axis controllers, that is, u_α and u_β, are equivalent to the $\alpha\beta$-frame components of the fundamental harmonic of the VSC AC-side terminal voltage.

7.3.4   Selection of DC-Bus Voltage Level

In the VSC system of Figure 7.3, the DC-bus voltage must be adequately large for proper operation of the VSC, under both steady-state and dynamic conditions. In this section, we develop an expression for the amplitude of the VSC AC-side voltage and also identify factors that must be considered in the selection of the DC-bus voltage.

Consider the VSC system of Figure 7.3 and that the AC system voltage at the PCC, that is, V_{sabc}, is given by (7.1)–(7.3). Let the VSC AC-side terminal voltage and current be

$$V_{ta}(t) = \widehat{V}_t \cos(\theta + \delta),$$

$$V_{tb}(t) = \widehat{V}_t \cos\left(\theta + \delta - \frac{2\pi}{3}\right),$$

$$V_{tc}(t) = \widehat{V}_t \cos\left(\theta + \delta - \frac{4\pi}{3}\right), \tag{7.17}$$

and

$$i_a(t) = \widehat{i}\cos(\theta - \phi),$$

$$i_b(t) = \widehat{i}\cos\left(\theta - \phi - \frac{2\pi}{3}\right),$$

$$i_c(t) = \widehat{i}\cos\left(\theta - \phi - \frac{4\pi}{3}\right), \tag{7.18}$$

where $\theta = \omega_0 t + \theta_0$; δ and $-\phi$ are, respectively, the phase shifts of V_{tabc} and i_{abc} with respect to V_{sabc}. Under steady-state conditions, ϕ is the power-factor angle of the VSC system, in the conventional phasor analysis sense. Based on (4.2), the space phasors corresponding to V_{sabc} and i_{abc} are

$$\overrightarrow{V}_s(t) = \widehat{V}_s e^{j\theta}, \tag{7.19}$$

$$\overrightarrow{i}(t) = \widehat{i}e^{-j\phi}e^{j\theta}. \tag{7.20}$$

Substituting for $\overrightarrow{V}_s$ and $\overrightarrow{i}$ in (4.38) and (4.40), one obtains the real and reactive power delivered to the AC system as

$$P_s = \frac{3}{2}\widehat{i}\widehat{V}_s\cos\phi, \tag{7.21}$$

$$Q_s = \frac{3}{2}\widehat{i}\widehat{V}_s\sin\phi. \tag{7.22}$$

Equations (7.21) and (7.22) are rearranged as

$$\widehat{i}\cos\phi = P_s / \left(\frac{3}{2}\widehat{V}_s\right), \tag{7.23}$$

$$\widehat{i}\sin\phi = Q_s / \left(\frac{3}{2}\widehat{V}_s\right). \tag{7.24}$$

Based on (4.46), the α-axis components of V_{sabc}, V_{tabc}, and i_{abc} are

$$V_{s\alpha} = \widehat{V}_s\cos\theta, \tag{7.25}$$

$$V_{t\alpha} = \widehat{V}_t\cos(\theta + \delta) = \widehat{V}_t\cos\delta\cos\theta - \widehat{V}_t\sin\delta\sin\theta, \tag{7.26}$$

$$i_\alpha = \widehat{i}\cos(\theta - \phi) = \widehat{i}\cos\phi\cos\theta + \widehat{i}\sin\phi\sin\theta. \tag{7.27}$$

Substituting for $\widehat{i}\cos\phi$ and $\widehat{i}\sin\phi$ in (7.27), from (7.23) and (7.24), substituting for $V_{s\alpha}$, $V_{t\alpha}$, and i_α in (7.12), from (7.25), (7.26), and (7.27), and assuming $(R + r_{on}) \approx 0$, we obtain

$$\left(\frac{2L}{3\widehat{V}_s}\right)\frac{d}{dt}(P_s\cos\theta + Q_s\sin\theta) = \left(\widehat{V}_t\cos\delta - \widehat{V}_s\right)\cos\theta - \widehat{V}_t\sin\delta\sin\theta. \qquad (7.28)$$

We note that the assumption of $(R + r_{on}) \approx 0$ is reasonable since the AC-side interface circuit is predominantly inductive and its impedance is approximately equal to the reactance of the interface reactor, that is, $L\omega_0$ (in Example 7.1, $L\omega_0 \approx 23(R + r_{on})$).

Calculating the derivatives in (7.28) and equating the corresponding coefficients of $\cos\theta$ and $\sin\theta$ in the resultant, we obtain

$$\widehat{V}_t\cos\delta = \widehat{V}_s + \left(\frac{2L}{3\widehat{V}_s}\right)\left(\frac{dP_s}{dt} + \omega_0 Q_s\right), \qquad (7.29)$$

$$\widehat{V}_t\sin\delta = \left(\frac{2L}{3\widehat{V}_s}\right)\left(-\frac{dQ_s}{dt} + \omega_0 P_s\right), \qquad (7.30)$$

where $\omega_0 = d\theta/dt$ is the AC system frequency. Squaring both sides of (7.29) and (7.30), and adding the corresponding sides of the resultants, we deduce

$$\begin{aligned}
\widehat{V}_t(t) = \Bigg\{ & \widehat{V}_s^2 + \frac{4}{9}\left(\frac{L\omega_0}{\widehat{V}_s}\right)^2\left(P_s^2 + Q_s^2\right) + \left(\frac{4L\omega_0}{3}\right)Q_s \\
& + \frac{4}{9}\left(\frac{L\omega_0}{\widehat{V}_s}\right)^2\left[\left(\frac{1}{\omega_0}\frac{dP_s}{dt}\right)^2 + \left(\frac{1}{\omega_0}\frac{dQ_s}{dt}\right)^2\right] \\
& + \left[\frac{4L\omega_0}{3} + \frac{8}{9}\left(\frac{L\omega_0}{\widehat{V}_s}\right)^2 Q_s\right]\left(\frac{1}{\omega_0}\frac{dP_s}{dt}\right) \\
& + \left[-\frac{8}{9}\left(\frac{L\omega_0}{\widehat{V}_s}\right)^2 P_s\right]\left(\frac{1}{\omega_0}\frac{dQ_s}{dt}\right)\Bigg\}^{\frac{1}{2}}. \qquad (7.31)
\end{aligned}$$

Equation (7.31) expresses the amplitude of the VSC AC-side terminal voltages under steady-state and transient conditions; individual phase voltages are given by (7.17). Based on (5.10)–(5.12), the following modulating signal is required to generate the AC-side terminal voltage (7.17):

$$m_a(t) = (2/V_{DC})\,\widehat{V}_t(t)\cos\left(\omega_0 t + \theta_0 + \delta\right),$$

$$m_b(t) = (2/V_{DC})\,\widehat{V}_t(t)\cos\left(\omega_0 t + \theta_0 + \delta - \frac{2\pi}{3}\right),$$

$$m_c(t) = (2/V_{DC})\,\widehat{V}_t(t)\cos\left(\omega_0 t + \theta_0 + \delta - \frac{4\pi}{3}\right). \qquad (7.32)$$

It then follows from (7.32), subject to the constraints

$$-1 \leq m_a(t) \leq 1,$$
$$-1 \leq m_b(t) \leq 1,$$
$$-1 \leq m_c(t) \leq 1,$$

that the following inequality must be ensured:

$$\widehat{V}_t(t) \leq \frac{V_{DC}}{2}. \tag{7.33}$$

Inequality (7.33) indicates a substantial limitation of the PWM-controlled VSC; that is, if $|m| \leq 1$, the maximum fundamental AC-side voltage of the VSC cannot be larger than half the DC-bus voltage. If $|m| > 1$, the fundamental AC-side voltage is larger than $V_{DC}/2$ and the modulation process is called *overmodulation* [16]. The overmodulation results in low-order harmonics in the AC voltage spectrum, and thus in this text we confine our investigations to $|m| \leq 1$.

Based on (7.31), the maximum AC-side voltage of the VSC, $\widehat{V}_t(t)$, is composed of steady-state and transient components. The steady-state component is determined based on the steady-state values of P_s and Q_s, whereas the transient component is a function of dP_s/dt and dQ_s/dt. The worst-case scenario corresponds to the condition that dP_s/dt and dQ_s/dt are large (i.e., P_s and Q_s are rapidly changed) while the VSC system operates near its rated operating condition; this scenario corresponds to a large value of $\widehat{V}_t$. Under a steady state, the derivative terms are zero and (7.31) is simplified to

$$\widehat{V}_t = \sqrt{\widehat{V}_s^2 + \frac{4}{9}\left(\frac{L\omega_0}{\widehat{V}_s}\right)^2 (P_s^2 + Q_s^2) + \left(\frac{4L\omega_0}{3}\right)Q_s}. \tag{7.34}$$

Typically, $L\omega_0 \ll \widehat{V}_s$ and, therefore, (7.34) can be further simplified to

$$\widehat{V}_t \approx \sqrt{\widehat{V}_s^2 + \left(\frac{4L\omega_0}{3}\right)Q_s}. \tag{7.35}$$

Equation (7.35) suggests that, under steady-state conditions, the maximum amplitude of the VSC AC-side terminal voltage is primarily affected by the amount of the reactive power. Figure 7.7 plots the ratio $\widehat{V}_t/\widehat{V}_s$ as a function of P_s and Q_s, based on the parameters of Example 7.1. Figure 7.7(a) illustrates $\widehat{V}_t/\widehat{V}_s$ versus Q_s, when Q_s is varied from -1.0 to 1.0 MVAr, while $P_s \equiv 0$. In Figure 7.7(a), we refer to the result obtained based on (7.35) as *approximate* and to those of (7.34) as *exact*. Figure 7.7(a) shows that $\widehat{V}_t/\widehat{V}_s$ is unity for $Q_s = 0$ and approximately linearly proportional to Q_s. Figure 7.7(b) shows that $\widehat{V}_t$ is always negligibly larger than $\widehat{V}_s$ for a nonzero real power if $Q_s \equiv 0$. Thus, the ratio $\widehat{V}_t/\widehat{V}_s$ can be assumed to be independent of P_s.

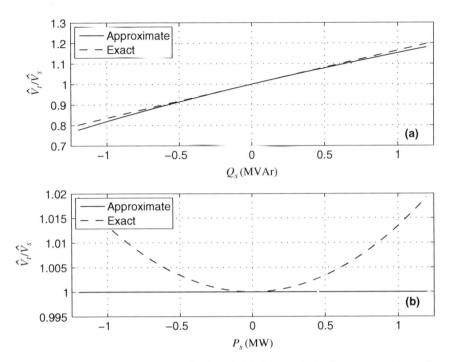

FIGURE 7.7   Variations of the normalized VSC AC-side terminal voltage versus (a) reactive power, when $P_s \equiv 0$, and (b) real power, when $Q_s \equiv 0$.

Equation (7.31) can also be simplified such that it includes the impact of transients of P_s and Q_s on the amplitude of the AC-side terminal voltage. Assuming $L\omega_0 \ll \widehat{V}_s$, (7.31) is simplified to

$$\widehat{V}_t(t) \approx \sqrt{\widehat{V}_s^2 + \left(\frac{4L\omega_0}{3}\right)Q_s + \left(\frac{4L}{3}\right)\frac{dP_s}{dt}}. \tag{7.36}$$

A comparison between (7.36) and (7.35) reveals that the transient component of $\widehat{V}_t$ is primarily due to the changes in P_s; the faster a change takes place, the larger is the transient component. Based on (7.36), if P_s does not change, $\widehat{V}_t$ is equal to its steady-state value, that is, $\sqrt{\widehat{V}_s^2 + \frac{4L\omega_0}{3}Q_s}$, as formulated by (7.35). Then, $\widehat{V}_t$ becomes larger (smaller) than its steady-state value if P_s starts to increase (decrease). Depending on the value of L and the rise time of $P_s(t)$, the deviation of $\widehat{V}_t$ from its steady-state value can be significant. For example, Figure 7.8 illustrates variations of $\widehat{V}_t$ in response to changes in P_s and Q_s, in the VSC system of Example 7.1. Figure 7.8 shows that until $t = 0.25$ s where $Q_s \equiv 0$, the steady-state value of $\widehat{V}_t$ is equal to $\widehat{V}_s = 391$ V. However, $\widehat{V}_t$ overshoots to about 725 V and then reverts to its predisturbance value when P_s rapidly changes from 0 to 1.0 MW at $t = 0.1372$ s. At $t = 0.2$ s, P_s rapidly decreases from 1.0 to -1.0 MW, and thus $\widehat{V}_t$ undergoes an

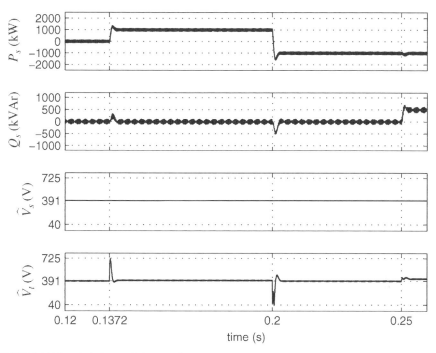

FIGURE 7.8 Variations of the amplitude of the AC-side terminal voltage in response to changes in real and reactive power of the VSC; Example 7.1.

undershoot down to about 40 V. Figure 7.8 also shows that, as predicted by (7.36), a rapid change in Q_s (from zero to 500 kVAr at $t = 0.25$ s) does not result in a transient overshoot or undershoot in $\widehat{V}_t$ but shifts the steady-state value of $\widehat{V}_t$ to a different (larger) value.

It is understood from the foregoing discussion that $\widehat{V}_t(t)$ can have a large peak if P_s is rapidly increased while Q_s is settled at a positive value. Usually, the instants at which Q_s and/or P_s are commanded to change are not known in advance. Therefore, in view of (7.32), the worst-case scenario corresponds to the instant when $\widehat{V}_t(t)$ peaks while $\cos(\omega_0 t + \theta_0 + \delta)$, $\cos(\omega_0 t + \theta_0 + \delta - \frac{2\pi}{3})$, or $\cos(\omega_0 t + \theta_0 + \delta - \frac{4\pi}{3})$ is equal or close to $+1$ or -1. To accommodate the worst-case scenario, V_{DC} must be selected in accordance with (7.33). In Example 7.1, as Figure 7.9 shows, $\widehat{V}_t(t)$ reaches its maximum value at about $t = 0.1372$ s when $\cos(\omega_0 t + \theta_0 + \delta) \approx 1$. Hence, $m_a(t)$ also reaches its maximum value of unity (Fig. 7.9).

7.3.5 Trade-Offs and Practical Considerations

As discussed in Section 7.3.4, since the DC-bus voltage must be larger than twice $\widehat{V}_t(t)$ (inequality (7.33)), it is desired to keep $\widehat{V}_t(t)$ small. To this end, as (7.36) suggests, L should be selected to be small and/or the rate of change of P_s must be limited. Practically, a small value for L requires a relatively large PWM switching frequency to ensure low harmonic distortion of the VSC AC-side current. However,

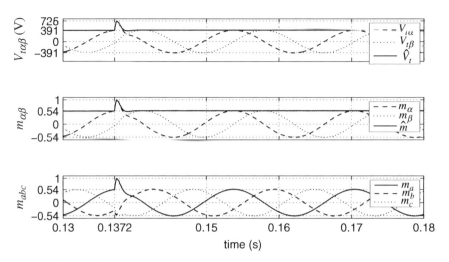

FIGURE 7.9   PWM modulating signals of the VSC system of Example 7.1.

the VSC loss increases as the switching frequency is increased (see (5.9)). Therefore, there is often a trade-off between the DC-bus voltage level and the PWM switching frequency.

On the other hand, to limit dP_s/dt, changes in P_{sref} must be gradual. This is in contradiction to fast control of P_s and Q_s. However, P_{sref} is often the output of an outer control loop of which the bandwidth is considerably smaller than that of the α- and β-axis current controllers. Thus, the outer control loop cannot exercise step changes in P_{sref}. If P_s is directly controlled in a particular application, P_{sref} can be passed through a rate limiter. However, applications similar to Example 7.1 in which P_{sref} and/or Q_{sref} are changed stepwise are not common in practice.

7.3.6   PWM with Third-Harmonic Injection

7.3.6.1   Principle of Operation   As discussed in Section 7.3.4, depending on the requirements for the VSC system steady-state and dynamic operations, relatively high levels of voltage may be required at the VSC AC-side terminals. Based on the developments of Chapters 2 and 6, the (averaged) AC-side terminal voltage of the half-bridge converter and the three-level half-bridge NPC is

$$V_t(t) = m(t)\frac{V_{DC}}{2}, \tag{7.37}$$

where the modulating signal $m(t)$ can be a function of time subject to $-1 \leq m(t) \leq 1$. Thus, the AC-side terminal voltage of each phase of the VSC is constrained by $-V_{DC}/2 \leq V_t(t) \leq V_{DC}/2$. The third-harmonic injected PWM is an approach to extend the range of $V_t(t)$ [70].

As presented in Chapters 5 and 6, the three-phase modulating signal of VSC can be expressed as

$$m_a(t) = \widehat{m}(t)\cos\left[\varepsilon(t)\right], \tag{7.38}$$

$$m_b(t) = \widehat{m}(t)\cos\left[\varepsilon(t) - \frac{2\pi}{3}\right], \tag{7.39}$$

$$m_c(t) = \widehat{m}(t)\cos\left[\varepsilon(t) - \frac{4\pi}{3}\right], \tag{7.40}$$

where $\widehat{m}(t)$ is the amplitude of the modulating signal $m_{abc}(t)$ and, in general, is a function of time. Under transient conditions, depending on the disturbances and the closed-loop bandwidth of the controllers, $\widehat{m}(t)$ can have large overshoots. Since the instant at which $\widehat{m}(t)$ reaches a peak is not known upfront, one must ensure that $\widehat{m}(t) \leq 1$, so that $-1 \leq m_{abc}(t) \leq 1$.

The method of third-harmonic injected PWM modifies the modulating signals (7.38)–(7.40) as

$$m_{aug\text{-}a}(t) = \widehat{m}(t)\cos\varepsilon - \frac{1}{6}\widehat{m}(t)\cos 3\varepsilon, \tag{7.41}$$

$$m_{aug\text{-}b}(t) = \widehat{m}(t)\cos\left(\varepsilon - \frac{2\pi}{3}\right) - \frac{1}{6}\widehat{m}(t)\cos 3\left(\varepsilon - \frac{2\pi}{3}\right), \tag{7.42}$$

$$m_{aug\text{-}c}(t) = \widehat{m}(t)\cos\left(\varepsilon - \frac{4\pi}{3}\right) - \frac{1}{6}\widehat{m}(t)\cos 3\left(\varepsilon - \frac{4\pi}{3}\right). \tag{7.43}$$

As indicated by (7.41)–(7.43), $m_{aug\text{-}b}$ and $m_{aug\text{-}c}$ have the same forms as $m_{aug\text{-}a}$ but are phase shifted, respectively, by $-2\pi/3$ and $-4\pi/3$ with respect to $m_{aug\text{-}a}$. Therefore, we focus on $m_{aug\text{-}a}$ to study the properties of $m_{aug\text{-}abc}$. Figure 7.10(a) and (b), respectively, illustrates the waveforms of m_a and $m_{aug\text{-}a}$, when $\widehat{m} = 1$. Under this condition, any further increase in $\widehat{m}$ would result in the overmodulation, if the conventional PWM was used (Fig. 7.10(a)). However, since the peak value of $m_{aug\text{-}a}$ is equal to 0.869 (Fig. 7.10(b)), $\widehat{m}$ can be increased to 1.15, that is, to 1/0.869, if $m_{aug\text{-}a}$ is used as the PWM modulating signal. Hence, the amplitude of the AC-side terminal voltage can be increased by about 15% for a given DC-bus voltage or, alternatively, the DC-bus voltage can be reduced by about 13% for a given amplitude of the AC-side terminal voltage.

Let us prove that if $m_{aug\text{-}abc}$ is employed as the PWM modulating signal, the VSC system dynamic and steady-state characteristics remain the same as those under the case where m_{abc} is used. Thus, with $m_{aug\text{-}abc}$ as the PWM modulating signal, based

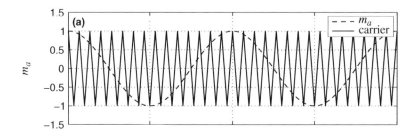

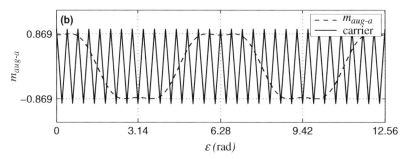

FIGURE 7.10   Modulating signal corresponding to (a) conventional sinusoidal PWM and (b) third-harmonic injected PWM.

on (7.37) the VSC AC-side terminal voltages are

$$V_{taug-a}(t) = \widehat{m}(t)\frac{V_{DC}}{2}\cos\varepsilon - \widehat{m}(t)\frac{V_{DC}}{12}\cos 3\varepsilon, \tag{7.44}$$

$$V_{taug-b}(t) = \widehat{m}(t)\frac{V_{DC}}{2}\cos\left(\varepsilon - \frac{2\pi}{3}\right) - \widehat{m}(t)\frac{V_{DC}}{12}\cos 3\varepsilon, \tag{7.45}$$

$$V_{taug-c}(t) = \widehat{m}(t)\frac{V_{DC}}{2}\cos\left(\varepsilon - \frac{4\pi}{3}\right) - \widehat{m}(t)\frac{V_{DC}}{12}\cos 3\varepsilon, \tag{7.46}$$

where $\widehat{m}(t) \leq 1.15$. Adding the corresponding sides of (7.44)–(7.46), one obtains

$$V_{taug-a} + V_{taug-b} + V_{taug-c} = -\widehat{m}(t)\frac{V_{DC}}{4}\cos 3\varepsilon. \tag{7.47}$$

With reference to the VSC system of Figure 7.3, adding the corresponding sides of (7.8)–(7.10), one deduces

$$L\frac{d}{dt}(i_a + i_b + i_c) = -(R + r_{on})(i_a + i_b + i_c)$$
$$+ \left(V_{taug-a} + V_{taug-b} + V_{taug-c}\right) - (V_{sa} + V_{sb} + V_{sc}) - 3V_{null}. \tag{7.48}$$

Since the VSC AC-side interface is a three-wire connection, $(i_a + i_b + i_c) \equiv 0$. Thus, substituting in (7.48) for V_{sabc} from (7.1)–(7.3), and for $(V_{taug-a} + V_{taug-b} + V_{taug-c})$ from (7.47), we deduce

$$V_{null} - \widehat{m}(t)\frac{V_{DC}}{12}\cos 3\varepsilon. \tag{7.49}$$

Equation (7.49) indicates that the third-harmonic component of the PWM modulating signal is amplified by $V_{DC}/2$ and appears at the AC-side neutral point. The same would be true if the modulating signal included any other triple-n harmonic component. As stated in Section 4.2.4, the space phasors corresponding to the triple-n harmonics of a balanced three-phase waveform are of zero length and referred to as *zero-sequence harmonics*. In general, zero-sequence components of modulating signal would be amplified by $V_{DC}/2$ and appear at the AC-side neutral point.

Substituting in (7.8)–(7.10) for $V_{taug-abc}$ from (7.44)–(7.46), and for V_{null} from (7.49), we obtain

$$L\frac{di_a}{dt} = -(R + r_{on})i_a + \widehat{m}(t)\frac{V_{DC}}{2}\cos\varepsilon - V_{sa}, \tag{7.50}$$

$$L\frac{di_b}{dt} = -(R + r_{on})i_b + \widehat{m}(t)\frac{V_{DC}}{2}\cos\left(\varepsilon - \frac{2\pi}{3}\right) - V_{sb}, \tag{7.51}$$

$$L\frac{di_c}{dt} = -(R + r_{on})i_c + \widehat{m}(t)\frac{V_{DC}}{2}\cos\left(\varepsilon - \frac{4\pi}{3}\right) - V_{sc}, \tag{7.52}$$

where $\widehat{m}(t) \leq 1.15$. Comparing (7.38)–(7.40) and (7.50)–(7.52), we deduce

$$L\frac{di_a}{dt} = -(R + r_{on})i_a + \frac{V_{DC}}{2}m_a - V_{sa}, \tag{7.53}$$

$$L\frac{di_b}{dt} = -(R + r_{on})i_b + \frac{V_{DC}}{2}m_b - V_{sb}, \tag{7.54}$$

$$L\frac{di_c}{dt} = -(R + r_{on})i_c + \frac{V_{DC}}{2}m_c - V_{sc}. \tag{7.55}$$

Equations (7.53)–(7.55) indicate that the third harmonic of the modulating signal $m_{aug-abc}$ has no impacts on the system transient and steady-state behavior. The space-phasor representation of (7.53)–(7.55) is

$$L\frac{d\vec{i}}{dt} = -(R + r_{on})\vec{i} + \frac{Vdc}{2}\vec{m} - \vec{V_s}, \tag{7.56}$$

corresponding to the following $\alpha\beta$-axis equations:

$$L\frac{di_\alpha}{dt} = -(R + r_{on})i_\alpha + \frac{V_{DC}}{2}m_\alpha - V_{s\alpha}, \tag{7.57}$$

$$L\frac{di_\beta}{dt} = -(R + r_{on})i_\beta + \frac{V_{DC}}{2}m_\beta - V_{s\beta}, \tag{7.58}$$

which are identical to (7.14)–(7.15), respectively. Thus, a VSC system that utilizes $m_{aug\text{-}abc}$ as the PWM modulating signal possesses the same control model as the one that assumes m_{abc}. However, with $m_{aug\ abc}$ the magnitude of $\overrightarrow{m}$, that is, $\hat{m}$, can be as high as 1.15, and thus the VSC is able to produce an AC-side voltage of 1.15 times that of the conventional PWM. Similarly, for a given AC-side voltage level, the VSC with the third-harmonic injected PWM can employ a DC-bus voltage level of 13% lower.

7.3.6.2 *Implementation*

To generate the PWM modulating signal for the third-order harmonic injection, values of $\varepsilon(t)$ and $\hat{m}(t)$ are required based on (7.41)–(7.43). However, in the VSC system of Figure 7.3, the controllers provide m_α and m_β, and then m_{abc} is generated based on the $\alpha\beta$- to abc-frame transformation. Therefore, it is imperative to formulate $m_{aug\text{-}abc}$ in terms of m_{abc}, m_α, and m_β.

Consider that $m_{aug\text{-}a}$ is given by (7.41). Using the identity $\cos 3\varepsilon = 4\cos^3\varepsilon - 3\cos\varepsilon$, (7.41) can be rewritten as

$$m_{aug\text{-}a}(t) = \frac{3}{2}\hat{m}(t)\cos\varepsilon - \frac{2}{3}\hat{m}(t)\cos^3\varepsilon. \tag{7.59}$$

Multiplying the second term of the right-hand side of (7.59) by $\hat{m}^2(t)/\hat{m}^2(t)$, we obtain

$$m_{aug\text{-}a}(t) = \frac{3}{2}\hat{m}(t)\cos\varepsilon - \frac{2}{3}\frac{\left[\hat{m}(t)\cos\varepsilon\right]^3}{\hat{m}^2(t)}. \tag{7.60}$$

Eliminating $\hat{m}(t)\cos\varepsilon$ between (7.38) and (7.60), and using $\hat{m}^2(t) = m_\alpha^2 + m_\beta^2$, we conclude that

$$m_{aug\text{-}a}(t) = \frac{3}{2}m_a(t) - \frac{2}{3}\frac{m_a^3(t)}{m_\alpha^2 + m_\beta^2}. \tag{7.61}$$

Similarly, $m_{aug\text{-}b}$ and $m_{aug\text{-}c}$ are derived as

$$m_{aug\text{-}b}(t) = \frac{3}{2}m_b(t) - \frac{2}{3}\frac{m_b^3(t)}{m_\alpha^2 + m_\beta^2} \tag{7.62}$$

$$m_{aug\text{-}c}(t) = \frac{3}{2}m_c(t) - \frac{2}{3}\frac{m_c^3(t)}{m_\alpha^2 + m_\beta^2}. \tag{7.63}$$

Figure 7.11 shows a block diagram of a signal transformer that receives m_α and m_β, and generates $m_{aug\text{-}abc}$. Figure 7.12 shows a schematic diagram of a real-/reactive-power controller that employs the third-harmonic injected PWM. The real-/reactive-power controller of Figure 7.12 is the same as that of Figure 7.3, with the exception that the $m_{\alpha\beta}$-to-m_{abc} block is now replaced by the block diagram of Figure 7.11. The third-harmonic injected PWM is equally applicable to the two-level VSC (Fig. 6.19) and the three-level NPC (Fig. 6.18). Therefore, for real-/reactive-power controllers of Figures 7.3 and 7.12, the VSC can be either a two-level VSC or a three-level NPC.

αβ- to abc-frame transformer with third-harmonic addition

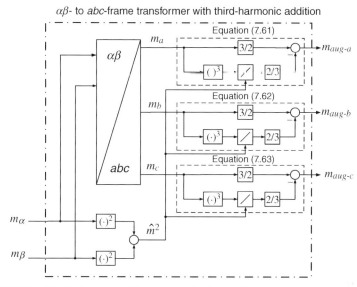

FIGURE 7.11   Block diagram of the $\alpha\beta$- to abc-frame signal transformer used for third-harmonic injected PWM.

As Figure 7.12 illustrates, for a VSC employing the third-harmonic injected PWM, each component of $m_{aug-abc}$ is limited to ± 1. This is equivalent to $|m_{abc}| < 1$ for a VSC that employs the conventional PWM strategy. Thus, based on the third-harmonic

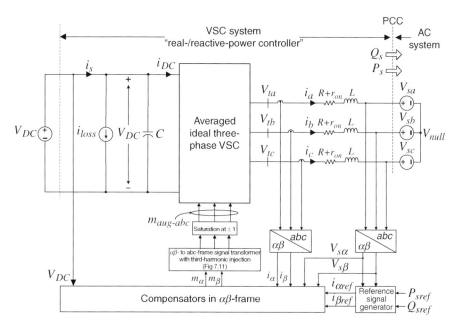

FIGURE 7.12   Schematic diagram of the real-/reactive-power controller adopting the third-harmonic injected PWM.

injected PWM, the VSC AC-side voltage can be as high as $\pm1.15(V_{DC}/2)$, instead of $\pm(V_{DC}/2)$ under the conventional PWM. Equivalently, if the third-harmonic injected PWM is employed, a lower DC-bus voltage can be adopted for the VSC, as the next example illustrates.

EXAMPLE 7.2   Third-Harmonic Injected PWM

Consider the VSC system of Example 7.1 that adopts the third-harmonic injected PWM. The VSC system parameters are the same as those of Example 7.1, and the system is subjected to the same sequence of disturbances. However, the DC-bus voltage is chosen to be $V_{DC} = 1250$ V in this example. Figure 7.13 illustrates the behavior of the VSC system in response to a step change in P_{sref} at $t = 0.1372$ s, from zero to 1.0 MW.

A comparison between Figures 7.13(a) and 7.9 indicates that m_{abc} in this case is larger than that in Example 7.1. In the VSC system of Example 7.1, following the disturbance, m_a transiently reaches its maximum value of unity (Fig. 7.9) and is not permitted to increase further. However, under the third-harmonic injected PWM, m_a can reach about 1.15 (Fig. 7.13(a)), while $m_{aug\text{-}a}$ reaches unity (Fig. 7.13(b)); this is acceptable since under the third-harmonic injected PWM, $m_{aug\text{-}abc}$ (and not m_{abc}) is constrained to ±1.

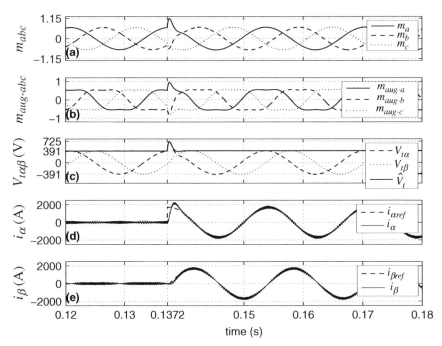

FIGURE 7.13 Dynamic response of the VSC system using third-harmonic injected PWM; Example 7.2.

A comparison between Figures 7.13(c) and 7.9 reveals that the third-harmonic injected PWM generates the same AC-side voltage as that in Example 7.1. Figures 7.13(d) and (e) indicate that the VSC system with the third-harmonic injected PWM exhibits the same dynamic and steady-state behavior as those shown in Figure 7.6 for the VSC system of Example 7.1. However, the DC-bus voltage is 13% smaller than that in Example 7.1.

7.4 REAL-/REACTIVE-POWER CONTROLLER BASED ON THREE-LEVEL NPC

The real-/reactive-power controllers of Figures 7.3 and 7.12 can utilize a three-level NPC instead of a two-level VSC. Thus, based on the unified dynamic model of Section 6.7.4, the same procedures described for the two-level VSC can be applied to the design of the α- and β-axis compensators of the VSC system. The only difference is that the three-level NPC also requires a DC-side voltage equalizing scheme as discussed in Section 6.7.2. The DC-side voltage equalizing scheme is designed based on the block diagram of Figure 6.17, the three-level NPC parameters, and the operating range. The design is, however, independent of the parameters of α- and β-axis current controllers. In this section, the procedure for designing the DC-side voltage equalizing scheme is presented.

Assume that the three-level NPC of Figure 7.14 is used as the VSC for the real-/reactive-power controller of Figure 7.3 or Figure 7.12. The control block diagram of the DC-side voltage equalizing scheme is the same as that of Figure 6.17, which, for ease of reference, is repeated here as Figure 7.15. Figure 7.15 indicates that the control plant is an integrator with a variable gain of $-\hat{i}\cos\gamma$ where, as discussed in Chapter 6, $\hat{i}$ is the amplitude of the AC-side current and γ is the angle of the AC-side terminal voltage minus that of the AC-side current, that is, the power-factor angle of the VSC. Both $\hat{i}$ and γ are functions of the operating point of the three-level NPC and, therefore, $-\hat{i}\cos\gamma$ can assume different (positive or negative) values. Consequently, a fixed-structure compensator, although stable for a specific operating point, may be unstable at another operating point. To make the loop gain independent of the operating point, the compensator output can be divided by $-\hat{i}\cos\gamma$ (Fig. 7.15). Since the converter system is controlled in $\alpha\beta$-frame, it is preferred to express the gain $-\hat{i}\cos\gamma$ in terms of the $\alpha\beta$-frame variables.

The modulating waveforms of the three-level NPC are

$$m_a(t) = m_0 + \widehat{m}(t)\cos\left[\varepsilon(t)\right] - \frac{1}{6}\widehat{m}(t)\cos\left[3\varepsilon(t)\right], \tag{7.64}$$

$$m_b(t) = m_0 + \widehat{m}(t)\cos\left[\varepsilon(t) - \frac{2\pi}{3}\right] - \frac{1}{6}\widehat{m}(t)\cos\left[3\varepsilon(t)\right], \tag{7.65}$$

$$m_c(t) = m_0 + \widehat{m}(t)\cos\left[\varepsilon(t) - \frac{4\pi}{3}\right] - \frac{1}{6}\widehat{m}(t)\cos\left[3\varepsilon(t)\right]. \tag{7.66}$$

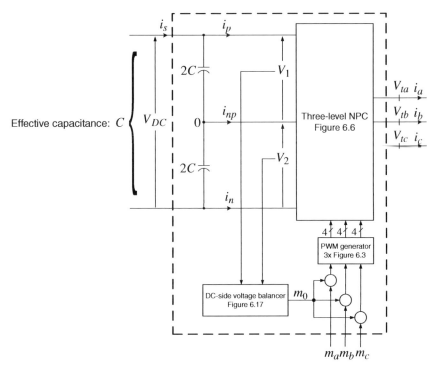

FIGURE 7.14   Block diagram of the three-level NPC.

These are the same modulating signals (6.46)–(6.48) that are augmented with the third-harmonic component. Based on (7.37), the AC-side terminal voltages are

$$V_{ta}(t) = m_0 \frac{V_{DC}}{2} + \widehat{m}(t) \frac{V_{DC}}{2} \cos\left[\varepsilon(t)\right] - \frac{1}{6}\widehat{m}(t) \frac{V_{DC}}{2} \cos\left[3\varepsilon(t)\right], \qquad (7.67)$$

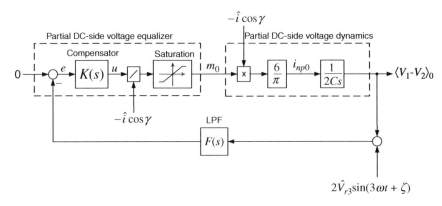

FIGURE 7.15   Control block diagram of the DC-side voltage equalizing scheme for the three-level NPC of Figure 7.14.

$$V_{tb}(t) = m_0 \frac{V_{DC}}{2} + \widehat{m}(t) \frac{V_{DC}}{2} \cos\left[\varepsilon(t) - \frac{2\pi}{3}\right] - \frac{1}{6}\widehat{m}(t) \frac{V_{DC}}{2} \cos[3\varepsilon(t)], \quad (7.68)$$

$$V_{tc}(t) = m_0 \frac{V_{DC}}{2} + \widehat{m}(t) \frac{V_{DC}}{2} \cos\left[\varepsilon(t) - \frac{4\pi}{3}\right] - \frac{1}{6}\widehat{m}(t) \frac{V_{DC}}{2} \cos[3\varepsilon(t)]. \quad (7.69)$$

Based on (4.2), the space-phasor equivalent of (7.67)–(7.69) is

$$\overrightarrow{V}_t(t) = \widehat{m}(t) \frac{V_{DC}}{2} e^{j\varepsilon(t)}. \quad (7.70)$$

Based on (6.35)–(6.37), the converter AC-side currents are

$$i_a(t) = \widehat{i} \cos\left[\varepsilon(t) - \gamma(t)\right], \quad (7.71)$$

$$i_b(t) = \widehat{i} \cos\left[\varepsilon(t) - \gamma(t) - \frac{2\pi}{3}\right], \quad (7.72)$$

$$i_c(t) = \widehat{i} \cos\left[\varepsilon(t) - \gamma(t) - \frac{4\pi}{3}\right], \quad (7.73)$$

which correspond to the space phasor

$$\overrightarrow{i}(t) = \widehat{i}(t) e^{j\varepsilon(t)} e^{-j\gamma(t)}. \quad (7.74)$$

Based on (4.38), the instantaneous real power leaving the converter AC-side terminals is

$$P_t(t) = Re\left\{\frac{3}{2}\overrightarrow{v}_t(t)\overrightarrow{i}^*(t)\right\} = \frac{3}{4}V_{DC}\widehat{m}\left(\widehat{i}\cos\gamma\right). \quad (7.75)$$

Dividing both sides of (7.75) by $(3/4)V_{DC}\widehat{m}$, we deduce

$$\widehat{i}\cos\gamma = \left(\frac{4}{3V_{DC}\widehat{m}}\right)P_t(t). \quad (7.76)$$

If the instantaneous power exchanged with the interface reactors is neglected, then $P_t(t) \approx P_s(t)$ and (7.76) can be written as

$$\widehat{i}\cos\gamma \approx \left(\frac{4}{3V_{DC}\widehat{m}}\right)P_s(t). \quad (7.77)$$

Equation (7.77) indicates that since V_{DC} and $\widehat{m}$ are always positive, the sign of $\widehat{i}\cos\gamma$ is the same as that of $P_s(t)$. In addition, V_{DC} is often regulated at a fairly constant value, and $\widehat{m}$ typically changes over a fairly narrow range, for example, from 0.7

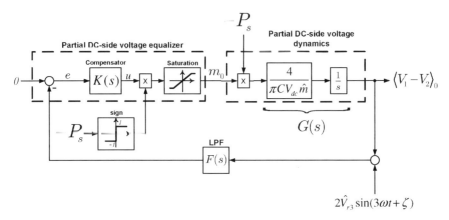

FIGURE 7.16   Modified control block diagram of the partial DC-side voltage equalizing scheme for the three-level NPC of Figure 7.14.

to 1.0.[2] Thus, $\widehat{i}\cos\gamma$ is approximately proportional to $P_s(t)$. Based on (7.77), the loop gain of the DC-side voltage equalizing system of Figure 7.15 is mainly a function of the real-power exchange between the three-level NPC and the AC system, and assumes opposite signs under inverting $(P_s(t) \geq 0)$ and rectifying $(P_s(t) \leq 0)$ modes of operation.

Figure 7.16 shows the control block diagram of Figure 7.15 in which the variable gain $-\widehat{i}\cos\gamma$ in the plant model has been replaced by its equivalent expression in terms of P_s, based on (7.77). However, only the sign of P_s is compensated in the control loop. Thus, the magnitude of the loop gain is proportional to the absolute value of P_s, while its sign always remains positive irrespective of the sign of P_s.

For the control loop of Figure 7.16, P_s can be considered as the disturbance input. In the steady state, this disturbance does not impact the output, that is, $\langle V_1 - V_2 \rangle_0$, even if $K(s)$ is a pure gain; the reason is that the control plant includes an integral term. The (compensator) gain can be selected for the converter-rated real power. Thus, the loop gain and as a result the bandwidth of the closed-loop system drop as P_s decreases. However, this is not an issue for the DC-side voltage equalizing system as the reference command, that is, the desired difference between the voltages of the two DC-side capacitors is normally set to zero and not changed.

As explained in Section 6.7.2, $V_1 - V_2$ includes a dominant third-harmonic component that must be attenuated by the filter $F(s)$. This is possible if a pair of complex-conjugate zeros at $s = \pm j(3\omega_0)$ is assigned to $F(s)$. Then, implementation considerations require $F(s)$ to have at least two poles,[3] that is, $F(s)$ should

---

[2] As we saw in Section 7.3.4, $\widehat{m}(t)$, under steady-state conditions, is a function of the reactive power, but indicates large excursions from its steady-state value if the real power quickly changes.
[3] This renders $F(s)$ as a notch filter with stop frequency of $3\omega_0$.

be a proper transfer function.[4] The following example demonstrates the design procedure.

EXAMPLE 7.3   Real-/Reactive-Power Controller Based on the Three-level NPC

Consider the VSC system of Figure 7.12 that utilizes the three-level NPC of Figure 7.14 and adopts the third-harmonic injected PWM strategy. The system parameters and controllers are the same as those in Example 7.1, except that $2C = 19250$ μF, $r_{on} = 0.44$ mΩ, and $V_{DC} = 1.25$ kV. The rated power of the system of Figure 7.12 is $P_s = 1.0$ MW. Based on $\widehat{m} = 2\widehat{V}_t/V_{DC}$ and assuming $\widehat{V}_t \approx \widehat{V}_s = 0.391$ kV, we obtain $\widehat{m} \approx 0.63$. Thus, the plant transfer function, in the closed-loop system of Figure 7.16, is

$$G(s) = \left(\frac{4}{\pi C V_{DC}\widehat{m}}\right)\frac{1}{s} = \frac{168}{s} \quad [(\text{kA})^{-1}].$$

Let $F(s)$ be

$$F(s) = \frac{s^2 + (3\omega_0)^2}{(s + 3\omega_0)^2} = \frac{s^2 + 1131^2}{s^2 + 2262s + 1131^2},$$

where $\omega_0 = 377$ rad/s. Then, (i) the third-harmonic component of $V_1 - V_2$ is suppressed, (ii) $F(s)$ has a unity DC gain, and (iii) the loop gain continues to roll off for frequencies beyond $3\omega_0$. If a pure gain is considered for the compensator, that is, $K(s) = k$, and $|P_s| = 1$ MW, then the loop gain is

$$\ell(s) = K(s)G(s)F(s)|P_s| = (168k)\frac{s^2 + 1131^2}{s\left(s^2 + 2262s + 1131^2\right)}.$$

For $\omega \leq \omega_0/10$, the phase of $\ell(j\omega)$ is almost constant at $-90°$, corresponding to a phase margin of $90°$. For ω larger than $\omega_0/10$, due to the double real pole at $\omega = \omega_0$, the phase drops with a slope of approximately $-90°/$dec. Thus, if we need a phase margin of, for example, $70°$, we should select the loop gain crossover frequency, that is, ω_c, to be about 1.93 decade larger than $\omega_0/10$, that is, $\omega_c = 218$ rad/s. Substituting for $\omega_c = 218$ rad/s in equation $|\ell(j\omega_c)| = 1$, we find $k = 1.40$ (kV)$^{-1}$, for which the exact phase margin is $68°$, and the closed-loop poles are located at $s = -1910$ rad/s and $s = -294 \pm j267$ rad/s. Figure 7.17 shows the frequency responses of the loop gain and closed-loop transfer function. It is observed that the bandwidth of the closed-loop system is about $\omega_b = 400$ rad/s. The closed-loop transfer function is $G_{cl}(s) = |P_s|K(s)G(s)/(1 + \ell(s))$.

---

[4]By definition, a proper transfer function is one whose denominator has a degree equal to or higher than that of its numerator.

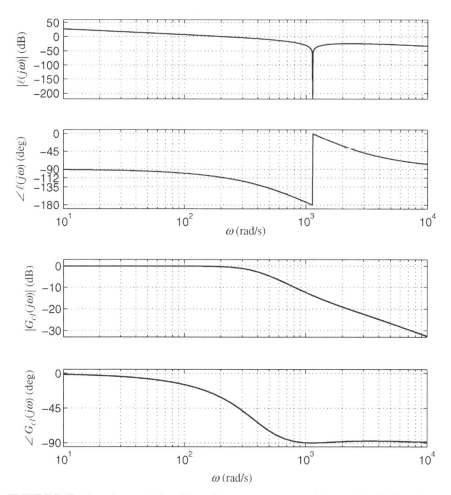

FIGURE 7.17   Open-loop and closed-loop frequency responses of the partial DC-side voltage balancer of Figure 7.16.

Figures 7.18 and 7.19 illustrate the dynamic behavior of the real-/reactive-power controller in response to step changes in real- and reactive-power commands. Based on the unified model of the VSC, presented in Section 6.7.4, we anticipate no differences between the dynamic responses of the real-/reactive-power controller of this example (using a three-level NPC) and that of Example 7.1 (utilizing a two-level VSC). A comparison of Figures 7.18 and 7.5 confirms this. Figures 7.19(a)–(c) show the modulating signals, that is, m_{abc}, the DC offset added to the modulating signals to equalize the partial DC-side voltages, that is, m_0, and the third-harmonic injected modulating signals, that is, $m_{aug-abc}$. Figure 7.19(b) shows that during transients m_0 is changed by the closed-loop controller of Figure 7.16 to maintain the DC-side voltages V_1 and V_2.

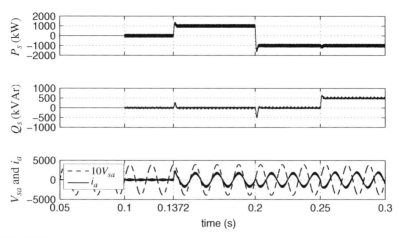

FIGURE 7.18 Dynamic responses of real and reactive power for the three-level NPC of Example 7.3. Compare the results with Figure 7.5 of Example 7.1.

Figure 7.19(d) illustrates that V_1 and V_2 preserve their equal DC components during the transients, but their third-harmonic components change according to the operating condition. As explained in Chapter 6, the third-harmonic component of the midpoint current of the three-level NPC increases as the converter

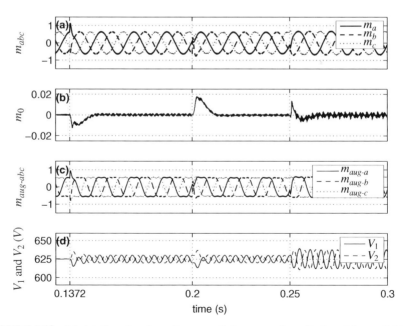

FIGURE 7.19 Modulating signals, voltage equalizing corrective offset, and the partial DC-side voltages of the converter system of Example 7.3.

power factor decreases. Thus, as Figure 7.19(d) illustrates, the (third-harmonic) ripples of V_1 and V_2 increase after $t = 0.25$ s when the three-level NPC starts to supply reactive power.

7.4.1 Midpoint Current of Three-level NPC Based on Third-Harmonic Injected PWM

Section 6.6.3 formulated the midpoint current of the three-level NPC and concluded that based on the conventional PWM strategy, the midpoint current can be approximated by the following third-harmonic component:

$$i_{np}(t) \approx \left(0.764\widehat{mi}\right) \sqrt{1 - 0.555 \cos^2\gamma} \cdot \cos\left(3\varepsilon + \zeta\right), \tag{7.78}$$

where γ is the power-factor angle of the three-level NPC and $\zeta = \pi - \tan^{-1}(1.5 \tan \gamma)$. Then, based on (7.78), it was shown that the peak of the third harmonic of the midpoint current resides within the following range:

$$0.51\widehat{mi} \leq \widehat{i}_{np} \leq 0.76\widehat{mi}, \tag{7.79}$$

where the lower bound corresponds to unity power factor, whereas the upper bound corresponds to zero power factor. Adopting the third-harmonic injected PWM strategy, the midpoint current of the three-level NPC can be expressed as

$$i_{np}(t) \approx \left(0.709\widehat{mi}\right) \sqrt{1 - 0.934 \cos^2\gamma} \cdot \cos\left(3\varepsilon + \zeta\right), \tag{7.80}$$

where γ is the power-factor angle of the converter and $\zeta = \tan^{-1}(3.9 \tan \gamma)$. Equation (7.80) is derived by substituting for m_a from (7.64) in (6.31) and following the procedures detailed in Section 6.6.3. As (7.80) indicates, based on the third-harmonic injected PWM strategy, the amplitude of the midpoint current is subject to

$$0.18\widehat{mi} \leq \widehat{i}_{np} \leq 0.71\widehat{mi}, \tag{7.81}$$

where, similar to (7.79), the lower bound corresponds to unity power factor and the upper bound corresponds to zero power factor. A comparison between (7.81) and (7.79) indicates that if the three-level NPC operates at a high power factor, that is, the converter is mainly a real-power processor, the amplitude of the midpoint current and the third-harmonic ripples of the partial DC-side voltages are considerably lower when the third-harmonic injected PWM is used. This can be regarded as another advantage of the third-harmonic injected PWM strategy for the three-level NPC. As the power factor of the converter decreases, the reduction in the midpoint current under the third-harmonic injected PWM strategy becomes less significant.

Figure 7.20 compares the third-harmonic voltage ripples of DC-side voltages of the three-level NPC of Example 7.3 under the conventional and the third-harmonic

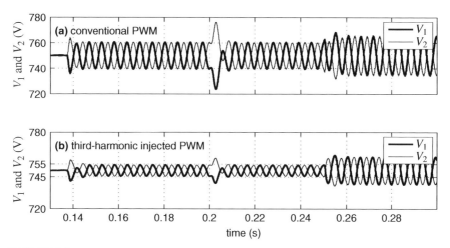

FIGURE 7.20   Third-harmonic ripples on the partial DC-side voltages of the three-level NPC using (a) conventional PWM and (b) third-harmonic injected PWM.

injected PWM strategies, when the net DC-bus voltage is 1500 V. Figure 7.20 illustrates that the magnitude of ripples is reduced to about half when the third-harmonic injected PWM strategy is utilized.

It should be noted that the use of the third-harmonic injected PWM has been proposed in the technical literature as a means to eliminate the third-harmonic component of the midpoint current and, therefore, those of the DC capacitor voltages [71]. As described in Ref. [71], to eliminate the third harmonic of the midpoint current, the PWM modulating signals must be augmented with a third-harmonic component whose amplitude and phase angle are functions of γ; for any γ, the amplitude and the phase angle can be obtained from a look-up table [71]. Since the augmented modulating signals do not have the form described by (7.41)–(7.43), the PWM strategy proposed in Ref. [71] does not result in the maximum DC-to-AC voltage gain for the VSC.[5] The method of Ref. [71] is primarily intended to mitigate the third-harmonic voltage ripple, the associated stress imposed on the DC-side capacitors, and the low-order AC-side voltage harmonics generated as a result of the ripple.

7.5   CONTROLLED DC-VOLTAGE POWER PORT

The objective of the real-/reactive-power controller, as discussed with respect to the configuration of Figures 7.3 and 7.12, is to control the real and reactive power that the VSC system exchanges with the AC system. In the real-/reactive-power controller, the VSC DC side is connected to an ideal DC voltage source that dictates the DC-bus

---

[5]As discussed in Section 7.3.6, the DC-to-AC voltage gain is $V_{DC}/2$ for conventional PWM strategy and $1.15V_{DC}/2$ for third-harmonic injected PWM.

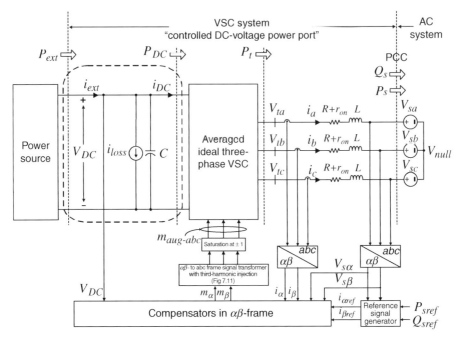

FIGURE 7.21   Schematic diagram of the controlled DC-voltage power port.

voltage. Thus, the VSC system acts as a bidirectional power-flow path between the AC system and the DC voltage source. However, in many applications, for example, photovoltaic (PV) systems and fuel-cell systems. The VSC DC side is not interfaced with a voltage source; rather, it is connected to a (DC) power source that needs to be interfaced and exchange (real) power with the AC system. Thus, the DC-bus voltage is not imposed and, therefore, needs to be regulated. This scenario is illustrated in Figure 7.21.

The VSC system of Figure 7.21 is conceptually the same as that of Figure 7.12, except that the DC voltage source is replaced by a (variable) DC power source. The power source typically represents a power-electronic unit (or a cluster of units) with the prime source of energy behind it, for example, a PV array, a variable-speed wind turbine-generator set, a fuel-cell unit, and a gas turbine-generator set. We consider it as a black box and assume that it exchanges a time-varying power $P_{ext}(t)$ with the VSC DC side. Thus, the VSC system of Figure 7.21 enables a bidirectional power exchange between the power source (black box) and the AC system. Hereinafter, we refer to the VSC system of Figure 7.21 as the controlled DC-voltage power port. The controlled DC-voltage power port is employed as an integral part of the STATCOM, the back-to-back HVDC converter system, and variable-speed wind-power units; these are discussed in Chapters 11, 12, and 13, respectively.

7.5.1   Model of Controlled DC-Voltage Power Port

Since there is no DC voltage source in the VSC system of Figure 7.21, any power imbalance within the outlined area of Figure 7.21 results in excursions (and potentially instability) of the DC-bus voltage, V_{DC}. Thus, the controlled DC-voltage power port of Figure 7.21 requires that V_{DC} be regulated. Normally, $P_{ext}(t)$ is an exogenous signal that cannot be controlled by the VSC system. Therefore, to ensure the balance of power, P_{DC} must be controlled via the VSC system. In the VSC system of Figure 7.21, the power balance is formulated as

$$P_{ext} - P_{loss} - \frac{d}{dt}\left(\frac{1}{2}CV_{DC}^2\right) = P_{DC}, \tag{7.82}$$

where $P_{loss} = V_{DC}\,i_{loss}$, $P_{DC} = V_{DC}\,i_{DC}$, and the third term at the left-hand side of (7.82) corresponds to the rate of change in the energy of the DC-bus capacitor. It should be noted that, based on the unified dynamic model of Section 6.7.4, (7.82) is valid for both the two-level VSC and the three-level NPC. However, the effective capacitor of the DC bus for a three-level NPC is half each of its DC-side capacitors. Substituting for $P_{DC} = P_t$ and rearranging the resultant, we deduce

$$\left(\frac{C}{2}\right)\frac{dV_{DC}^2}{dt} = P_{ext} - P_{loss} - P_t, \tag{7.83}$$

where P_t is the VSC AC-side terminal power. Equation (7.83) represents a dynamic system in which V_{DC}^2 is the state variable and the output, P_t is the control input, and P_{ext} and P_{loss} are the disturbance inputs.

Since the VSC system of Figure 7.21 enables to control P_s and Q_s, we express the control input P_t in terms of P_s and Q_s. Thus, by multiplying both sides of (7.11) with $(3/2)\,\vec{i}\,^*$, equating $\vec{i}\,\vec{i}\,^*$ to $\hat{i}^2$, and applying the real-part operator to both sides of the resultant, we obtain

$$\frac{3L}{2}Re\left\{\frac{d\vec{i}}{dt}\,\vec{i}\,^*\right\}$$
$$= -\frac{3}{2}(R + r_{on})\hat{i}^2 + Re\left\{\frac{3}{2}\vec{V}_t\,\vec{i}\,^*\right\} - Re\left\{\frac{3}{2}\vec{V}_s\,\vec{i}\,^*\right\}. \tag{7.84}$$

Based on (4.38), the second and third terms of the right-hand side of (7.84) are, respectively, equal to P_t and P_s. Thus,

$$\frac{3L}{2}Re\left\{\frac{d\vec{i}}{dt}\,\vec{i}\,^*\right\} = -\frac{3}{2}(R + r_{on})\hat{i}^2 + P_t - P_s. \tag{7.85}$$

Equation (7.85) can be solved for P_t as

$$P_t = P_s + \frac{3}{2}(R + r_{on})\widehat{i}^2 + \frac{3L}{2} Re\left\{\frac{d\overrightarrow{i}}{dt}\overrightarrow{i}^*\right\}.$$ (7.86)

As shown in Example 4.5, the term $\frac{3L}{2} Re\left\{\frac{d\overrightarrow{i}}{dt}\overrightarrow{i}^*\right\}$ is the instantaneous power absorbed by the three-phase inductor bank L (see (4.41)) and $\frac{3}{2}(R + r_{on})\widehat{i}^2$ is the power dissipated by the resistance of the three-phase inductor.

Practically, $(R + r_{on})$ is a small resistance and its associated power is negligible compared to P_t and P_s. However, the power absorbed by the three-phase inductor can be significant during transients. The reason is that, since in a high-power VSC the switching frequency is limited by power loss considerations, L must be adequately large to suppress the switching harmonics. Furthermore, since the $\alpha\beta$-frame current controllers are fast, $\overrightarrow{i}$ can undergo rapid phase and amplitude changes, during the real-/reactive-power command tracking process. Substituting for $\overrightarrow{i} = i_\alpha + ji_\beta$ in (4.41), we obtain

$$\frac{3L}{2} Re\left\{\frac{d\overrightarrow{i}}{dt}\overrightarrow{i}^*\right\} = \frac{3L}{4}\frac{d}{dt}\left(i_\alpha^2 + i_\beta^2\right) = \frac{3L}{4}\frac{d\widehat{i}^2}{dt}.$$ (7.87)

Based on (4.40), we deduce

$$P_s + jQ_s = \frac{3}{2}\overrightarrow{V}_s\overrightarrow{i}^*.$$ (7.88)

Applying the complex-conjugate operator to (7.88) and multiplying the resultant by (7.88), we deduce

$$P_s^2 + Q_s^2 = \frac{9}{4}\widehat{V}_s^2\widehat{i}^2.$$ (7.89)

Substituting for $\widehat{i}^2$ from (7.89) in (7.87), and assuming that $\widehat{V}_s$ is constant, we obtain

$$\frac{3L}{2} Re\left\{\frac{d\overrightarrow{i}}{dt}\overrightarrow{i}^*\right\} = \left(\frac{L}{3\widehat{V}_s^2}\right)\frac{dP_s^2}{dt} + \left(\frac{L}{3\widehat{V}_s^2}\right)\frac{dQ_s^2}{dt}.$$ (7.90)

Thus, (7.86) can be rewritten as

$$P_t \approx P_s + \left(\frac{2L}{3\widehat{V}_s^2}\right)P_s\frac{dP_s}{dt} + \left(\frac{2L}{3\widehat{V}_s^2}\right)Q_s\frac{dQ_s}{dt}.$$ (7.91)

Substituting for P_t from (7.91) in (7.83), we have

$$\frac{dV_{DC}^2}{dt} = \frac{2}{C}P_{ext} - \frac{2}{C}P_{loss} - \frac{2}{C}\left[P_s + \left(\frac{2LP_s}{3\widehat{V}_s^2}\right)\frac{dP_s}{dt}\right]$$

$$+ \frac{2}{C}\left[\left(\frac{2LQ_s}{3\widehat{V}_s^2}\right)\frac{dQ_s}{dt}\right]. \qquad (7.92)$$

Equation (7.92) describes dynamics of V_{DC}^2. Based on (7.92), V_{DC}^2 is the output, P_s is the control input, and P_{ext}, P_{loss}, and Q_s are the disturbance inputs. Thus, to control V_{DC}^2, one can form the control scheme shown in Figure 7.22, which consists of an inner control loop nested inside an outer loop. The outer loop compares V_{DC}^2 with the reference V_{DCref}^2, processes the error by a compensator, and delivers P_{sref} to the inner control loop. The inner control loop is the real-power controller detailed in Section 7.3, which regulates P_s at its reference, P_{sref}.

In the system of Figure 7.22, Q_s can be independently controlled. If the command Q_{sref} is set to zero, the VSC system operates at unity power factor. However, Q_{sref} may be set to a positive or a negative value, depending on the required level of the reactive-power exchange with the AC system; this is the operational strategy of a STATCOM. If the AC system has a significant equivalent impedance, the voltage at the PCC is subject to variations depending on the load changes and/or level of the real-power exchange. In this case, the PCC voltage can be regulated by controlling Q_s, via a closed-loop system that monitors the PCC voltage and determines Q_{sref}. The STATCOM mode of operation and the AC voltage regulation strategy through reactive-power control are discussed in Chapter 11.

Figure 7.22 illustrates that an estimate of P_{ext} can be added as a feed-forward signal to the output of the compensator $K_v(s)$. Thus, a change in P_{ext} is rapidly reflected in P_{sref}, and its impact on V_{DC}^2 is mitigated. Such an estimate of P_{ext} is available in many applications. For example, if the power source connected to the VSC system of Figure 7.22 is another VSC system that controls an electric machine, P_{ext} is approximately equal to the product of the machine torque and speed, ignoring the power losses of the machine and VSC. Figure 7.22 also shows that the compensator output, augmented by the measure of P_{ext}, is passed through a saturation block before the real-power controller. This ensures that the VSC is protected from overcurrent conditions in case the DC-bus voltage is subjected to a significant deviation from its reference value, or in case there is a large excursion in $P_{ext}(t)$.

In the VSC system of Figure 7.22, the compensator $K_v(s)$ is designed according to (7.92), which indicates that P_{ext} and P_{loss} impact V_{DC}^2 under both transient and steady-state conditions. The impact of P_{ext} is, to a great extent, mitigated by the feed-forward compensation. However, P_{loss} cannot be measured or estimated with certainty to be compensated by the feed-forward compensation. Therefore, to eliminate the steady-state error of V_{DC}^2 due to P_{loss}, $K_v(s)$ must include an integral term. Since the control plant also includes an integral term, to ensure stability and an adequate phase margin, $K_v(s)$ must also include a zero.

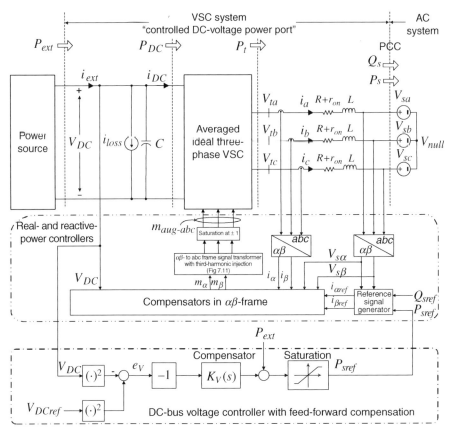

FIGURE 7.22 Schematic diagram of the controlled DC-voltage power port with DC-bus voltage regulator.

Due to the presence of terms $P_s \frac{dP_s}{dt}$ and $Q_s \frac{dQ_s}{dt}$, the control plant described by (7.92) is nonlinear. Thus, to design $K_v(s)$, we first linearize (7.92) about a steady-state operating point, which is computed by replacing all the derivatives in (7.92) by zero. Thus,

$$P_{s0} = P_{ext0} - P_{loss} \approx P_{ext0}, \qquad (7.93)$$

and (7.92) is linearized as

$$\frac{d\tilde{V}_{DC}^2}{dt} = \frac{2}{C}\tilde{P}_{ext} - \frac{2}{C}\left[\tilde{P}_s + \left(\frac{2LP_{s0}}{3\widehat{V}_s^2}\right)\frac{d\tilde{P}_s}{dt}\right]$$

$$+ \frac{2}{C}\left[\left(\frac{2LQ_{s0}}{3\widehat{V}_s^2}\right)\frac{d\tilde{Q}_s}{dt}\right], \qquad (7.94)$$

where the symbol $\sim$ denotes small-signal perturbations. Transformation of (7.94) into Laplace domain results in a transfer function from $\tilde{P}_s$ to $\tilde{V}_{DC}^2$ as

$$G_v(s) = \tilde{V}_{DC}^2(s)/\tilde{P}_s(s) = -\left(\frac{2}{C}\right)\frac{\tau s + 1}{s}, \tag{7.95}$$

where the time constant τ is

$$\tau = \frac{2L P_{s0}}{3\widehat{V}_s^2} = \frac{2L P_{ext0}}{3\widehat{V}_s^2}. \tag{7.96}$$

Equation (7.96) indicates that τ is proportional to the (steady-state) real-power flow P_{ext0} (or P_{s0}). Equation (7.96) suggests that if P_{ext0} is small, τ is insignificant and the plant is predominantly an integrator. As P_{ext} increases, τ becomes larger and brings about a shift in the phase of $G_v(s)$. In the inverting mode of operation where P_{ext0} is positive, τ is positive and adds to the phase of $G_v(s)$. However, in the rectifying mode of operation, that is, when P_{ext0} is negative, τ is negative and reduces the phase of $G_v(s)$; a larger absolute value of P_{ext0} results in a smaller phase of $G_v(s)$. Based on (7.95), the plant zero is $z = -1/\tau$. Therefore, a negative τ corresponds to a zero on the right-half plane (RHP), and the controlled DC-voltage power port represents a nonminimum-phase system in the rectifying mode of operation [72]. The phase reduction, associated with the nonminimum-phase zero, has a detrimental impact on the closed-loop stability, as discussed in Section 7.5.3.

7.5.2 DC-Bus Voltage Control in Controlled DC-Voltage Power Port

Figure 7.23 shows a block diagram of the DC-bus voltage control scheme for the controlled DC-voltage power port of Figure 7.22. The control scheme is composed of $K_v(s)$, the real-power controller $G_p(s)$, and the control plant $G_v(s)$ described by (7.95). As Figures 7.22 and 7.23 indicate, $K_v(s)$ is multiplied by -1 to compensate for the negative sign of $G_v(s)$. Thus, the loop gain is

$$\ell(s) = -K_v(s)G_p(s)G_v(s). \tag{7.97}$$

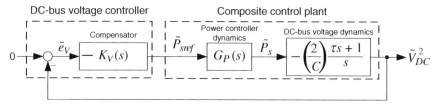

FIGURE 7.23 Control block diagram of the DC-bus voltage regulation loop based on the linearized model.

Let $K_v(s)$ be

$$K_v(s) = \left(\frac{C}{2}\right)\frac{H(s)}{s}, \tag{7.98}$$

where $H(s)$ is a proper transfer fraction with no zero at $s = 0$. Substituting in (7.97) for $K_v(s)$ and $G_v(s)$, respectively, from (7.98) and (7.95), we obtain

$$\ell(s) = G_p(s)H(s)\frac{\tau s + 1}{s^2}. \tag{7.99}$$

The objective is to select $H(s)$ such that $\ell(j\omega)$ crosses the 0 dB axis with the slope of about -20 dB/dec at $\omega = \omega_c$, and the phase of $\ell(j\omega_c)$ is larger than $-180°$ by a reasonable phase margin. This must be achieved with respect to the facts that (i) $G_p(j\omega)$ has a negative phase at high frequencies that can reduce the phase margin, and (ii) depending on the level of the real power, τ assumes different values.

To ensure that the phase delay due to $G_p(j\omega_c)$ is negligible, ω_c should be chosen adequately lower than the bandwidth of $G_p(j\omega)$, such that $G_p(j\omega_c) \approx 1 + j0$. Thus, ω_c is selected to be about 0.1–0.5 times the bandwidth of $G_p(j\omega)$. This should not be considered as a major limitation on the DC-bus voltage controller. The reason is that a very fast DC-bus voltage controller is often neither necessary nor even desired; in a high-power VSC system, particularly if a three-level NPC is used, the DC-bus capacitor is typically large and consequently the DC-bus voltage cannot be rapidly changed.

Assuming $G_p(j\omega_c) \approx 1$, (7.99) is rewritten as

$$\ell(j\omega_c) \approx H(j\omega_c)\frac{j\tau\omega_c + 1}{-\omega_c^2}. \tag{7.100}$$

$H(j\omega_c)$ is then selected to ensure that (i) $|\ell(j\omega_c)| = 1$ and (ii) $\angle\ell(j\omega_c)$ is larger than $-180°$ by a certain phase margin. The next example details the design process.

EXAMPLE 7.4   Design of the DC-Bus Voltage Controller

Consider the controlled DC-voltage power port of Figure 7.22. The system parameters and all controllers are the same as those in Example 7.1, and $C = 9625$ μF. The rated power of the converter system is $P_s = \pm 1$ MW. Therefore, based on (7.96), τ can vary from -0.43 to 0.43 ms. As in Example 7.1, $\omega_b = 3820$ rad/s for the $\alpha\beta$-axis controllers. Thus, we select the gain crossover frequency of the DC-bus voltage control loop to be $\omega_c = 700$ rad/s.

Let us first adjust the gain of $H(s)$ to achieve $|\ell(j\omega_c)| = 1$. Based on (7.100), and since $(\tau\omega_c)^2 \ll 1$, if $H(j\omega)$ is a pure gain equal to ω_c^2, then the loop gain is unity at $\omega = \omega_c$. This is shown in the plots of Figure 7.24 (solid lines) as the *uncompensated* case. The Bode plots of Figure 7.24 correspond to the positive rated real-power, the zero real-power, and the negative rated real-power

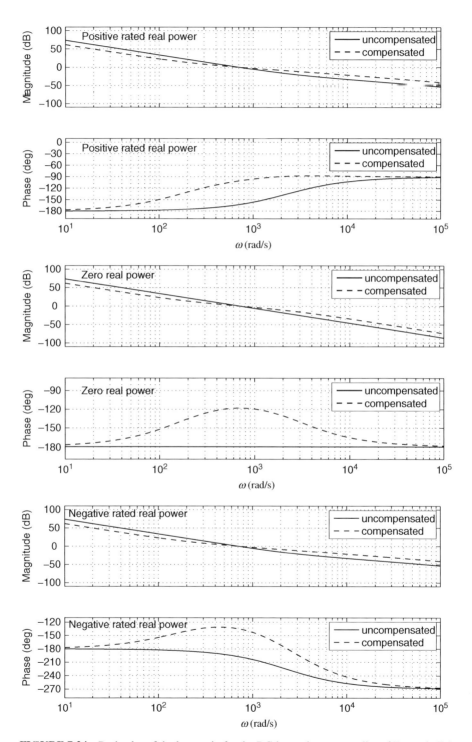

FIGURE 7.24 Bode plot of the loop gain for the DC-bus voltage controller of Example 7.4.

operating points. The three operating points correspond to $\tau = 0.43$ ms, $\tau = 0$, and $\tau = -0.43$ ms, respectively.

Figure 7.24 illustrates that the magnitude plot is similar for all three operating points and $|\ell(j700)| = 1$. Moreover, the closed-loop system is stable for the positive rated real power, with a phase margin of about 18°. However, the loop is unstable under the zero-power and the negative rated power operating points, as $\angle\ell(j700)$ is correspondingly equal to 180° and $-197°$. Based on parameters of this specific example, the closed-loop system is stable, although poorly, for the positive rated power; however, it becomes unstable as the power becomes increasingly negative [72].

To design the compensator, let us consider the worst-case scenario that corresponds to the negative rated real-power operating point, for which $\angle\ell(j700) \approx -197°$. Thus, if a phase margin of, for example, 45° is required, 62° must be added to $\angle\ell(j700)$. This can be achieved by means of a lead filter, as explained in Example 3.6. Let $H(s)$ be the lead filter

$$H(s) = h\frac{s + (p_1/\alpha)}{s + p_1}, \tag{7.101}$$

where p_1 is the filter pole and $\alpha > 1$ is a real constant. The maximum phase of the lead filter is

$$\delta_m = \sin^{-1}\left(\frac{\alpha - 1}{\alpha + 1}\right), \tag{7.102}$$

which corresponds to the frequency

$$\omega_m = \frac{p_1}{\sqrt{\alpha}}. \tag{7.103}$$

In our example, $\delta_m = 62°$ and $\omega_m = \omega_c = 700$ rad/s. Hence, based on (7.102) and (7.103), we obtain $\alpha = 16.1$ and $p_1 = 2808$ rad/s, and (7.101) becomes

$$H(s) = h\frac{s + 174.4}{s + 2808}. \tag{7.104}$$

Solving for h based on $|H(j700)| = \omega_c^2$, one finds $h = 1,965,666$ s^{-2}. Substituting for $H(s)$ from (7.104) in (7.98), one concludes

$$K_v(s) = 9459\frac{s + 174.4}{s(s + 2808)} \quad [\Omega^{-1}]. \tag{7.105}$$

Dashed lines in Figure 7.24 illustrate the Bode plots of the compensated loop gain. Figure 7.24 shows that for all the three operating points, the magnitude of $\ell(j\omega)$ remains similar to that of the uncompensated loop, and

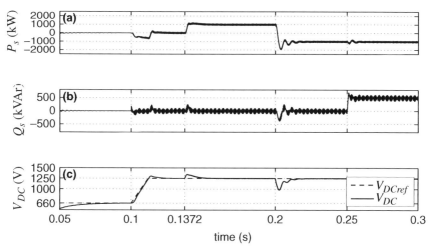

FIGURE 7.25 Behavior of the DC-bus voltage of the controlled DC-voltage power port, in response to changes in P_s and Q_s, without feed-forward compensation; Example 7.4.

$|\ell(j700)| = 1$. Moreover, the phase characteristic of the loop gain is improved such that $\angle\ell(j700)$ is equal to $-101°$, $-117°$, and $-135°$, respectively, for the positive rated power, zero-power, and negative rated power operating points. This, in turn, corresponds to phase margins of $79°$, $63°$, and $45°$, respectively.

Figure 7.25 illustrates the time response of the controlled DC-voltage power port of Figure 7.22 when the feed-forward compensation is disabled in the DC-bus voltage control loop. In addition, initially the DC power source is inactive, that is, $P_{ext} = 0$, the VSC gating pulses are blocked, and the controllers are inactive. However, the AC system (source) is interfaced with three-phase reactors of the VSC through three 0.5 Ω resistors, one in series with each inductor. Thus, the DC-bus capacitor is gradually charged to about 660 V via the antiparallel diodes of VSC switch cells (Fig. 7.25(c)). At $t = 0.1$ s, the three (0.5 Ω) start-up resistors are bypassed, the gating pulses are unblocked, and the controllers are activated. To limit the charging current of the DC-bus capacitor and also to avoid large overshoots or ringings in V_{DC}, from $t = 0.1$ s on, V_{DCref} is changed from 660 V to the nominal value of 1250 V at the rate of 50 V/ms, as Figure 7.25(c) illustrates. To track V_{DCref}, the DC-bus voltage controller forces P_s to become negative, as Figure 7.25(a) indicates. Figures 7.25(c) and (a) illustrate that before about $t = 0.135$, V_{DC} is settled at V_{DCref} and P_s assumes a small value corresponding to P_{loss}.

At $t = 0.1372$ s, P_{ext} of the DC power source is changed stepwise from zero to 1.0 MW. This results in an overshoot in V_{DC}, as Figure 7.25(c) shows. However, the DC-bus voltage controller increases P_s to maintain the balance of power (Fig. 7.25(a)). At $t = 0.2$ s, P_{ext} is step changed from 1.0 to -1.0 MW. Thus, the DC-bus voltage controller reduces P_s from about 1.0 to about -1.0

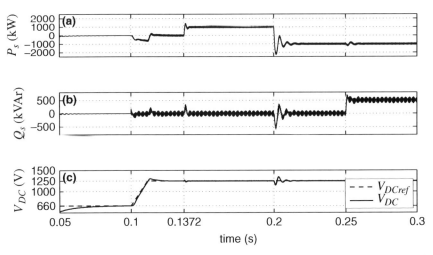

FIGURE 7.26   Behavior of the DC-bus voltage of the controlled DC-voltage power port, in response to changes in P_s and Q_s, with feed-forward compensation; Example 7.4.

MW (Fig. 7.25(a)), and V_{DC} is regulated at V_{DCref} following an undershoot as shown in Figure 7.25(c). At $t = 0.25$ s, Q_{sref} assumes a step change from zero to 500 kVAr (Fig. 7.25(b)). The disturbance, however, has an insignificant impact on V_{DC}, as Figure 7.25(c) illustrates. The reason is that, based on (7.92), the contribution of Q_s in dV_{DC}^2/dt is weighted by the term $2L/(3\widehat{V}_s^2)$, which typically is a small value.

Figure 7.26 illustrates the response of the controlled DC-voltage power port of Figure 7.22, to the same sequence of events as described for Figure 7.25, with the exception that the output of $K_v(s)$ is augmented with the measure of P_{ext} as the feed-forward signal. A comparison between Figures 7.26(c) and 7.25(c) indicates that deviations of V_{DC} from its reference value are considerably smaller in the control system with the feed-forward compensation. Due to the feed-forward compensation, the changes in P_{ext} are rapidly transferred to P_{sref}, which, as Figure 7.26(a) indicates, result in a faster response of P_s compared to the case without the feed-forward compensation (Fig. 7.25(a)).

7.5.3   Simplified and Accurate Models

In the dynamic model of the DC-bus voltage control developed in Section 7.5.1, we employed the principle of power balance, taking into account the instantaneous power of the VSC AC-side interface RL branches (reactors), that is, (7.91). However, traditionally, in the technical literature the instantaneous power of the interface reactors is ignored and $P_t = P_s$ is assumed [73]–[76]. Let us refer to this model as the *simplified model*. Replacing P_s by P_t in (7.83), we deduce the simplified model as

$$\left(\frac{C}{2}\right)\frac{dV_{DC}^2}{dt} = P_{ext} - P_{loss} - P_s, \tag{7.106}$$

which describes a linear model with V_{DC}^2 as the output, P_s as the control input, and P_{ext} and P_{loss} as the disturbance inputs. Based on (7.106), the transfer function from P_s to V_{DC}^2 is

$$G_v(s) = V_{DC}^2(s)/P_s(s) = -\left(\frac{2}{C}\right)\frac{1}{s}. \tag{7.107}$$

As (7.107) indicates, the simplified model is equivalent to the accurate model of (7.95), if the VSC system exchanges zero power with the AC system or equivalently if the time constant τ is zero. This, in turn, is equivalent to the fact that the compensator design based on the simplified model of (7.107) is as if $K_v(s)$ is designed based on the accurate model of (7.95) for the zero real-power operating point. As discussed in Example 7.4, the zero real-power operating point is not the worst-case scenario in terms of the compensator design since the loop-gain phase further drops in the rectifying mode of operation. The simplified model also fails to reveal the phase roll-off of the loop gain under the rectifying mode of operation. Consequently, the compensator design based on the simplified model of (7.107) may result in poor dynamic performance or even instabilities [72], as further highlighted in Example 7.5.

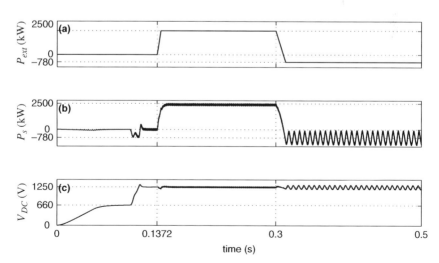

FIGURE 7.27   Instability of the DC-bus voltage controller in the rectifying mode of operation; Example 7.5.

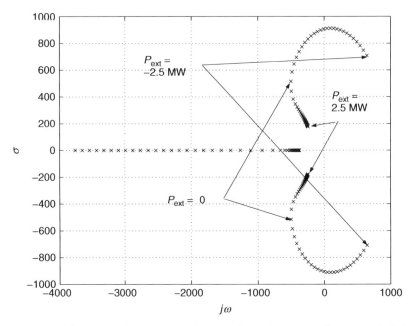

FIGURE 7.28   Root loci of the closed-loop DC-bus voltage controller; Example 7.5.

EXAMPLE 7.5   Instability in the Rectifying Mode of Operation

Consider the controlled DC-voltage power port of Figure 7.22 with parameters $L = 200\,\mu\text{H}$, $R = 2.38\,\text{m}\Omega$, $r_{on} = 0.88\,\text{m}\Omega$, $V_d = 1.0\,\text{V}$, $V_{DC} = 1250\,\text{V}$, and $f_s = 1620\,\text{Hz}$. The line-to-line rms voltage of the AC system is 480 V and its frequency is 377 rad/s. The α- and β-axis compensators of the real-/reactive-power controller are

$$K_\alpha(s) = K_\beta(s) = 2516\left(\frac{s + 16.34}{s^2 + 377^2}\right)\left(\frac{s + 966}{s + 5633}\right)\left(\frac{s + 2}{s + 0.05}\right) \quad [\Omega]. \quad (7.108)$$

Each compensator provides a closed-loop bandwidth of about $\omega_b = 3820$ rad/s (see Example 3.6 for details) for the corresponding loop. The transfer function of the feed-forward filter is $G_{ff}(s) = 1/(8 \times 10^{-6}s + 1)$. In this example, the PWM switching frequency is smaller than that of the VSC system of Example 7.4. Therefore, to achieve low harmonic distortion for the AC-side current, inductances of the interface reactors are larger than those of the VSC system of Example 7.4. However, since the switching frequency is reduced, the VSC system of this example can operate at a relatively higher power level.

We design the DC-bus voltage controller based on the simplified model of (7.107). Thus, the (simplified) plant is an integrator and has the phase $-90°$ regardless of the operating point. Assuming that a phase margin of $45°$ is required at a gain crossover frequency of $\omega_c = 700$ rad/s, following the procedure of Example 7.4 we deduce

$$K_v(s) = 5700 \frac{s + 290}{s(s + 1688)} \quad [\Omega^{-1}]. \tag{7.109}$$

Since the simplified model of (7.107) does not show any dependence on the system operating point, one expects the closed-loop system to remain stable over the entire system operating range. This, however, is not the case. Figure 7.27 illustrates that the system is initially stable and V_{DC} is regulated when $P_{ext} = 2500$ kW. However, the system becomes oscillatory and unstable when P_{ext} is changed from 2500 to -780 kW. The reason can be explained by the accurate model of (7.95). Figure 7.28 illustrates loci of the closed-loop poles when the accurate model of (7.95) is employed in the analysis. Figure 7.28 illustrates that as P_{ext} varies from 2500 to -2500 kW, the real parts of a pair of complex-conjugate poles move from the left-half plane (LHP) to the RHP. Therefore, the closed-loop system becomes unstable beyond a certain negative power. In this example, while the controlled DC-voltage power port can transfer 2.5 MW in the inverting mode of operation, it cannot reach more than 30% of its rated capacity in the rectifying mode of operation.

8 Grid-Imposed Frequency VSC System: Control in *dq*-Frame

8.1 INTRODUCTION

Chapter 5 presented dynamic models for the two-level VSC in $\alpha\beta$-frame and *dq*-frame and briefly discussed its control based on generic block diagrams of Figures 5.5 and 5.7. Chapter 6 introduced the three-level NPC as an extension of the two-level VSC and established that the dynamic model of the three-level NPC is identical to that of the two-level VSC, except that the three-level NPC requires a DC-side voltage equalizing system to maintain DC-side capacitor voltages, each at half the net DC-side voltage. Thus, Chapter 6 presented a unified model for the three-level NPC and the two-level VSC (Fig. 6.18 and 6.19). Chapter 7 introduced a class of VSC systems referred to as *grid-imposed frequency VSC systems*. On the basis of the unified model of Chapter 6, Chapter 7 presented $\alpha\beta$-frame models and controls for two members of the family of the grid-imposed frequency VSC systems, namely, the *real-/reactive-power controller* and the *controlled DC-voltage power port*. In parallel with Chapter 7, this chapter presents *dq*-frame models and controls for the real-/reactive-power controller and the controlled DC-voltage power port.

As discussed in Chapter 7, compared to the *abc*-frame control, the $\alpha\beta$-frame control of a grid-imposed frequency VSC system reduces the number of plants to be controlled from three to two. Moreover, instantaneous decoupled control of the real and reactive power, exchanged between the VSC system and the AC system, is possible in $\alpha\beta$-frame. However, the control variables, that is, feedback signals, feed-forward signals, and control signals are sinusoidal functions of time. It is shown in this chapter that the *dq*-frame control of a grid-imposed VSC system features all merits of the $\alpha\beta$-frame control, in addition to the advantage that the control variables are DC quantities in the steady state. This feature it remarkably facilitates the compensator design, especially in variable-frequency scenarios.

To achieve zero steady-state error in the $\alpha\beta$-frame control, the bandwidth of the closed-loop system must be adequately larger than the AC system frequency; alternatively, the compensators can include complex-conjugate pairs of poles at the AC system frequency and other frequencies of interest, to increase the loop gain. In the *dq*-frame control, however, zero steady-state error is readily achieved by including

Voltage-Sourced Converters in Power Systems, by Amirnaser Yazdani and Reza Iravani
Copyright © 2010 John Wiley & Sons, Inc.

integral terms in the compensators since the control variables are DC quantities [77]. The *dq*-frame representation and control of a grid-imposed VSC system is also consistent with the approach used for the dynamic analysis of the large power system. The small-signal dynamics of the power system is conventionally modeled and analyzed in *dq*-frame [4?]

Compared to the $\alpha\beta$-frame control, the *dq*-frame control requires a synchronization mechanism that is usually achieved through the *phase-locked loop* (PLL); this requirement can be regarded a demerit of the *dq*-frame control.

8.2 STRUCTURE OF GRID-IMPOSED FREQUENCY VSC SYSTEM

Figure 8.1 shows a schematic diagram of a grid-imposed frequency VSC system. The VSC represents either a three-level NPC with a DC-side voltage equalizing scheme or a two-level VSC. In either case, the VSC is modeled by a lossless power processor including an equivalent DC-bus capacitor, a current source representing the VSC switching power loss, and series on-state resistances at the AC side representing the VSC conduction power loss, as Figure 8.1 shows. The DC side of the VSC may be interfaced with a DC voltage source or a DC power source. Each phase of the VSC is interfaced with the AC system via a series *RL* branch.

In this chapter, as an approximation we consider an infinitely stiff AC system. Thus, the AC system is modeled by an ideal three-phase voltage source, V_{sabc}.[1] It is also assumed that V_{sabc} is balanced, sinusoidal, and of a relatively constant frequency. The VSC system of Figure 8.1 exchanges the real- and reactive-power components

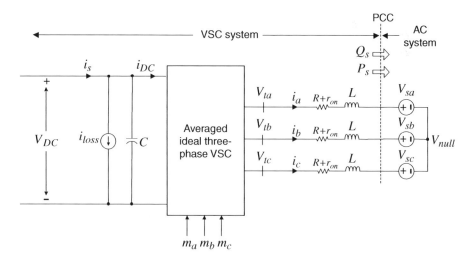

FIGURE 8.1   Schematic diagram of a grid-imposed frequency VSC system.

[1] In Chapter 11, we investigate the dynamics of a VSC system under nonstiff AC system conditions.

$P_s(t)$ and $Q_s(t)$ with the AC system, at the point of common coupling (PCC). Depending on the control strategy, the VSC system of Figure 8.1 is used as either a real-/reactive-power controller or a controlled DC-voltage power port. In Chapter 12, we employ the real-/reactive-power controller as part of a back-to-back HVDC converter system. The controlled DC-voltage power port is employed as part of the static compensator (STATCOM), the back-to-back HVDC converter system, and variable-speed wind-power units, in Chapters 11, 12, and 13, respectively.

8.3   REAL-/REACTIVE-POWER CONTROLLER

The grid-imposed frequency VSC system of Figure 8.1 can be employed as a real-/reactive-power controller. As such, the VSC DC side is connected in parallel with a DC voltage source and the objective is to control the instantaneous real and reactive power that the VSC system exchanges with the AC system, that is, $P_s(t)$ and $Q_s(t)$.

8.3.1   Current-Mode Versus Voltage-Mode Control

Two main methods exist for controlling P_s and Q_s in the VSC system of Figure 8.1. The first approach that is known as *voltage-mode control* and illustrated in Figure 8.2 has been dominantly utilized in high-voltage-/-power applications such as in FACTS controllers [44, 45], although its industrial applications have also been reported [47]. Figure 8.2 illustrates that in a voltage-controlled VSC system, the real and reactive

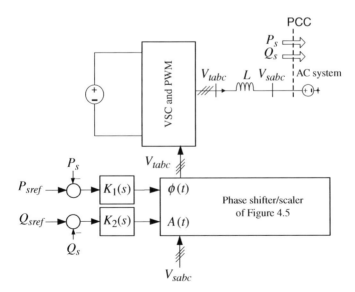

FIGURE 8.2   Schematic diagram of a voltage-controlled real-/reactive-power controller.

power are controlled, respectively, by the phase angle and the amplitude of the VSC AC-side terminal voltage relative to the PCC voltage [46]. If the amplitude and phase angle of V_{tabc} are close to those of V_{sabc}, the real and reactive power are almost decoupled and two independent compensators can be employed for their control (Fig. 8.2). The voltage-mode control is simple and has a low number of control loops. However, the main shortcoming of the voltage-mode control is that there is no control loop closed on the VSC line current. Consequently, the VSC is not protected against overcurrents, and the current may undergo large excursions if the power commands are rapidly changed or faults take place in the AC system.

The second approach to the control of the real and reactive power in the VSC system of Figure 8.1 is referred to as the *current-mode control*. In this approach, the VSC line current is tightly regulated by a dedicated current-control scheme, through the VSC AC-side terminal voltage. Then, the real and reactive power are controlled by the phase angle and the amplitude of the VSC line current with respect to the PCC voltage. Thus, due to the current regulation scheme, the VSC is protected against overcurrent conditions. Other advantages of the current-mode control include robustness against variations in parameters of the VSC system and the AC system, superior dynamic performance, and higher control precision [68]. We demonstrated the basics of the current-mode control strategy in Chapter 3 and will exclusively focus on this method throughout the rest of the book.

Figure 8.3 shows a schematic diagram of a current-controlled real-/reactive-power controller, illustrating that the control is performed in *dq*-frame. Thus, P_s and Q_s are

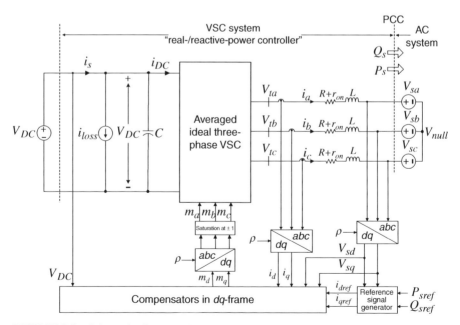

FIGURE 8.3 Schematic diagram of a current-controlled real-/reactive-power controller in *dq*-frame.

controlled by the line current components i_d and i_q. The feedback and feed-forward signals are first transformed to the dq-frame and then processed by compensators to produce the control signals in dq-frame. Finally, the control signals are transformed to the abc-frame and fed to the VSC (Fig. 8.3). To protect the VSC, the reference commands i_{dref} and i_{qref} are limited by the corresponding saturation blocks (not shown in the figure). It is noted that the block diagram of Figure 8.3 is a special case of the general block diagram of Figure 4.27. In Chapter 12, we employ the real-/reactive-power controller as part of the back-to-back HVDC converter system.

8.3.2   Representation of Space Phasors in dq-Frame

In this chapter, we need to express space phasors in dq-frame. The transformation and its inverse were extensively discussed in Chapter 4. However, they are briefly reviewed in this section, for ease of reference.

Consider the space phasor $\overrightarrow{f}(t) = f_\alpha + jf_\beta$. The dq- to $\alpha\beta$-frame transformation is defined as

$$f_d + jf_q = \overrightarrow{f}(t)e^{-j\rho(t)} = (f_\alpha + jf_\beta)e^{-j\rho(t)}, \qquad (8.1)$$

which is a phase shift in $\overrightarrow{f}(t)$ by $-\rho(t)$. The angle $\rho(t)$ can be chosen arbitrarily. However, if, for example, $\overrightarrow{f}(t) = \widehat{f}e^{j(\omega t+\theta_0)}$, then choosing $\rho(t)$ to be equal to ωt results in the space phasor

$$f_d + jf_q = \underbrace{\widehat{f}e^{j(\omega t+\theta_0)}}_{\overrightarrow{f}(t)}e^{-j\omega t} = \widehat{f}e^{j\theta_0},$$

which is no longer time-varying and, therefore, f_d and f_q are DC quantities. The inverse transformation is

$$\overrightarrow{f}(t) = f_\alpha + jf_\beta = (f_d + jf_q)e^{j\rho(t)}. \qquad (8.2)$$

8.3.3   Dynamic Model of Real-/Reactive-Power Controller

Assume that the AC system voltage in the VSC system of Figure 8.3 is expressed as

$$V_{sa}(t) = \widehat{V}_s \cos(\omega_0 t + \theta_0),$$

$$V_{sb}(t) = \widehat{V}_s \cos\left(\omega_0 t + \theta_0 - \frac{2\pi}{3}\right),$$

$$V_{sc}(t) = \widehat{V}_s \cos\left(\omega_0 t + \theta_0 - \frac{4\pi}{3}\right), \qquad (8.3)$$

where $\widehat{V}_s$ is the peak value of the line-to-neutral voltage, ω_0 is the AC system (source) frequency, and θ_0 is the source initial phase angle. Based on (4.2), the space-phasor equivalent of V_{s-abc} is

$$\overrightarrow{V_s}(t) = \widehat{V}_s e^{j(\omega_0 t + \theta_0)}. \tag{8.4}$$

Dynamics of the AC side of the VSC system of Figure 8.3 are described by the following space-phasor equation (refer to (7.11) for details):

$$L\frac{d\overrightarrow{i}}{dt} = -(R + r_{on})\overrightarrow{i} + \overrightarrow{V_t} - \overrightarrow{V_s}. \tag{8.5}$$

Substituting for $\overrightarrow{V_s}$ from (8.4) in (8.5), we deduce

$$L\frac{d\overrightarrow{i}}{dt} = -(R + r_{on})\overrightarrow{i} + \overrightarrow{V_t} - \widehat{V}_s e^{j(\omega_0 t + \theta_0)}. \tag{8.6}$$

Then, we use (8.2) to express (8.6) in a dq-frame. Thus, substituting for $\overrightarrow{i} = i_{dq}e^{j\rho}$ and $\overrightarrow{V_t} = V_{tdq}e^{j\rho}$ in (8.6), we deduce

$$L\frac{d}{dt}\left(i_{dq}e^{j\rho}\right) = -(R + r_{on})\left(i_{dq}e^{j\rho}\right) + \left(V_{tdq}e^{j\rho}\right) - \widehat{V}_s e^{j(\omega_0 t + \theta_0)}, \tag{8.7}$$

where $f_{dq} = f_d + jf_q$. Equation (8.7) can be rewritten as

$$L\frac{d}{dt}\left(i_{dq}\right) = -j\left(L\frac{d\rho}{dt}\right)i_{dq} - (R + r_{on})i_{dq} + V_{tdq} - \widehat{V}_s e^{j(\omega_0 t + \theta_0 - \rho)}. \tag{8.8}$$

Decomposing (8.8) into real and imaginary components, we deduce

$$L\frac{di_d}{dt} = \left(L\frac{d\rho}{dt}\right)i_q - (R + r_{on})i_d + V_{td} - \widehat{V}_s \cos(\omega_0 t + \theta_0 - \rho), \tag{8.9}$$

$$L\frac{di_q}{dt} = -\left(L\frac{d\rho}{dt}\right)i_d - (R + r_{on})i_q + V_{tq} - \widehat{V}_s \sin(\omega_0 t + \theta_0 - \rho). \tag{8.10}$$

Equations (8.9) and (8.10) are not in the standard state-space form. Thus, we introduce the new control variable ω to (8.9) and (8.10), where $\omega = d\rho/dt$. This yields

$$L\frac{di_d}{dt} = L\omega(t)i_q - (R + r_{on})i_d + V_{td} - \widehat{V}_s \cos(\omega_0 t + \theta_0 - \rho), \qquad (8.11)$$

$$L\frac{di_q}{dt} = -L\omega(t)i_d - (R + r_{on})i_q + V_{tq} - \widehat{V}_s \sin(\omega_0 t + \theta_0 - \rho), \qquad (8.12)$$

$$\frac{d\rho}{dt} = \omega(t). \qquad (8.13)$$

In (8.11)–(8.13), i_d, i_q, and ρ are the state variables, and V_{td}, V_{tq}, and ω are the control inputs. The system described by (8.11)–(8.13) is nonlinear due to the presence of the terms ωi_d, ωi_q, $\cos(\omega_0 t + \theta_0 - \rho)$, and $\sin(\omega_0 t + \theta_0 - \rho)$.

To further investigate (8.11)–(8.13), assume that ρ has a zero initial condition and $\omega(t) \equiv 0$. Consequently, ρ remains zero at all times, and (8.11) and (8.12) assume the forms

$$L\frac{di_d}{dt} = -(R + r_{on})i_d + V_{td} - \widehat{V}_s \cos(\omega_0 t + \theta_0), \qquad (8.14)$$

$$L\frac{di_q}{dt} = -(R + r_{on})i_q + V_{tq} - \widehat{V}_s \sin(\omega_0 t + \theta_0). \qquad (8.15)$$

Equations (8.14) and (8.15) describe two, decoupled, first-order systems that are excited by inputs $-\widehat{V}_s \cos(\omega_0 t + \theta_0)$ and $-\widehat{V}_s \sin(\omega_0 t + \theta_0)$, respectively. Thus, the superposition principle requires that i_d and i_q also include sinusoidal components, irrespective of V_{td} and V_{tq}. This result is expected since if $\rho = 0$, then based on (8.1) the dq-frame is the same as the $\alpha\beta$-frame in which the signals are sinusoidal functions of time. In other words, (8.14) and (8.15) represent the VSC system in $\alpha\beta$-frame; comparison of (8.14) and (8.15), respectively, with (7.12) and (7.13) confirms this conclusion.

The foregoing discussion shows that the usefulness of the dq-frame depends on proper selection of ω and ρ. For the VSC system of Figure 8.3, if $\omega = \omega_0$ and $\rho(t) = \omega_0 t + \theta_0$, then (8.11) and (8.12) take the forms

$$L\frac{di_d}{dt} = L\omega_0 i_q - (R + r_{on})i_d + V_{td} - \widehat{V}_s, \qquad (8.16)$$

$$L\frac{di_q}{dt} = -L\omega_0 i_d - (R + r_{on})i_q + V_{tq}, \qquad (8.17)$$

which describe a second-order linear system that is excited by the constant input $\widehat{V}_s$. Thus, if V_{td} and V_{tq} are DC variables, i_d and i_q are also DC variables in the steady state. The mechanism to ensure $\rho(t) = \omega_0 t + \theta_0$ is referred to as the PLL. The following section presents the structure, model, and stabilization of the PLL.

8.3.4 Phase-Locked Loop (PLL)

Substituting for $\vec{V}_s(t)$ from (8.4) in (8.1), we deduce

$$V_{sd} = \widehat{V}_s \cos(\omega_0 t + \theta_0 - \rho), \tag{8.18}$$

$$V_{sq} = \widehat{V}_s \sin(\omega_0 t + \theta_0 - \rho). \tag{8.19}$$

Thus, (8.11)–(8.13) can be rewritten as

$$L\frac{di_d}{dt} = L\omega(t)i_q - (R + r_{on})i_d + V_{td} - V_{sd}, \tag{8.20}$$

$$L\frac{di_q}{dt} = -L\omega(t)i_d - (R + r_{on})i_q + V_{tq} - V_{sq}, \tag{8.21}$$

$$\frac{d\rho}{dt} = \omega(t). \tag{8.22}$$

Based on (8.19), $\rho(t) = \omega_0 t + \theta_0$ corresponds to $V_{sq} = 0$. Therefore, we devise a mechanism to regulate V_{sq} at zero. This can be achieved based on the following feedback law:

$$\omega(t) = H(p)V_{sq}(t), \tag{8.23}$$

where $H(p)$ is a linear transfer function (compensator) and $p = d(\cdot)/dt$ is a differentiation operator. Substituting for V_{sq} from (8.19) in (8.23), and substituting for ω from (8.23) in (8.22), we deduce

$$\frac{d\rho}{dt} = H(p)\widehat{V}_s \sin(\omega_0 t + \theta_0 - \rho). \tag{8.24}$$

Equation (8.24) describes a nonlinear dynamic system, which is referred to as PLL [49], [78–80]. The function of the PLL is to regulate ρ at $\omega_0 t + \theta_0$. However, in view of its nonlinear characteristic, the PLL can exhibit unsatisfactory behavior under certain conditions. For example, if the PLL starts from an initial condition corresponding to $\rho(0) = 0$ and $\omega(0) = 0$, then the term $\widehat{V}_s H(p) \sin(\omega_0 t + \theta_0 - \rho)$ in (8.24) is a sinusoidal function of time with frequency ω_0. Then, if $H(s)$ has a low-pass frequency response, the right-hand side of (8.24) and also $d\rho/dt$ exhibit small sinusoidal perturbations about zero, the PLL falls in a limit cycle, and ρ does not track $\omega_0 t + \theta_0$. To prevent the limit cycle from taking place, the control law can be modified as

$$\omega(t) = H(p)V_{sq}(t), \quad \omega(0) = \omega_0 \text{ and } \omega_{min} \leq \omega \leq \omega_{max}, \tag{8.25}$$

where $\omega(t)$ has the initial value $\omega(0) = \omega_0$ and is limited to the lower and upper limits of, respectively, ω_{min} and ω_{max}. ω_{min} and ω_{max} are selected to be close to ω_0 and thus to define a narrow range of variations for $\omega(t)$. On the other hand, the range of

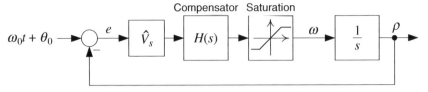

FIGURE 8.4   Control block diagram of the PLL.

variations should be selected adequately wide to permit excursions of $\omega(t)$ during transients. If the PLL tracks $\omega_0 t + \theta_0$, the term $\omega_0 t + \theta_0 - \rho$ is close to zero and $\sin(\omega_0 t + \theta_0 - \rho) \approx (\omega_0 t + \theta_0 - \rho)$. Therefore, (8.24) can be simplified to

$$\frac{d\rho}{dt} = \widehat{V}_s H(p)(\omega_0 t + \theta_0 - \rho). \tag{8.26}$$

Equation (8.26) represents a classical feedback control loop in which $\omega_0 t + \theta_0$ is the reference input, ρ is the output, and $\widehat{V}_s H(s)$ is the transfer function of the effective compensator, as shown in the block diagram of Figure 8.4.

Figure 8.5 illustrates a schematic diagram of the PLL based on (8.19), (8.22), and (8.23). Figure 8.5 shows that the PLL transforms V_{sabc} to V_{sdq} (based on (4.73)) and adjusts the rotational speed of the dq-frame, that is, ω, such that V_{sq} is forced to zero in the steady state. The end result is that $\rho = \omega_0 t + \theta_0$ and $V_{sd} = \widehat{V}_s$. It should be pointed out that in the block diagram of Figure 8.5, the integrator of (8.22) is realized by means of a voltage-controlled oscillator (VCO). The VCO can be regarded as a resettable integrator whose output, ρ, is reset to zero whenever it reaches 2π.

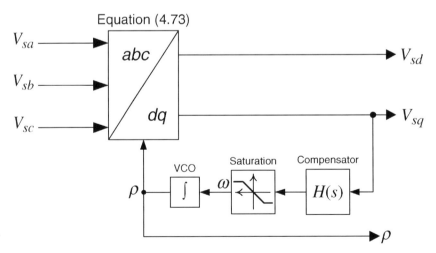

FIGURE 8.5   Schematic diagram of the PLL.

8.3.5   Compensator Design for PLL

Dynamic performance of the PLL is highly influenced by the compensator $H(s)$. Consider the block diagram of Figure 8.4 indicating that the reference signal, $\omega_0 t + \theta_0$, is composed of a constant component, that is, θ_0, and a ramp function, that is, $\omega_0 t$. Since the loop gain includes an integral term, ρ tracks the constant component of the reference signal with zero steady-state error. However, to ensure a zero steady-state error for the ramp component, the loop gain must include at least two integrators. Therefore, $H(s)$ must include at least one integral term, that is, one pole at $s = 0$. The other poles and zeros of $H(s)$ are determined mainly on the basis of the closed-loop bandwidth of the PLL and stability indices such as phase margin and gain margin.

Another consideration in designing $H(s)$ is the issue of unbalanced and/or harmonically distorted three-phase voltages. Assume that V_{sabc} represents an unbalanced voltage with a negative-sequence fundamental component and a fifth-order harmonic component [81], as

$$V_{sa}(t) = \widehat{V}_s \cos(\omega_0 t + \theta_0) + k_1 \widehat{V}_s \cos(\omega_0 t + \theta_0)$$
$$+ k_5 \widehat{V}_s \cos(5\omega_0 t + \phi_5),$$
$$V_{sb}(t) = \widehat{V}_s \cos\left(\omega_0 t + \theta_0 - \frac{2\pi}{3}\right) + k_1 \widehat{V}_s \cos\left(\omega_0 t + \theta_0 - \frac{4\pi}{3}\right)$$
$$+ k_5 \widehat{V}_s \cos\left(5\omega_0 t + \phi_5 - \frac{4\pi}{3}\right),$$
$$V_{sc}(t) = \widehat{V}_s \cos\left(\omega_0 t + \theta_0 - \frac{4\pi}{3}\right) + k_1 \widehat{V}_s \cos\left(\omega_0 t + \theta_0 - \frac{2\pi}{3}\right)$$
$$+ k_5 \widehat{V}_s \cos\left(5\omega_0 t + \phi_5 - \frac{2\pi}{3}\right), \quad (8.27)$$

where k_1 and k_5 are the amplitudes of the negative-sequence (fundamental) and fifth-order harmonic components, respectively, relative to the amplitude of the positive-sequence (fundamental) component. Based on (4.2), the space phasor corresponding to V_{sabc} is

$$\overrightarrow{V_s} = \widehat{V}_s e^{j(\omega_0 t + \theta_0)} + k_1 \widehat{V}_s e^{-j(\omega_0 t + \theta_0)} + k_5 \widehat{V}_s e^{-j(5\omega_0 t + \phi_5)}. \quad (8.28)$$

If the PLL of Figure 8.5 is under a steady-state operating condition, that is, $\rho = \omega_0 t + \theta_0$, then based on (8.1) V_{sd} and V_{sq} are

$$V_{sd} = \widehat{V}_s + k_1 \widehat{V}_s \cos(2\omega_0 t + 2\theta_0) + k_5 \widehat{V}_s \cos(6\omega_0 t + \theta_0 + \phi_5), \quad (8.29)$$
$$V_{sq} = -k_1 \widehat{V}_s \sin(2\omega_0 t + 2\theta_0) - k_5 \widehat{V}_s \sin(6\omega_0 t + \theta_0 + \phi_5). \quad (8.30)$$

Equations (8.29) and (8.30) indicate that, in addition to DC components, V_{sd} and V_{sq} include sinusoidal components with frequencies $2\omega_0$ and $6\omega_0$. Typical values of k_1 and k_5 are assumed to be 0.01 and 0.025, respectively [81]. However, under single-phase to ground faults, k_1 can be as large as 0.5. The sinusoidal components of V_{sq} must be attenuated by $H(s)$. Otherwise, ω and ρ also exhibit fluctuations that are modulated with feedback and control signals, through abc- to dq-frame and dq-to abc-frame transformations, and result in generation of undesirable voltage/current distortions in the VSC system.

Between the two AC components of V_{sq}, the component with frequency $2\omega_0$ is more important. The reason is that (i) the frequency of this component is three times lower than that of the other component and (ii) the magnitude of this component, k_1, can be significantly larger than that of the other component, for example, during a fault. One approach to attenuate the double-frequency component of V_{sq} is to ensure that $H(s)$ exhibits a strong low-pass characteristic. However, this method may compromise the PLL closed-loop bandwidth. Alternatively, one can include in $H(s)$ one pair of complex-conjugate zeros, at $s = \pm j2\omega_0$, to eliminate the double-frequency ripple of V_{sq}. The advantage of this technique is that the PLL closed-loop bandwidth is not sacrificed and can be selected to be arbitrarily large. Example 8.1 illustrates the second PLL design approach.

EXAMPLE 8.1 Compensator Design for the PLL

Consider the PLL of Figure 8.5 whose input is V_{sabc} defined by (8.27), where $\omega_0 = 2\pi \times 60$ rad/s and $\widehat{V}_s = 391$ V. The objective is to design the PLL compensator $H(s)$.

As explained in Section 8.3.5, $H(s)$ must include one pole at $s = 0$ and the complex-conjugate zeros $s = \pm j2\omega_0$. In addition, to ensure that the loop gain magnitude continues to drop with the slope of -40 dB/dec for $\omega > 2\omega_0$, a double real pole at $s = -2\omega_0$ is included in $H(s)$. Thus,

$$H(s) = \left(\frac{h}{\widehat{V}_{sn}} \right) \frac{s^2 + (2\omega_0)^2}{s\,(s + 2\omega_0)^2} F(s), \tag{8.31}$$

where $\widehat{V}_{sn}$ is the nominal value of $\widehat{V}_s$ and $F(s)$ is the proper transfer function with no zero at $s = 0$. Based on the block diagram of Figure 8.4, the loop gain is formulated as

$$\ell(s) = h \frac{s^2 + (2\omega_0)^2}{s^2(s + 2\omega_0)^2} F(s). \tag{8.32}$$

Let us assume that we need a gain crossover frequency of $\omega_c = 200$ rad/s and a phase margin of $60°$. If $hF(s) = 1$, it can be calculated that $\angle\ell(j200) = -210°$. Thus, to achieve the required phase margin, $F(j200)$ must add $90°$ to $\angle\ell(j200)$. As discussed in Example 3.6, a lead compensator can offer an optimum phase advance to the loop gain. In this example, the required phase advance is fairly large. Consequently, $F(s)$ can be composed of two cascaded lead compensators, each to provide $45°$ at 200 rad/s. Thus,

$$F(s) = \left(\frac{s + (p/\alpha)}{s + p}\right)\left(\frac{s + (p/\alpha)}{s + p}\right), \tag{8.33}$$

where

$$p = \omega_c\sqrt{\alpha} \tag{8.34}$$

$$\alpha = \frac{1 + \sin\delta_m}{1 - \sin\delta_m}, \tag{8.35}$$

and δ_m is the phase of each lead compensator at ω_c. If $\delta_m = 45°$, based on (8.33)–(8.35), we calculate $F(s)$ as

$$F(s) = \left(\frac{s + 83}{s + 482}\right)^2. \tag{8.36}$$

Substituting for $F(s)$ from (8.36) in (8.32), we deduce

$$\ell(s) = \frac{h\left(s^2 + 568{,}516\right)\left(s^2 + 166s + 6889\right)}{s^2\left(s^2 + 1508s + 568{,}516\right)\left(s^2 + 964s + 232{,}324\right)}. \tag{8.37}$$

It then follows from $|\ell(j200)| = 1$ and $\widehat{V}_{sn} = 391$ V that $h = 2.68 \times 10^5$. Therefore, $h/\widehat{V}_{sn} = 685.42$ and the final compensator is

$$H(s) = \frac{685.42\left(s^2 + 568{,}516\right)\left(s^2 + 166s + 6889\right)}{s\left(s^2 + 1508s + 568{,}516\right)\left(s^2 + 964s + 232{,}324\right)} \quad [(\text{rad/s})/\text{V}]. \tag{8.38}$$

Figure 8.6 depicts the frequency response of $\ell(j\omega)$ based on the compensator of (8.38). It is observed that $|\ell(j\omega)|$ drops with the slope of -40 dB/dec, for $\omega \ll \omega_c = 200$. However, around ω_c the slope of $|\ell(j\omega)|$ reduces to about -20 dB/dec and $\angle\ell(j\omega)$ rises to about $-120°$ at $\omega = \omega_c$, corresponding to a phase margin of $60°$. Figure 8.6 also illustrates that $|\ell(j\omega)|$ continues to drop with a slope of -40 dB/dec for $\omega > \omega_c$. This characteristic is desired as the AC components of V_{sq} due to the harmonic distortion of V_{sabc} are attenuated. In particular, at $\omega = 6\omega_0$, $|\ell(j\omega)|$ is about -30 dB.

Figure 8.7 illustrates the start-up transient of the PLL. Figure 8.7 shows that, from $t = 0$ to $t = 0.07$ s, the compensator output is saturated at $\omega_{min} = 2\pi \times 55$ rad/s and, therefore, V_{sd} and V_{sq} vary with time. At about $t = 0.07$ s, V_{sq} crosses zero and intends to become negative. Thus, $H(s)$ increases ω to regulate V_{sq} at zero. Figure 8.7 indicates that V_{sq} is regulated at zero within 0.15 s. It should be noted that if ω_{min} is selected closer to ω_0, the start-up transient period becomes shorter. However, ω_{min} cannot be selected too close to ω_0 since the PLL would not be able to quickly react to other types of disturbance.

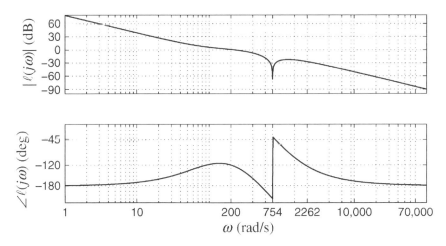

FIGURE 8.6   Open-loop frequency response of the PLL of Example 8.1.

Figure 8.8 illustrates the dynamic response of the PLL to a sudden imbalance in V_{sabc}. Initially, the PLL is in a steady state. At $t = 0.05$ s, the AC system voltage V_{sabc} becomes unbalanced such that $\hat{V}_s$ and k_1 undergo step changes, respectively, from 391 to 260 V and from zero to 0.5, and at $t = 0.15$ s, V_{sabc}

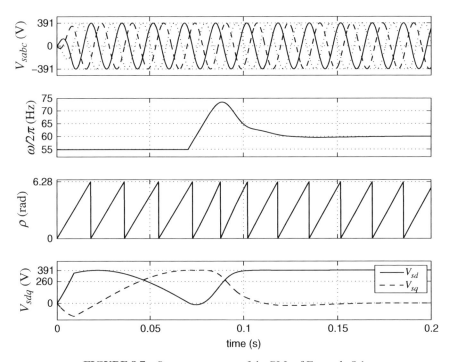

FIGURE 8.7   Start-up response of the PLL of Example 8.1.

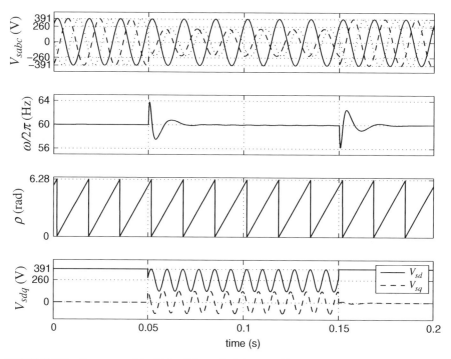

FIGURE 8.8   Response of the PLL of Example 8.1 to a sudden AC system voltage imbalance.

reverts to its balanced predisturbance condition. In response to the voltage imbalance, $H(s)$ transiently changes ω, as Figure 8.8 shows, to maintain the DC component of V_{sq} at zero. Figure 8.8 also shows that V_{sq} (and V_{sd}) includes a 120-Hz sinusoidal ripple due to the negative-sequence component of V_{sabc}. The ripple is, however, suppressed by $H(s)$, and ω and ρ remain free of distortion.

Figure 8.9 depicts the dynamic response of the PLL to two stepwise changes in ω_0, the first one from $2\pi \times 60 = 377$ rad/s to $2\pi \times 63 = 396$ rad/s at $t = 0.05$ s, and the other from 396 rad/s to $2\pi \times 57 = 358$ rad/s at $t = 0.1$ s. As Figure 8.9 shows, V_{sq} is rapidly regulated at zero and ω tracks the changes.

Equation (8.31) denotes that $H(s)$ is normalized such that the constant gain of the loop gain h is independent of $\widehat{V}_{sn}$. Thus, in subsequent chapters when we need a PLL, we will employ the compensator of (8.38), but modify its constant gain, that is, $h/\widehat{V}_{sn}$, according to $\widehat{V}_{sn}$ for the specific problem in hand, based on $h = 2.68 \times 10^5$.

8.4   CURRENT-MODE CONTROL OF REAL-/REACTIVE-POWER CONTROLLER

With reference to the real-/reactive-power controller of Figure 8.3, based on (4.83) and (4.84), the real and reactive power delivered to the AC system at

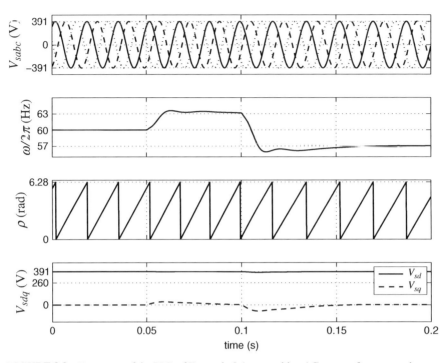

FIGURE 8.9   Response of the PLL of Example 8.1 to a sudden AC system frequency change.

the PCC are

$$P_s(t) = \frac{3}{2} \left[V_{sd}(t)i_d(t) + V_{sq}(t)i_q(t) \right],$$ (8.39)

$$Q_s(t) = \frac{3}{2} \left[-V_{sd}(t)i_q(t) + V_{sq}(t)i_d(t) \right],$$ (8.40)

where V_{sd} and V_{sq} are the AC system *dq*-frame voltage components and cannot be controlled by the VSC system. As described in Section 8.3.4, if the PLL is in a steady state, $V_{sq} = 0$ and (8.39) and (8.40) can be rewritten as

$$P_s(t) = \frac{3}{2} V_{sd}(t)i_d(t),$$ (8.41)

$$Q_s(t) = -\frac{3}{2} V_{sd}(t)i_q(t).$$ (8.42)

Therefore, based on (8.41) and (8.42), $P_s(s)$ and $Q_s(s)$ can be controlled by i_d and i_q, respectively. Let us introduce

$$i_{dref}(t) = \frac{2}{3V_{sd}} P_{sref}(t), \tag{8.43}$$

$$i_{qref}(t) = -\frac{2}{3V_{sd}} Q_{sref}(t). \tag{8.44}$$

Then, if the control system can provide fast reference tracking, that is, $i_d \approx i_{dref}$ and $i_q \approx i_{qref}$, then $P_s \approx P_{sref}$ and $Q_s \approx Q_{sref}$, that is, $P_s(t)$ and $Q_s(t)$, can be independently controlled by their respective reference commands. Since V_{sd} is a DC variable (in the steady state), i_{dref} and i_{qref} are also DC variables if P_{sref} and Q_{sref} are constant signals. Thus, as expected, the control system in dq-frame deals with DC variables, unlike the control system in $\alpha\beta$-frame that deals with sinusoidal signals.

8.4.1   VSC Current Control

The dq-frame control of the real-/reactive-power controller of Figure 8.3 is based on (8.11) and (8.12). Assuming a steady-state operating condition and substituting for $\omega(t) = \omega_0$ in (8.11) and (8.12), we deduce

$$L\frac{di_d}{dt} = L\omega_0 i_q - (R + r_{on})i_d + V_{td} - V_{sd}, \tag{8.45}$$

$$L\frac{di_q}{dt} = -L\omega_0 i_d - (R + r_{on})i_q + V_{tq} - V_{sq}, \tag{8.46}$$

in which, based on (5.22) and (5.23), V_{td} and V_{tq} are

$$V_{td}(t) = \frac{V_{DC}}{2} m_d(t), \tag{8.47}$$

$$V_{tq}(t) = \frac{V_{DC}}{2} m_q(t). \tag{8.48}$$

Equations (8.47) and (8.48) represent the VSC model in dq-frame. The model is applicable to both the two-level VSC and the three-level NPC. In (8.45) and (8.46), i_d and i_q are state variables, V_{td} and V_{tq} are control inputs, and V_{sd} and V_{sq} are disturbance inputs. Due to the presence of $L\omega_0$ terms in (8.45) and (8.46), dynamics of i_d and i_q are coupled. To decouple the dynamics, we determine m_d and m_q as

$$m_d = \frac{2}{V_{DC}} (u_d - L\omega_0 i_q + V_{sd}), \tag{8.49}$$

$$m_q = \frac{2}{V_{DC}} (u_q + L\omega_0 i_d + V_{sq}), \tag{8.50}$$

where u_d and u_q are two new control inputs [69, 82]. Substituting for m_d and m_q in (8.47) and (8.48), respectively, from (8.49) and (8.50), and substituting for V_{td} and

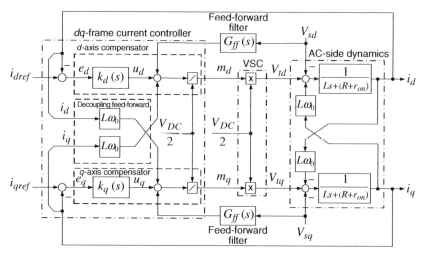

FIGURE 8.10   Control block diagram of a current-controlled VSC system.

V_{tq} from the resultant in (8.45) and (8.46), we deduce

$$L\frac{di_d}{dt} = -(R + r_{on})i_d + u_d, \tag{8.51}$$

$$L\frac{di_q}{dt} = -(R + r_{on})i_q + u_q. \tag{8.52}$$

Equations (8.51) and (8.52) describe two decoupled, first-order, linear systems. Based on (8.51) and (8.52), i_d and i_q can be controlled by u_d and u_q, respectively. Figure 8.10 shows a block representation of the d- and q-axis current controllers of the VSC system in which u_d and u_q are the outputs of two corresponding compensators. The d-axis compensator processes $e_d = i_{dref} - i_d$ and provides u_d. Then, based on (8.49), u_d contributes to m_d. Similarly, the q-axis compensator processes $e_q = i_{qref} - i_q$ and provides u_q that, based on (8.50), contributes to m_q. The VSC then amplifies m_d and m_q by a factor of $V_{DC}/2$ and generates V_{td} and V_{tq} that, in turn, control i_d and i_q based on (8.45) and (8.46). On the basis of the above-mentioned control process, one can sketch the simplified control block diagram of Figure 8.11, which is equivalent to the control system of Figure 8.10. It should be noted that in the control system of Figure 8.10, all the control, feed-forward, and feedback signals are DC quantities in the steady state.

Figure 8.11 indicates that the control plants in both d- and q-axis current-control loops are identical. Therefore, the corresponding compensators can also be identical. Consider the d-axis control loop. Unlike the $\alpha\beta$-frame control where the compensators are fairly difficult to optimize and typically are of high dynamic orders, $k_d(s)$ can be a simple proportional-integral (PI) compensator to enable tracking of a DC reference

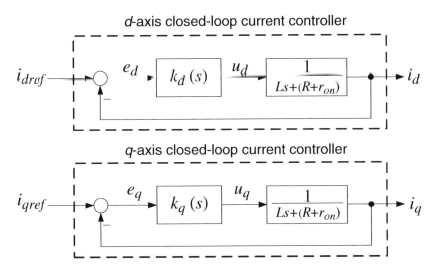

FIGURE 8.11 Simplified block diagram of the current-controlled VSC system of Figure 8.10.

command. Let

$$k_d(s) = \frac{k_p s + k_i}{s}, \tag{8.53}$$

where k_p and k_i are proportional and integral gains, respectively. Thus, the loop gain is

$$\ell(s) = \left(\frac{k_p}{Ls}\right) \frac{s + k_i/k_p}{s + (R + r_{on})/L}. \tag{8.54}$$

It is noted that due to the plant pole at $s = -(R + r_{on})/L$, which is fairly close to the origin, the magnitude and the phase of the loop gain start to drop from a relatively low frequency. Thus, the plant pole is first canceled by the compensator zero $s = -k_i/k_p$, and the loop gain assumes the form $\ell(s) = k_p/(Ls)$. Then, the closed-loop transfer function, that is, $\ell(s)/(1 + \ell(s))$, becomes

$$\frac{I_d(s)}{I_{dref}(s)} = G_i(s) = \frac{1}{\tau_i s + 1}, \tag{8.55}$$

if

$$k_p = L/\tau_i, \tag{8.56}$$

$$k_i = (R + r_{on})/\tau_i. \tag{8.57}$$

where τ_i is the time constant of the resultant closed-loop system.

Equation (8.55) indicates that, if k_p and k_i are selected based on (8.56) and (8.57), the response of $i_d(t)$ to $i_{dref}(t)$ is based on a first-order transfer function whose time constant τ_i is a design choice. τ_i should be made small for a fast current-control response but adequately large such that $1/\tau_i$, that is, the bandwidth of the closed-loop control system, is considerably smaller, for example, 10 times, than the switching frequency of the VSC (expressed in rad/s). Depending on the requirements of a specific application and the converter switching frequency, τ_i is typically selected in the range of 0.5–5 ms. The same compensator as $k_d(s)$ can also be adopted for the q-axis compensator $k_q(s)$. Example 8.2 demonstrates the design procedures.

EXAMPLE 8.2   Dynamic Performance of Real-/Reactive-Power Controller

Consider the real-/reactive-power controller of Figure 8.3 with parameters $L = 100$ μH, $R = 0.75$ mΩ, $r_{on} = 0.88$ mΩ, $V_d = 1.0$ V, $V_{DC} = 1250$ V, and $f_s = 3420$ Hz. The AC system frequency and line-to-line rms voltage are $\omega_0 = 377$ rad/s and 480 V (i.e., $V_{sd} = 391$ V), respectively. The transfer function of the feed-forward filter is $G_{ff}(s) = 1/(8 \times 10^{-6}s + 1)$. The PLL of Example 8.1 is used to synchronize the dq-frame to the AC system voltage.

Assuming a closed-loop time constant of $\tau_i = 2.0$ ms, based on (8.56) and (8.57), we deduce the following d- and q-axis compensators:

$$k_d(s) = k_q(s) = \frac{0.05s + 0.815}{s} \quad [\Omega].$$

The system is subjected to the following sequence of events: until $t = 0.15$ s, the gating pulses are blocked and the controllers are inactive. This permits the PLL to reach its steady state. At $t = 0.15$ s, the gating pulses are unblocked and the controllers are activated, while $P_{sref} = Q_{sref} \equiv 0$. At $t = 0.20$ s, P_{sref} is subjected to a step change from 0 to 2.5 MW. At $t = 0.30$ s, P_{sref} is subjected to another step change from 2.5 to -2.5 MW. At $t = 0.35$ s, Q_{sref} is subjected to a step change from 0 to 1.0 MVAr.

Figure 8.12 illustrates the time responses of the VSC system to the start-up process and the disturbances. Figure 8.12 illustrates that P_s and Q_s rapidly track P_{sref} and Q_{sref}, respectively. Figure 8.12 also shows that the responses of P_s and Q_s are decoupled when either of them is changed. Figure 8.12 also illustrates the AC system phase-a voltage waveform, that is, V_{sa}, and the converter phase-a current waveform, that is, i_a. Figure 8.12 shows that i_a is (i) in phase with V_{sa} when $(P_s, Q_s) = (2.5 \text{ MW}, 0)$, (ii) 180° behind V_{sa} when $(P_s, Q_s) = (-2.5 \text{ MW}, 0)$, and (iii) 158° behind V_{sa} when $(P_s, Q_s) = (-2.5 \text{ MW}, 1.0 \text{ MVAr})$.

Figure 8.13 provides a close-up of i_d and i_q around $t = 0.20$ s. Figure 8.13 verifies that the step response of i_d is that of a first-order exponential function that reaches its final value at about $t = 0.21$ s, that is, after about 10 ms. It should be noted that such an inspective verification is not readily possible for $\alpha\beta$-frame current controllers of Chapter 7. The reason is that compensators in $\alpha\beta$-frame are essentially of high dynamic orders, and so are the resultant closed-loop

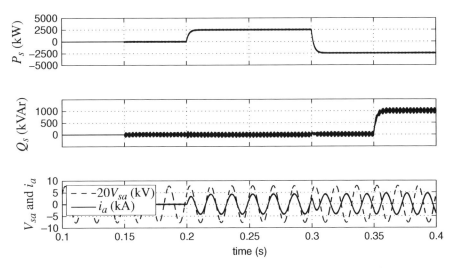

FIGURE 8.12   Dynamic responses of real and reactive power; Example 8.2.

systems. To design $\alpha\beta$-frame compensators, we adopted the frequency response method (Bode plots) that usually does not offer a quantitative insight into the time-response characteristics of the closed-loop system, unless the closed-loop system is predominantly a first-order or a second-order system. Figure 8.13 also confirms that i_d and i_q are well decoupled; it is observed that i_q remains regulated at zero while i_d is changing from zero to 4.26 kA. Ripples on the waveforms of i_d and i_q are due to the pulse-width modulation (PWM) switching side-band harmonics of VSC AC-side currents, which are modulated by 60 Hz via the *abc*- to *dq*-frame transformation.

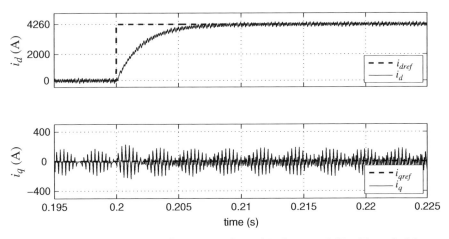

FIGURE 8.13   A close-up of responses of i_d and i_q about $t = 0.20$ s; Example 8.2.

8.4.2 Selection of DC-Bus Voltage Level

As discussed in Sections 7.3.4, 7.3.5, and 7.3.6, the DC-bus voltage of the real-/reactive-power controller of Figure 8.3 must satisfy the following criteria:

$$V_{DC} \geq 2\widehat{V}_t, \quad \text{PWM}, \tag{8.58}$$

$$V_{DC} \geq 1.74\widehat{V}_t, \quad \text{PWM with third-harmonic injection.} \tag{8.59}$$

Thus, one must properly evaluate $\widehat{V}_t$ under the worst-case operating condition. Since the VSC system controls P_s and Q_s, $\widehat{V}_t$ should also be expressed in terms of P_s and Q_s. Based on (8.45) and (8.46), and under the assumptions that $V_{sq} = 0$ and $(R + r_{on}) \approx 0$, we deduce

$$V_{td} = L\frac{di_d}{dt} - L\omega_0 i_q + V_{sd}, \tag{8.60}$$

$$V_{tq} = L\frac{di_q}{dt} + L\omega_0 i_d. \tag{8.61}$$

Substituting for i_d and i_q from (8.41) and (8.42) in (8.60) and (8.61), and assuming that V_{sd} is constant, we obtain

$$V_{td} = \left(\frac{2L}{3V_{sd}}\right)\frac{dP_s}{dt} + \left(\frac{2L\omega_0}{3V_{sd}}\right)Q_s + V_{sd}, \tag{8.62}$$

$$V_{tq} = -\left(\frac{2L}{3V_{sd}}\right)\frac{dQ_s}{dt} + \left(\frac{2L\omega_0}{3V_{sd}}\right)P_s. \tag{8.63}$$

Based on (4.77), the amplitude of the AC-side terminal voltage is

$$\widehat{V}_t = \sqrt{V_{td}^2 + V_{tq}^2}. \tag{8.64}$$

Furthermore, the amplitude of the modulating signal is

$$\widehat{V}_t = \widehat{m}\frac{V_{DC}}{2}. \tag{8.65}$$

As discussed in Section 7.3.6, if the conventional PWM is employed, $\widehat{m}$ can assume a value up to unity, whereas with the PWM with third-harmonic injection, $\widehat{m}$ can be as large as 1.15.

To calculate the maximum of $\widehat{V}_t$, consider the following worst-case scenario. Initially, the system is under a steady-state condition, that is, $P_s = P_{sref} = P_{s0}$ and $Q_s = Q_{sref} = Q_{s0}$. At $t = t_0$, P_{sref} and Q_{sref} are subjected to step changes from P_{s0} to $P_{s0} + \Delta P_s$, and Q_{s0} to $Q_{s0} + \Delta Q_s$, respectively. As discussed in Section 8.4.1,

P_s and Q_s respond to step changes in their corresponding reference commands as

$$P_s(t) = (P_{s0} + \Delta P_s) - \Delta P_s e^{-(t-t_0)/\tau_i}, \tag{8.66}$$

$$Q_s(t) = (Q_{s0} + \Delta Q_s) - \Delta Q_s e^{-(t-t_0)/\tau_i}, \tag{8.67}$$

for $t \geq t_0$. Substituting for $P_s(t)$ and $Q_s(t)$ in (8.62) and (8.63), from (8.66) and (8.67), we deduce

$$V_{td} = V_{sd} + \left(\frac{2L\omega_0}{3V_{sd}}\right)(Q_{s0} + \Delta Q_s)$$

$$+ \left(\frac{2L\omega_0}{3V_{sd}}\right)\left(\frac{\Delta P_s}{\omega_0 \tau_i} - \Delta Q_s\right)e^{-(t-t_0)/\tau_i}, \tag{8.68}$$

$$V_{tq} = \left(\frac{2L\omega_0}{3V_{sd}}\right)(P_{s0} + \Delta P_s) - \left(\frac{2L\omega_0}{3V_{sd}}\right)\left(\frac{\Delta Q_s}{\omega_0 \tau_i} + \Delta P_s\right)e^{-(t-t_0)/\tau_i}. \tag{8.69}$$

Equation (8.68) indicates that at $t = t_0$, V_{td} jumps from the initial value of $V_{sd} + (\frac{2L\omega_0}{3V_{sd}})Q_{s0}$ to $V_{sd} + (\frac{2L\omega_0}{3V_{sd}})Q_{s0} + (\frac{2L}{3\tau_i V_{sd}})\Delta P_s$ and then exponentially approaches the final value of $V_{sd} + (\frac{2L\omega_0}{3V_{sd}})(Q_{s0} + \Delta Q_s)$. Equation (8.69) indicates that V_{tq} jumps from the initial value of $(\frac{2L\omega_0}{3V_{sd}})P_{s0}$ to $(\frac{2L\omega_0}{3V_{sd}})P_{s0} - (\frac{2L}{3\tau_i V_{sd}})\Delta Q_s$ at $t = t_0$ and then exponentially approaches the final value of $(\frac{2L\omega_0}{3V_{sd}})(P_{s0} + \Delta P_s)$. The worst-case scenario corresponds to $t = t_0^+$ (immediately after $t = t_0$) where the jumps in both P_s and Q_s coincide, and

$$V_{td}(t_0^+) = V_{sd} + \left(\frac{2L\omega_0}{3V_{sd}}\right)Q_{s0} + \left(\frac{2L}{3\tau_i V_{sd}}\right)\Delta P_s, \tag{8.70}$$

$$V_{tq}(t_0^+) = \left(\frac{2L\omega_0}{3V_{sd}}\right)P_{s0} - \left(\frac{2L}{3\tau_i V_{sd}}\right)\Delta Q_s. \tag{8.71}$$

Depending on the steady-state power flow and the values of ΔP_s and ΔQ_s, $V_{td}(t_0^+)$ and $V_{tq}(t_0^+)$ can be estimated based on (8.70) and (8.71). The maximum AC-side terminal voltage, $\widehat{V}_t(t_0^+)$, is then calculated from (8.64), based on $V_{td}(t_0^+)$ and $V_{tq}(t_0^+)$. Finally, the minimum required DC-bus voltage is calculated based on (8.58) or (8.59), depending on the PWM strategy adopted. These calculations are demonstrated in Example 8.3.

EXAMPLE 8.3   Selection of DC-Bus Voltage Level

Consider the real-/reactive-power controller of Example 8.2, in which $V_{sd} = 0.391$ kV, $L = 100$ μH, $\tau_i = 2.0$ ms, and $V_{DC} = 1.250$ kV. Assume that for this system the worst-case scenario corresponds to $P_{s0} = 0$, $\Delta P_s = 2.5$ MW, $Q_{s0} = 0$, and $\Delta Q_s = 0$. Thus, based on (8.70), (8.71), and (8.64), $V_{td}(t_0^+) = 0.604$ kV, $V_{tq}(t_0^+) = 0$, and $\widehat{V}_t(t_0^+) = 0.604$ kV. If the conventional sinusoidal

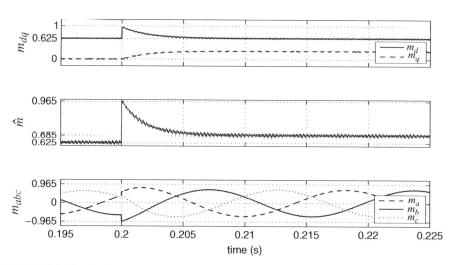

FIGURE 8.14 Steady-state and dynamic responses of the modulating signals to step change in P_{sref}; Example 8.3.

PWM is employed, V_{DC} must be larger than 1.208 kV (equation (8.58)) to avoid overmodulation. However, if the third-harmonic injected PWM is employed, V_{DC} can be lowered to about 1.050 kV (equation (8.59)). For the VSC system of Example 8.2, $V_{DC} = 1.250$ kV was selected since the conventional PWM was employed.

Figure 8.14 illustrates the waveforms of m_d, m_q, and $\widehat{m}$ for the VSC system of Example 8.2. Figure 8.14 illustrates that at $t_0 = 0.2$ s, $\widehat{m}$ jumps to 0.965, corresponding to $\widehat{V}_t = 0.604$ kV. Figure 8.14 also indicates that in this specific example, the instant when the disturbance takes place coincides with the instant when $m_b(t)$ reaches its negative peak; this corresponds to the worst-case scenario. However, since the DC-bus voltage is adequately large, neither $\widehat{m}$ nor $|m_b(t_0)|$ exceed unity, and the VSC does not experience overmodulation.

8.4.3 AC-Side Equivalent Circuit

Traditionally, balanced three-phase linear circuits have been analyzed based on their corresponding phasor diagrams and single-phase equivalent circuits. In the conventional phasor analysis, which is restricted to steady-state conditions, the voltages and currents are represented by phasors, and the passive elements are represented by impedances. This section first presents a space-phasor diagram, analogous to the conventional phasor diagram, for the AC side of the real-/reactive-power controller of Figure 8.3. Then, the relationships between the magnitude/phase-angle of an AC-side variable and the d-/q-axis components of the variable are identified. It is also demonstrated that, under steady-state conditions, the space-phasor differential equations of the real-/reactive-power controller become equivalent to the algebraic equations derived based on the conventional phasor-domain analysis. Finally, based on the steady-state phasor model, a simplified equivalent circuit is presented for the real-/reactive-power controller of Figure 8.3.

8.1.3.1 Space-Phasor Diagram of the AC Side   With reference to the real-/reactive-power controller of Figure 8.3, V_{sabc}, V_{tabc}, and i_{abc} are

$$V_{sa}(t) = \widehat{V}_s \cos(\theta),$$

$$V_{sb}(t) = \widehat{V}_s \cos\left(\theta - \frac{2\pi}{3}\right),$$

$$V_{sc}(t) = \widehat{V}_s \cos\left(\theta - \frac{4\pi}{3}\right), \tag{8.72}$$

$$V_{ta}(t) = \widehat{V}_t \cos(\theta + \delta),$$

$$V_{tb}(t) = \widehat{V}_t \cos\left(\theta + \delta - \frac{2\pi}{3}\right),$$

$$V_{tc}(t) = \widehat{V}_t \cos\left(\theta + \delta - \frac{4\pi}{3}\right), \tag{8.73}$$

$$i_a(t) = \widehat{i} \cos(\theta - \phi),$$

$$i_b(t) = \widehat{i} \cos\left(\theta - \phi - \frac{2\pi}{3}\right),$$

$$i_c(t) = \widehat{i} \cos\left(\theta - \phi - \frac{4\pi}{3}\right), \tag{8.74}$$

where $\theta = \omega_0 t + \theta_0$, and δ and $-\phi$ are the phase shifts of V_{tabc} and i_{abc} with respect to V_{sabc}, respectively. V_{sabc}, V_{tabc}, and i_{abc} are equivalently expressed by the space phasors

$$\overrightarrow{V_s} = \widehat{V}_s e^{j\theta}, \tag{8.75}$$

$$\overrightarrow{V_t} = \left(\widehat{V}_t e^{j\delta}\right) e^{j\theta}, \tag{8.76}$$

$$\overrightarrow{i} = \left(\widehat{i} e^{-j\phi}\right) e^{j\theta}. \tag{8.77}$$

The space phasors $\overrightarrow{V_t}$ and $\overrightarrow{i}$ are phase shifted with respect to $\overrightarrow{V_s}$ by angles δ and $-\phi$, respectively. Under transient conditions, in addition to δ and ϕ, the magnitudes of $\overrightarrow{V_t}$ and $\overrightarrow{i}$ (i.e., $\widehat{V}_t$ and $\widehat{i}$) can also change with time. In a steady state, however,

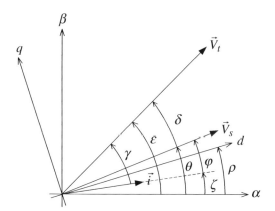

FIGURE 8.15   Space-phasor diagram for AC side of real-/reactive-power controller of Figure 8.3.

δ, ϕ, $\widehat{V}_t$, and $\widehat{i}$ are constant values, and $\overrightarrow{V}_s$, $\overrightarrow{V}_t$, and $\overrightarrow{i}$ assume constant lengths and rotate with the constant angular frequency ω_0.

Figure 8.15 illustrates the space phasors $\overrightarrow{V}_s$, $\overrightarrow{V}_t$, and $\overrightarrow{i}$ on the $\alpha\beta$-plane. Figure 8.15 also shows a *dq*-frame whose *d*-axis makes an angle ρ with respect to the α-axis. The *d* or *q* component of each space phasor is the projection of the space phasor on the corresponding axis. Therefore, if $d\rho/dt = \omega_0$, that is, the *dq*-frame rotates with the angular speed ω_0, then V_{sdq}, V_{tdq}, and i_{dq} settle at constant values in steady state. As discussed earlier in this chapter, the PLL not only guarantees $d\rho/dt = \omega_0$ but also ensures that $\rho = \theta$; the latter implies that $V_{sq} = 0$ and $V_{sd} = \widehat{V}_s$, as perceived from Figure 8.15.

To relate the lengths and phase angles of the space phasors to their *d*- and *q*-axis components, we use the space-phasor to *dq*-frame transformation of (8.1), with $\rho = \theta$. This yields

$$V_{sd} + jV_{sq} = \widehat{V}_s, \tag{8.78}$$

$$V_{td} + jV_{tq} = \widehat{V}_t e^{j\delta} = \left(\widehat{V}_t \cos \delta\right) + j\left(\widehat{V}_t \sin \delta\right), \tag{8.79}$$

$$i_d + ji_q = \widehat{i}e^{-j\phi} = \left(\widehat{i}\cos\phi\right) + j\left(-\widehat{i}\sin\phi\right). \tag{8.80}$$

It follows from (8.79) and (8.80) that

$$\delta = \tan^{-1}\left(V_{tq}/V_{td}\right), \tag{8.81}$$

$$\phi = -\tan^{-1}\left(i_q/i_d\right). \tag{8.82}$$

The angles of $\overrightarrow{V}_t$ and $\overrightarrow{i}$ with respect to the α-axis are identified as ε and ζ, respectively. Figure 8.15 illustrates that $\varepsilon = \theta + \delta$ and $\zeta = \theta - \phi$. Thus,

$$\varepsilon = \theta + \tan^{-1}\left(V_{tq}/V_{td}\right), \tag{8.83}$$

$$\zeta = \theta + \tan^{-1}\left(i_q/i_d\right). \tag{8.84}$$

Figure 8.15 also shows that $\gamma = \varepsilon - \zeta$ is the angle of $\overrightarrow{V_t}$ with respect to $\overrightarrow{i}$, that is, γ is the power-factor angle of the three-phase circuit seen from the VSC AC-side terminals. Based on (8.83) and (8.84), we deduce

$$\gamma = \tan^{-1}\left(V_{tq}/V_{td}\right) - \tan^{-1}\left(i_q/i_d\right). \tag{8.85}$$

8.4.3.2 AC-Side Steady-State Equivalent Circuit The AC-side dynamics of the real-/reactive-power controller of Figure 8.3 are described by (8.8). If the PLL is under a steady-state condition, then $\rho = \omega_0 t + \theta_0$ and (8.8) can be rewritten as

$$V_{tdq} - V_{sdq} - L\frac{d}{dt}i_{dq} = \left[jL\omega_0 + (R + r_{on})\right]i_{dq}, \tag{8.86}$$

where $f_{dq} = f_d + jf_q$ and $\omega_0 = d\rho/dt$. In a steady state, the time derivative is zero and we obtain

$$V_{tdq} - V_{sdq} = \underbrace{\left[jL\omega_0 + (R + r_{on})\right]}_{\underline{Z}} i_{dq} = \underline{Z}i_{dq}, \tag{8.87}$$

which is identical to the conventional phasor-domain equation for an equivalent single-phase circuit. Although (8.87) is valid under steady-state conditions, it may also be employed for analysis and control design purposes, if a quasi-steady-state condition is assumed. In this case, i_d and i_q are not constant quantities, but change relatively slowly with time. Therefore, di_d/dt and di_q/dt are insignificant and can be ignored in the analysis.

Figure 8.16(a) illustrates a time-domain equivalent circuit for the AC side of the real-/reactive-power controller of Figure 8.3. Based on (8.86), the circuit of Figure 8.16(a) can be represented by the space-phasor-domain equivalent circuit of Figure 8.16(b). In the circuit of Figure 8.16(b), all the time-domain variables of the original circuit are represented by the corresponding space phasors. Thus, the equivalent circuit is valid under both dynamic and steady-state operating conditions. If a quasi-steady-state condition is assumed, based on (8.87) the circuit of Figure 8.16(a) can be represented by the steady-state phasor-domain circuit of Figure 8.16(c).

Substituting for $\overrightarrow{V_s}$ and $\overrightarrow{i}$, from (8.75) and (8.77), in (4.40), we deduce

$$P_s = \frac{3}{2}\widehat{V_s}\widehat{i}\cos\phi, \tag{8.88}$$

$$Q_s = \frac{3}{2}\widehat{V_s}\widehat{i}\sin\phi. \tag{8.89}$$

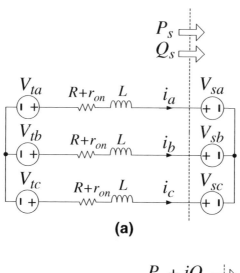

(a)

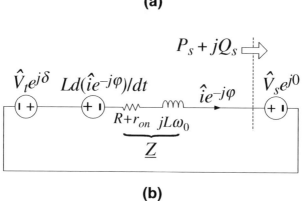

(b)

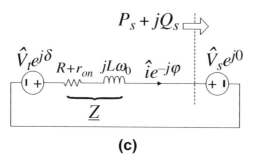

(c)

FIGURE 8.16   Equivalent circuits for AC side of the real-/reactive-power controller of Figure 8.3: (a) time-domain equivalent circuit; (b) dynamic space-phasor-domain equivalent circuit; (c) quasi-steady-state space-phasor-domain equivalent circuit.

Equations (8.88) and (8.89) exhibit the same forms as their counterparts in the conventional phasor-domain analysis. However, they are also valid for dynamic operating regimes where $\widehat{V}_s$, $\hat{i}$, and ϕ can all be functions of time.

8.4.4 PWM with Third-Harmonic Injection

In Section 7.3.6, we explained the need for the third-harmonic injected PWM as a means for extending the VSC permissible voltage range. We then formulated the third-harmonic injected PWM and presented the block diagram of Figure 7.11 for its implementation in $\alpha\beta$-frame. Figure 8.17 shows a block diagram equivalent to that of Figure 7.11, for the third-harmonic injected PWM in dq-frame.

As explained in Section 7.3.6, the modulating signals for the third-harmonic injected PWM are constructed by m_{abc}, based on (7.61)–(7.63). Thus, as shown in Figure 8.17, we obtain m_{abc} from the dq- to abc-frame transformation of m_d and m_q. The third-harmonic injected PWM also requires $\hat{m}^2$, as indicated by (7.61)–(7.63). Therefore, we express $\hat{m}^2$ in terms of m_d and m_q as $\hat{m} = \sqrt{m_d^2 + m_q^2}$ (Fig. 8.17).

Figure 8.18 shows a schematic diagram of a real-/reactive-power controller that employs the third-harmonic injected PWM. The real-/reactive-power controller of

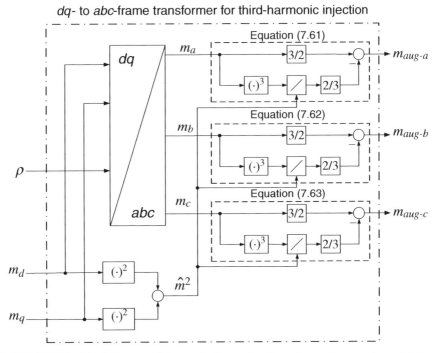

FIGURE 8.17   Block diagram of dq- to abc-frame signal transformer to generate modulating signals for third-harmonic injected PWM.

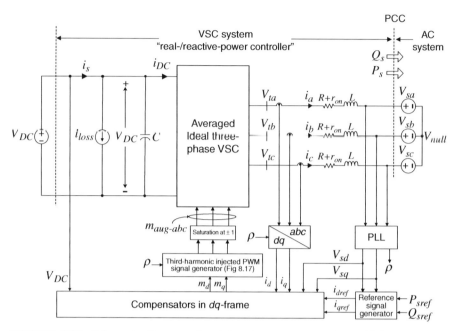

FIGURE 8.18 Schematic diagram of real-/reactive-power controller utilizing the third-harmonic injected PWM.

Figure 8.18 is the same as the real-/reactive-power controller of Figure 8.3 in which the m_{dq}-to-m_{abc} block is replaced by the block diagram of Figure 8.17. The VSC employed in the real-/reactive-power controller of Figure 8.18 can be a two-level VSC or a three-level NPC. Figure 8.18 also illustrates that for the VSC system with the third-harmonic injected PWM, $m_{aug-abc}$ is limited to ± 1, which corresponds to the limit of ± 1.15 for m_{abc}. Thus, using the third-harmonic injected PWM, the VSC AC-side terminal voltage can reach up to $\pm 1.15(V_{DC}/2)$, instead of $\pm(V_{DC}/2)$ under the conventional PWM.

8.5   REAL-/REACTIVE-POWER CONTROLLER BASED ON THREE-LEVEL NPC

The real-/reactive-power controllers of Figures 8.3 and 8.18 can also utilize the three-level NPC (Fig. 8.19) as the power processor. Based on the unified dynamic model of Section 6.7.4 presented for the two-level VSC and the three-level NPC, the *dq*-frame model and control design procedures presented in Sections 8.3 and 8.4 are equally applicable to both the two-level VSC and the three-level NPC. However, as shown in Figure 8.19, the three-level NPC also requires a DC-side voltage equalizing scheme, as discussed in Section 6.7.2.

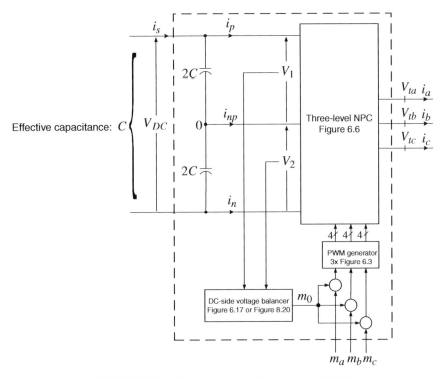

FIGURE 8.19   Block diagram of the three-level NPC.

Figure 8.20 illustrates a control block diagram of the DC-side voltage equalizing scheme. Figure 8.20 indicates that the control plant is an integrator whose gain is proportional to $-P_s$, that is, the real power that the VSC system exchanges with the AC system. Thus, the output of the compensator $K(s)$ is multiplied by -1 if P_s is

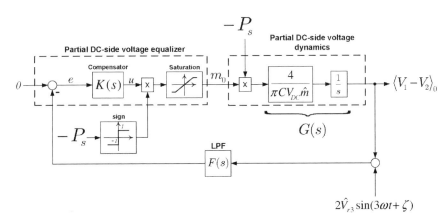

FIGURE 8.20   Control block diagram of the partial DC-side voltage equalizing scheme for the three-level NPC.

positive, to ensure a negative feedback irrespective of the direction of the power flow. P_s can be readily calculated using $P_s = (3/2)V_{sd}i_d$ or, assuming a fast d-axis current controller, approximated by P_{sref}. Figure 8.20 also shows that the difference between the partial DC-side voltages is fed back through a filter, $F(s)$. The filter is required to attenuate the third-order harmonic component of the measured signal and prevent it from distorting the corrective offset m_0. Details of modeling, derivation of the block diagram of Figure 8.20, and the compensator design approach for the DC-side voltage equalizing scheme are given in Section 7.4.

8.6   CONTROLLED DC-VOLTAGE POWER PORT

Previous sections presented the model and controls of the real-/reactive-power controller (Figs. 8.3 and 8.18), whose function is to control the real and reactive power that is exchanged with the AC system. In the real-/reactive-power controller, the VSC DC-bus voltage is impressed by an ideal, DC, voltage source, and the VSC system acts as a bidirectional energy exchanger between the AC system and the DC voltage source. However, in many applications, for example, photovoltaic (PV) systems and fuel-cell systems, the VSC DC side is not interfaced with a voltage source; rather, it is connected to a (DC) power source that needs to be interfaced and exchange (real) power with the AC system. Thus, the DC-bus voltage is not imposed and, therefore, needs to be regulated. This scenario is illustrated in Figure 8.21.

The VSC system of Figure 8.21 is conceptually the same as that of Figure 8.18, except that the DC voltage source is replaced by a (variable) DC power source. The power source typically represents a power-electronic unit (or a cluster of them) with a prime source of energy, for example, a PV array, a variable-speed wind turbine-generator set, a fuel-cell unit, or a gas turbine-generator set, behind it, and is considered as a black box in our investigations. The power source is assumed to exchange a time-varying power, $P_{ext}(t)$, with the VSC DC side. Thus, the VSC system of Figure 8.21 enables a bidirectional power exchange between the power source (black box) and the AC system. We refer to the VSC system of Figure 8.21 as controlled DC-voltage power port, which is employed as an integral part of the STATCOM, the back-to-back HVDC converter system, and variable-speed wind-power units; these are discussed in Chapters 11, 12, and 13, respectively.

The main control objective for the controlled DC-voltage power port is to regulate the DC-bus voltage V_{DC}. As Figure 8.21 illustrates, the kernel of the controlled DC-voltage power port is the real-/reactive-power controller of Figure 8.18 by which P_s and Q_s can be independently controlled. Therefore, to regulate the DC-bus voltage, a feedback mechanism compares V_{DC} with its reference command and accordingly adjusts P_s, such that the net power exchanged with the DC-bus capacitor is kept at zero. However, the reactive power Q_s can be independently controlled. In many applications, Q_s is regulated at zero, that is, the VSC system operates at unity power factor. Alternatively, Q_s may be controlled in a closed-loop mechanism to regulate the PCC voltage, as discussed in Chapter 11.

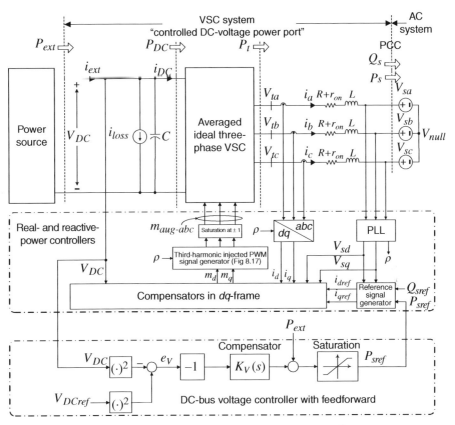

FIGURE 8.21   Schematic diagram of the controlled DC-voltage power-port.

8.6.1   Model of Controlled DC-Voltage Power Port

The main control requirement of the controlled DC-voltage power port of Figure 8.21 is to regulate the DC-bus voltage, V_{DC}. Equivalently, as discussed in Section 7.5.1, we choose to regulate V_{DC}^2 rather than V_{DC}. Based on (7.92), dynamics of V_{DC}^2 are described by

$$\frac{dV_{DC}^2}{dt} = \frac{2}{C}P_{ext} - \frac{2}{C}P_{loss} - \frac{2}{C}\left[P_s + \left(\frac{2LP_s}{3V_{sd}^2}\right)\frac{dP_s}{dt}\right]$$

$$+ \frac{2}{C}\left[\left(\frac{2LQ_s}{3V_{sd}^2}\right)\frac{dQ_s}{dt}\right], \tag{8.90}$$

where $\widehat{V}_s$ of (7.92) is replaced by V_{sd}. Based on the unified dynamic model of the two-level VSC and the three-level NPC that was presented in Section 6.7.4, (8.90) is valid for both VSC configurations. Based on (8.90), V_{DC}^2 is the output, P_s is the control input, and P_{ext}, P_{loss}, and Q_s are the disturbance inputs. As shown in Figure 8.21, V_{DC}^2 is compared with V_{DCref}^2, the error signal is processed by the compensator $K_v(s)$, and the command P_{sref} is issued for the real-power controller. The real-power controller, in turn, regulates P_s at P_{sref}, while Q_s can be independently controlled. Q_{sref} can be set to a nonzero value if an exchange of reactive power with the AC system is required. In an AC system with a large impedance, the PCC voltage is subject to variations as P_s changes with time (i.e., due to the changes of P_{ext}). In this case, the PCC voltage can be regulated by controlling Q_s in a closed-loop system that feeds the PCC voltage back and commands Q_{sref}; this reactive-power control strategy is discussed in Chapter 11.

To derive the transfer function $G_p(s) = P_s(s)/P_{sref}(s)$, we note that

$$I_d(s) = G_i(s)I_{dref}(s), \tag{8.91}$$

where $G_i(s)$ is given by (8.55). Assuming that V_{sd} is constant, multiplying both sides of (8.91) by $(3/2)V_{sd}$, we obtain

$$P_s(s) = G_i(s)P_{sref}(s). \tag{8.92}$$

Therefore, $G_p(s) = G_i(s)$ and based on (8.55), we have

$$\frac{P_s(s)}{P_{sref}(s)} = G_p(s) = \frac{1}{\tau_i s + 1}. \tag{8.93}$$

The form of (8.93) is intuitively expected as real power in dq-frame is proportional to i_d. The control plant described by (8.90) is nonlinear due to $P_s\frac{dP_s}{dt}$ and $Q_s\frac{dQ_s}{dt}$ terms. The linearized plant is provided by (7.94), which is repeated here as (8.94), in which $\hat{V}_s$ is substituted by V_{sd}.

$$\frac{d\tilde{V}_{DC}^2}{dt} = \frac{2}{C}\tilde{P}_{ext} - \frac{2}{C}\left[\tilde{P}_s + \left(\frac{2LP_{s0}}{3V_{sd}^2}\right)\frac{d\tilde{P}_s}{dt}\right]$$
$$+ \frac{2}{C}\left[\left(\frac{2LQ_{s0}}{3V_{sd}^2}\right)\frac{d\tilde{Q}_s}{dt}\right], \tag{8.94}$$

where superscripts $\sim$ and 0 represent, respectively, small-signal perturbations and steady-state values of the variables. Applying Laplace transform to (8.94), we deduce the transfer function $G_v(s) = \tilde{V}_{DC}^2/\tilde{P}_s$ as

$$G_v(s) = \tilde{V}_{DC}^2(s)/\tilde{P}_s(s) = -\left(\frac{2}{C}\right)\frac{\tau s + 1}{s}, \tag{8.95}$$

where the time constant τ is

$$\tau = \frac{2LP_{s0}}{3V_{sd}^2} = \frac{2LP_{ext0}}{3V_{sd}^2}.$$ (8.96)

Equation (8.96) indicates that τ is proportional to the (steady-state) real-power flow P_{ext0} (or P_{s0}). Thus, if P_{ext0} is small, τ is insignificant and the plant is predominantly an integrator. As P_{ext} increases, τ becomes larger and causes a shift in the phase of $G_v(s)$. In the inverting mode of operation where P_{ext0} is positive, τ is also positive and adds to the phase of $G_v(s)$. However, in the rectifying mode of operation, that is, where P_{ext0} is negative, τ is negative and results in reduction in the phase of $G_v(s)$; the phase drops further as the absolute value of P_{ext0} becomes larger. Based on (8.95), the plant zero is given by $z = -1/\tau$. Therefore, a negative τ corresponds to a zero on the right-half plane (RHP). Consequently, the controlled DC-voltage power port is a non-minimum-phase system in the rectifying mode of operation [72]. As discussed in Section 8.6.3, this non-minimum-phase property has a detrimental impact on the system stability and must be accounted for in the control design process [72].

8.6.2   Control of Controlled DC-Voltage Power Port

Figure 8.22 shows a block diagram of the DC-bus voltage controller for the controlled DC-voltage power port of Figure 8.21. The closed-loop system is composed of the compensator $K_v(s)$, real-power controller $G_p(s)$, and control plant $G_v(s)$, which is described by (8.95). Figures 8.21 and 8.22 indicate that $K_v(s)$ is multiplied by -1 to compensate for the negative sign of $G_v(s)$. The closed-loop system of Figure 8.22 is identical to that of Figure 7.23 for which the design guidelines have been provided in Section 7.5.2 and are, therefore, equally applicable to the closed-loop system of Figure 8.22. As described in Section 7.5.2, $K_v(s)$ should include an integral term and a lead transfer function. The lead transfer function compensates for the plant phase lag and ensures an adequate phase margin at the gain crossover frequency. Based on (8.95) and (8.96), $G_v(s)$ has the largest phase lag when P_{ext} is at its rated negative value. If an adequate phase margin can be guaranteed at this operating point, the closed-loop system remains stable for other operating points.

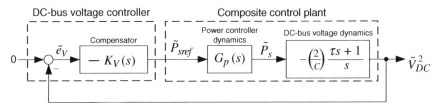

FIGURE 8.22   Control block diagram of DC-bus voltage controller based on the linearized model.

As outlined in Section 7.5.2, to design $K_v(s)$, we first select the gain crossover, ω_c, to be adequately smaller than the bandwidth of $G_p(s)$, such that one can assume $G_p(j\omega_c) \approx 1 + j0$. Then, $K_v(s)$ is designed for an adequately large phase margin under the worst-case operating condition. The design method presented in Section 7.5.2 was based on frequency response. The reason was that based on the $\alpha\beta$-frame control $G_p(s)$ typically is a high-order transfer function and primarily characterized by its bandwidth rather than its pole/zero map. Here, however, $G_p(s)$ (as given by (8.93)) is a first-order transfer function and the root-locus design method is also an option. The advantage of the root-locus method is that performance indices, for example, maximum overshoot and settling time, are related to the pole/zero loci in a more straightforward manner and can be readily taken into account in the design process.

EXAMPLE 8.4   Design of DC-Bus Voltage Controller in dq-Frame

Consider the controlled DC-voltage power port of Figure 8.21 that employs the three-level NPC of Figure 8.19. Parameters of the system are $2C = 19,250\ \mu F$, $L = 200\ \mu H$, $R = 2.38\ m\Omega$, $r_{on} = 0.88\ m\Omega$, $V_d = 1.0\ V$, $V_{DC} = 2500\ V$, $f_s = 1680\ Hz$, $V_{sd} = 391\ V$, and $\omega_0 = 377\ rad/s$. The rated power of the VSC system is $P_s = \pm 2.5\ MW$, and the third-harmonic injected PWM strategy is adopted.

With reference to Figure 8.20, the controllers of the DC-side voltage equalizing scheme are

$$K(s) = 0.0007 \quad [V^{-1}],$$

$$F(s) = \frac{s^2 + (3\omega_0)^2}{(s + 3\omega_0)^2} = \frac{s^2 + 1131^2}{s^2 + 2262s + 1131^2}.$$

From (8.56) and (8.57), for $\tau_i = 1.0$ ms parameters of the dq-frame current controllers must be $k_p = 0.2\ \Omega$ and $k_i = 3.26\ \Omega/s$, which correspond to

$$G_p(s) = G_i(s) = \frac{1000}{s + 1000}. \tag{8.97}$$

The DC-bus voltage controller is designed based on the block diagram of Figure 8.22. In Figure 8.22, $G_v(s)$ is a function of the operating point (see equations (8.95) and (8.96)). Therefore, $K_v(s)$ is designed for the worst-case operating point in the rectification mode, corresponding to $P_{ext0} = -2.5\ MW$. Equation (8.97) indicates that the bandwidth of $G_p(s)$ is 1000 rad/s. Thus, for the control loop of Figure 8.22, we choose ω_c to be about one-fifth of the bandwidth of $G_p(s)$, that is, 200 rad/s, to avoid excessive phase lag in the loop.

Based on Figure 8.22, the loop gain is

$$\ell(s) = -K_v(s)G_p(s)G_v(s), \tag{8.98}$$

where $G_v(s)$ and $G_p(s)$ are given by (8.95) and (8.97), respectively. To ensure zero steady-state errors, $K_v(s)$ must include an integral term. Let $K_v(s)$ be

$$K_v(s) - N(s)\frac{k_0}{s}, \tag{8.99}$$

where $N(s)$ is a proper transfer function with no zero at $s = 0$, and k_0 is a constant gain. Substituting for $G_v(s)$ and $K_v(s)$ in (8.98), respectively, from (8.95) and (8.99), we obtain

$$\ell(s) = N(s)k_0\left(\frac{2}{C}\right)\frac{\tau s + 1}{s^2\,(0.001s + 1)}. \tag{8.100}$$

If $N(s) = 1$, then $k_0 = 180$ yields $|\ell(j200)| = 1$ and

$$\ell(s) = 37423\frac{\tau s + 1}{s^2\,(0.001s + 1)}. \tag{8.101}$$

We refer to (8.101) as the *uncompensated* loop gain.

Figure 8.23 illustrates the magnitude and phase plots of the uncompensated loop gain, for $P_{ext0} = 2.5$ MW, $P_{ext0} = 0$, and $P_{ext0} = -2.5$ MW. Figure 8.23 shows that the magnitude response of the uncompensated loop gain is similar for all three operating points, and $|\ell(j200)| = 1$. However, $\angle\ell(j200)$ is $-168°$, $-191°$, and $-215°$, corresponding to $P_{ext0} = 2.5$, 0, and -2.5 MW, respectively. Therefore, the closed-loop system is poorly stable for $P_{ext0} = 2.5$ MW, and unstable for $P_{ext0} = 0$ and $P_{ext0} = -2.5$ MW. To ensure a stable closed-loop system for all operating points, we correct $\angle\ell(j200)$ by letting $N(s)$ in (8.100) be the lead filter

$$N(s) = n_0\frac{s + (p/\alpha)}{s + p_1}, \tag{8.102}$$

where p is the filter pole, $\alpha\ (> 1)$ is a real constant, and n_0 is the filter gain. The maximum phase of the filter is given by

$$\delta_m = \sin^{-1}\left(\frac{\alpha - 1}{\alpha + 1}\right), \tag{8.103}$$

which corresponds to the frequency

$$\omega_m = \frac{p}{\sqrt{\alpha}}. \tag{8.104}$$

Thus, if a phase margin of, for example, $45°$ is desired for $P_{ext} = -2.5$ MW, then $\angle N(j200)$ is required to be $80°$. Solving for α, p, and n_0, with $\delta_m = 80°$,

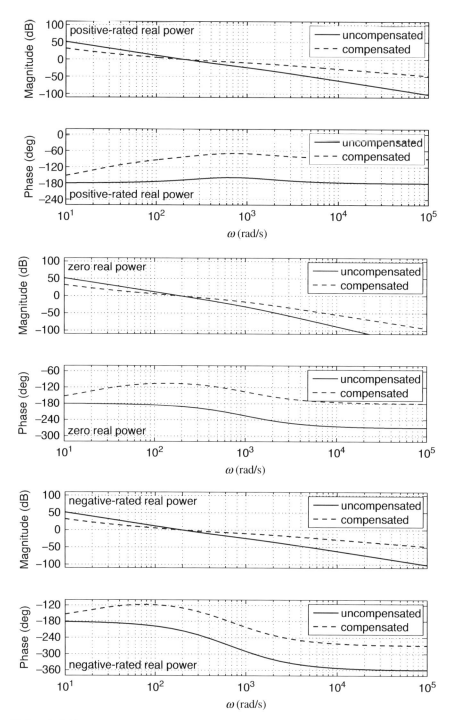

FIGURE 8.23 Bode plot of the open-loop gain of the DC-bus voltage controller; Example 8.4.

$\omega_m = 200$ rad/s, and $|N(j200)| = 1$, we obtain

$$N(s) = 10.38 \frac{s + 19}{s + 2077}. \tag{8.105}$$

Substituting for $N(s)$ in (8.99) and (8.100), from (8.105), we obtain

$$\ell(s) = 388455 \left(\frac{s + 19}{s + 2077} \right) \left(\frac{\tau s + 1}{s^2 (0.001s + 1)} \right), \tag{8.106}$$

$$K_v(s) = 1868 \frac{s + 19}{s (s + 2077)} \quad [\Omega^{-1}]. \tag{8.107}$$

We refer to the loop gain of (8.106) as the *compensated* loop gain. Figure 8.23 also shows the magnitude and phase plots of the compensated loop gain, for $P_{ext0} = 2.5$, 0, and -2.5 MW. Figure 8.23 illustrates that $|\ell(j200)| = 1$ for all three operating points. Moreover, $\angle \ell(j200)$ is $-89°$, $-112°$, and $-135°$, corresponding to $P_{ext0} = 2.5$, 0, and -2.5 MW, respectively. Thus, the closed-loop system is stable for the three operating points with a phase margin ranging from $45°$ to $91°$.

Figure 8.24 illustrates the response of the controlled DC-voltage power port of Figure 8.21 to the start-up process as well as stepwise changes in P_{ext}. The

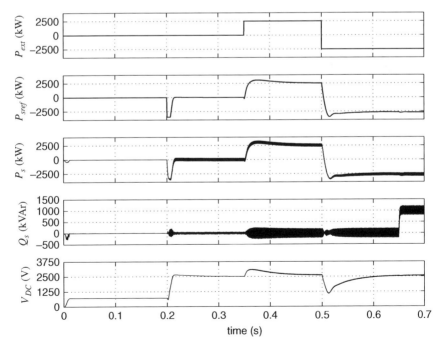

FIGURE 8.24 Dynamic performance of the controlled DC-voltage power port of Example 8.4 when feed-forward compensation is not in service.

results of Figure 8.24 are obtained under the condition that the feed-forward compensation in the DC-bus voltage control loop is disabled, and the VSC system is subjected to the following sequence of events.

Initially, $P_{ext} = 0$, the VSC gating signals are blocked, and the controllers are inactive. However, the DC-side capacitors of the VSC are charged via antiparallel diodes of VSC switch cells, and V_{DC} increases to about 700 V. At $t = 0.20$ s, gating signals are unblocked, all controllers are activated, and V_{DCref} is changed stepwise from 700 to 2500 V. Consequently, to move V_{DC} up, $K_v(s)$ commands a negative P_{sref} to import real power from the AC system to the VSC DC side; P_{sref} is saturated to its negative limit for a brief period. At about $t = 0.30$ s, V_{DC} is regulated at $V_{DCref} = 2500$ V, and P_{sref} and P_s assume small values corresponding to the VSC power loss. Figure 8.24 also shows that P_{ext} changes stepwise from 0 to 2.5 MW, at $t = 0.35$ s, which entails an overshoot in V_{DC}. The compensator reacts to this disturbance and increases P_{sref} (and thus P_s increases) to bring V_{DC} back to 2500 V. At $t = 0.50$ s, P_{ext} changes stepwise from 2.5 to -2.5 MW. Consequently, V_{DC} undergoes an undershoot until the compensator reacts and reduces P_{sref}. It should be noted that the pattern of the undershoot at $t = 0.50$ s is different from that of the overshoot at $t = 0.35$ s. The reason is that, as Figure 8.23 illustrates, the phase margin (and frequency response) is considerably different for these two operating points. Therefore, the system response to disturbances is also different for the two operating points. At $t = 0.65$ s, Q_{sref} assumes a step change from 0 to 1.0 MVAr. This disturbance, however, has no significant impact on V_{DC}, as Figure 8.24 illustrates. The reason is that, based on (8.90), the contribution of Q_s to dV_{DC}^2/dt is weighted by the term $2L/(3V_{sd}^2)$, which typically is a small value.

Figure 8.25 illustrates the response of the controlled DC-voltage power port of Figure 8.21 to the same disturbances as described above, but with the feed-forward compensation of the DC-bus voltage control loop enabled (i.e., a measure of P_{ext} is added to the output of $K_v(s)$, Fig. 8.21). A comparison between Figures 8.25 and 8.24 indicates that deviations of V_{DC} from V_{DCref} are considerably smaller when the feed-forward compensation is employed. The reason is that any change in P_{ext} is rapidly communicated to P_{sref}, and the balance of power is quickly regained.

8.6.3   Simplified and Accurate Models

The DC-bus voltage dynamics, described by (8.90), are nonlinear; the nonlinearity is due to the presence of the instantaneous power of VSC interface reactors. Thus, in the linearized model of (8.95), the time constant τ is a function of the operating point. Based on (8.96), τ is negative in the rectifying mode of operation and results in excessive phase lag in the loop gain. This phase lag can lead to unsatisfactory performance or even instabilities if it is not taken into account in the compensator design.

In the technical literature, the instantaneous power of the interface reactors is often ignored [73–76], that is, it is assumed that $L \approx 0$ and $P_t = P_s$. We refer to this model

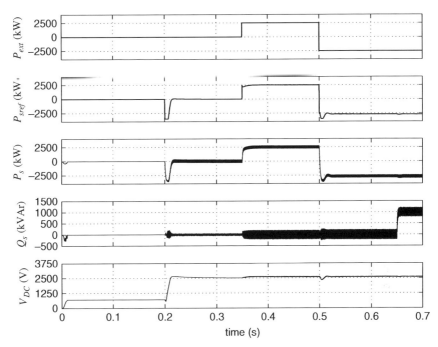

FIGURE 8.25   Dynamic performance of the controlled DC-voltage power port of Example 8.4 when feed-forward compensation is enabled.

as the *simplified model*, which can then be derived from (8.95) by substituting for $\tau = 0$.

$$G_v(s) = V_{DC}^2(s)/P_s(s) = -\left(\frac{2}{C}\right)\frac{1}{s}. \qquad (8.108)$$

The transfer function (8.108) indicates that the simplified model corresponds to the accurate model of (8.95) for the zero real-power operating point, that is, $P_{ext0} = 0$. However, as demonstrated in Example 8.4, the zero real-power operating point does not correspond to the worst-case scenario in terms of the compensator design, since the loop gain phase continues to drop in the rectifying mode of operation. Consequently, compensator design based on the simplified model of (8.108) may result in poor performance or even instabilities [72]. This is further highlighted in Example 8.5.

EXAMPLE 8.5   Instability in Rectifying Mode of Operation

Consider the controlled DC-voltage power port of Example 8.4 for which $G_p(s)$ is given by (8.97). Assume that we have to design a PI compensator, for the closed-loop system of Figure 8.22, based on the simplified model of (8.108). Thus, the loop gain includes a double integrator and a negative real pole (i.e.,

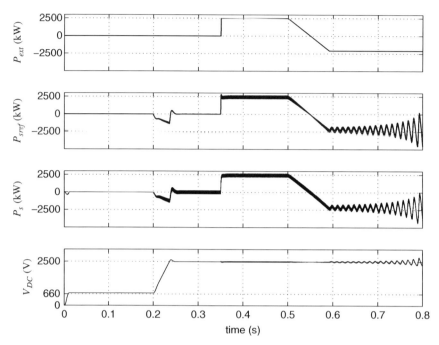

FIGURE 8.26   Instability of the DC-bus voltage controller in the rectifying mode of operation; Example 8.5.

the pole of $G_p(s)$); the compensator design process requires to identify the zero and the gain of the PI compensator. For a loop gain that possesses two integrators (including that of the PI compensator) and one first-order lag, the method of *symmetrical optimum* can be effectively employed to determine the compensator zero [43]. Based on the symmetrical optimum method, one obtains the following compensator that yields a phase margin of 45° and a crossover frequency of $\omega_c = 415$ rad/s:

$$K_v(s) = 1.996\frac{s + 172}{s} \quad [\Omega^{-1}]. \tag{8.109}$$

Since the simplified model of (8.108) does not exhibit any dependence on the operating point, one would expect that the closed-loop system remains stable over the entire power range. This is, however, not the case. Figure 8.26 illustrates that while the closed-loop system is stable for $P_{ext} = 2.5$ MW, it becomes oscillatory and unstable when P_{ext} drops from 2.5 to about -2.1 MW. The reason is that the actual control plant, described by (8.95) rather than (8.108), exhibits a non-minimum-phase zero when P_{ext} becomes negative. For this example, two of the three closed-loop poles lie on the RHP when P_{ext} becomes smaller than about -2.1 MW. These two poles are $s = 4.42 \pm j535$ rad/s and correspond to the observed unstable oscillatory response.

9 Controlled-Frequency VSC System

9.1 INTRODUCTION

Chapters 7 and 8 discussed control and operation of the grid-imposed frequency VSC system in which the operating frequency was predetermined and imposed by the AC system. These chapters implicitly translated the control of the grid-imposed frequency VSC system into the control of real and reactive power that the VSC system exchanges with the AC system, through a current-mode control strategy. This chapter investigates a class of VSC systems in which the operating frequency is not imposed by the AC system, but it is controlled by the VSC system itself. We refer to this class as *controlled-frequency VSC system*, in which the voltage and frequency at the point of common coupling (PCC) are controlled; thus, the real and reactive power that the VSC system exchanges with the AC system are the by-products.

Typical scenarios where a controlled-frequency VSC system is encountered include

- an electronically coupled distributed generation (DG) or distributed energy storage (DES) unit[1] that supplies a dedicated load, or a cluster of loads, under an islanded (off-grid) condition;
- a VSC-based HVDC converter system that supplies a passive or weak AC system; and
- an uninterruptible power supply (UPS) system that adopts a VSC system as its kernel to regulate the frequency and voltage of a sensitive load, for example, under emergency conditions.

In this chapter, we first present a dq-frame model for the controlled-frequency VSC system and then introduce a control strategy that does not require prior knowledge of the load model. We achieve this objective through a feed-forward compensation technique that can effectively decouple the VSC system dynamics from those of the load. Finally, we investigate transitions of the VSC system from the controlled-frequency mode to the grid-imposed frequency mode, and vice versa.

---

[1] These are collectively referred to as distributed energy resource (DER) units.

Voltage-Sourced Converters in Power Systems, by Amirnaser Yazdani and Reza Iravani
Copyright © 2010 John Wiley & Sons, Inc.

9.2   STRUCTURE OF CONTROLLED-FREQUENCY VSC SYSTEM

Figure 9.1 shows a schematic diagram of a controlled-frequency VSC system whose kernel is a current-controlled VSC [83]. The VSC can be either a two-level VSC or a three-level NPC. If the VSC is a three-level NPC, a partial DC-side voltage balancer is also required for its operation (see Sections 6.7, 7.4, and 8.5 for details). Irrespective of the converter type, based on the unified dynamic model of Section 6.7.4, the VSC can be considered as the composition of an ideal (lossless) VSC, an equivalent DC-bus capacitor, a parallel DC-side current source, and three series, one per each phase, AC-side resistances.

In this chapter, we assume that the net DC-side voltage, V_{DC}, is supported by a DC voltage source. Practically, the DC voltage source can be a DC energy source, for example, a battery bank, a photovoltaic (PV) array, etc., or it can be a representation of an electronically-interfaced AC or DC prime source of energy, for example, a wind turbine-generator set or a fuel cell. The DC voltage source can also be realized by the controlled DC-voltage power port of Figure 8.21.

In the VSC system of Figure 9.1, each phase of the VSC AC-side terminals is interfaced with a three-phase load via an interface RLC filter. Thus, the control objective is to regulate the amplitude and frequency of the load voltage, V_{sabc}, in the presence of disturbances in the load current, i_{Labc}. The RLC filter is composed of a series RL branch and a shunt capacitor, C_f. In this chapter, we develop the model and controllers of the controlled-frequency VSC system of Figure 9.1 with no specific assumptions regarding the load configuration or dynamics. Therefore, we need C_f to ensure that the RL branch is terminated to a node with some degree of voltage support. C_f provides a low-impedance path for switching current harmonics generated by the VSC and thus

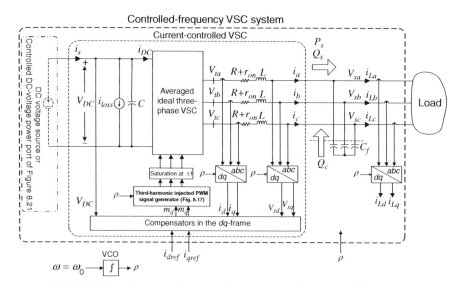

FIGURE 9.1   Schematic diagram of the controlled-frequency VSC system.

prevents them from penetrating into the load. Without C_f, harmonic distortion of load voltage, V_{sabc}, will significantly depend on load impedance at the switching frequency and its harmonics. The current-controlled VSC and the filter capacitor deliver the real and reactive power (P_s, Q_s) and $(0, Q_c)$, respectively. Therefore, the effective power output of the controlled-frequency VSC system of Figure 9.1, delivered to the load, is $(P_s, Q_s + Q_c)$.

Figure 9.1 illustrates that the controlled-frequency VSC system is controlled in a dq-frame. Thus, the feedback and feed-forward signals are transformed to the dq-frame and processed by the corresponding control loops to produce the required control signals in dq-frame. Finally, the control signals are transformed to the abc-frame and delivered to the pulse-width modulation (PWM) scheme of the VSC. The abc- to dq-frame transformation and its inverse are defined, respectively, by (4.73) and (4.75), while the angle for transformations, ρ, is obtained from the output of a voltage-controlled oscillator (VCO) whose input is $d\rho/dt = \omega(t)$. $\omega(t)$ can be set to a constant value, for example, the nominal frequency of the power system, ω_0, or it can be dynamically controlled in a variable-frequency environment [84]. The variable-frequency scenario is typically encountered where the VSC system is required to be capable of operating in both the grid-imposed frequency mode and the controlled-frequency mode; thus, $\omega(t)$ is controlled by a phase-locked loop (PLL) for synchronization to the AC system voltage [84]. An example of such an application is also provided in Section 9.4.1.

9.3   MODEL OF CONTROLLED-FREQUENCY VSC SYSTEM

The outlined part in the schematic diagram of Figure 9.1 represents a current-controlled VSC system, introduced in Section 8.4.1, whose detailed control block diagram is shown in Figure 9.2. The block diagram of Figure 9.2 is the same as that of Figure 8.10, which is repeated in this section for ease of reference. With reference to Figure 9.2, compensators $k_d(s)$ and $k_q(s)$ are of proportional-integral (PI) type, as

$$k_d(s) = k_q(s) = \frac{k_p s + k_i}{s}. \tag{9.1}$$

If parameters k_p and k_i are selected as

$$k_p = L/\tau_i, \tag{9.2}$$

$$k_i = (R + r_{on})/\tau_i, \tag{9.3}$$

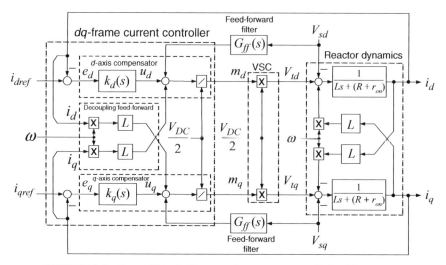

FIGURE 9.2   Control block diagram of a current-controlled VSC system.

where τ_i is a design parameter, then d- and q-axis closed-loop transfer functions assume the forms

$$I_d(s) = G_i(s)I_{dref}(s) = \frac{1}{\tau_i s + 1}I_{dref}(s), \tag{9.4}$$

$$I_q(s) = G_i(s)I_{qref}(s) = \frac{1}{\tau_i s + 1}I_{qref}(s). \tag{9.5}$$

It is noted that τ_i turns out to be the time constant of the first-order, closed-loop, transfer functions. With reference to Figure 9.1, dynamics of the load voltage are described by state-space equations:

$$C_f\frac{dV_{sa}}{dt} = i_a - i_{La}, \tag{9.6}$$

$$C_f\frac{dV_{sb}}{dt} = i_b - i_{Lb}, \tag{9.7}$$

$$C_f\frac{dV_{sc}}{dt} = i_c - i_{Lc}. \tag{9.8}$$

Equations (9.6)–(9.8) constitute the space-phasor equation

$$C_f\frac{d\overrightarrow{V_s}}{dt} = \overrightarrow{i} - \overrightarrow{i_L}. \tag{9.9}$$

Expressing each space phasor in (9.9) in terms of its dq-frame components, using $\vec{f} = (f_d + jf_q)e^{j\rho(t)}$, we obtain

$$C_f \frac{d}{dt}\left[\left(V_{sd} + jV_{sq}\right)e^{j\rho}\right] = \left(i_d + ji_q\right)e^{j\rho} - \left(i_{Ld} + ji_{Lq}\right)e^{j\rho}. \qquad (9.10)$$

It then follows from (9.10) that

$$C_f \frac{dV_{sd}}{dt} = C_f(\omega V_{sq}) + i_d - i_{Ld} \qquad (9.11)$$

$$C_f \frac{dV_{sq}}{dt} = -C_f(\omega V_{sd}) + i_q - i_{Lq}, \qquad (9.12)$$

in which $d\rho/dt$ has been replaced by $\omega(t)$, as

$$\frac{d\rho}{dt} = \omega(t). \qquad (9.13)$$

The load current components, that is, i_{Ld} and i_{Lq}, can be regarded as outputs of a dynamic system whose inputs are V_{sd} and V_{sq}. Thus,

$$\begin{bmatrix} i_{Ld} \\ i_{Lq} \end{bmatrix} = \begin{bmatrix} g_1(x_1, x_2, \ldots, x_n, V_{sd}, V_{sq}, \omega, t) \\ g_2(x_1, x_2, \ldots, x_n, V_{sd}, V_{sq}, \omega, t) \end{bmatrix}, \qquad (9.14)$$

where $x_1(t), \ldots, x_n(t)$ are the state variables, and $g_1(\cdot)$ and $g_2(\cdot)$ are nonlinear functions of their arguments. Dynamics of the state variables are governed by the equation

$$\frac{d}{dt}\begin{bmatrix} x_1 \\ x_2 \\ \cdot \\ \cdot \\ \cdot \\ x_n \end{bmatrix} = \begin{bmatrix} f_1(x_1, x_2, \ldots, x_n, V_{sd}, V_{sq}, \omega, t) \\ f_2(x_1, x_2, \ldots, x_n, V_{sd}, V_{sq}, \omega, t) \\ \cdot \\ \cdot \\ \cdot \\ f_n(x_1, x_2, \ldots, x_n, V_{sd}, V_{sq}, \omega, t) \end{bmatrix}, \qquad (9.15)$$

where $f_1(\cdot), \ldots, f_n(\cdot)$ are nonlinear functions of their arguments.

Equations (9.4), (9.5), and (9.11)–(9.15) describe dynamics of V_{sd} and V_{sq}, as shown in the block diagram of Figure 9.3, illustrating that V_{sd} and V_{sq} are the system outputs whereas i_d, i_q, and ω are the control inputs; i_d and i_q are, in turn, the responses of the d- and q-axis current controllers of the VSC system to i_{dref} and i_{qref}, respectively. Equations (9.14) and (9.15) describe the dynamics of i_{Ld} and i_{Lq} in terms of V_{sd}, V_{sq}, and ω. The next two examples show that the complexity and

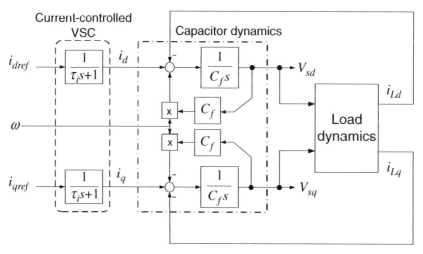

FIGURE 9.3 Block diagram of the load voltage dynamic model.

dynamic order of the load model depend on the load configuration and the number of storage elements[2].

EXAMPLE 9.1   Dynamic Model of a Series RL Load

Assume that the load in the system of Figure 9.1 is a three-phase series RL branch (Fig. 9.4). With reference to Figure 9.4, we have

$$L_1 \frac{di_{1a}}{dt} = -R_1 i_{1a} + V_{sa} - V_{n1},$$

$$L_1 \frac{di_{1b}}{dt} = -R_1 i_{1b} + V_{sb} - V_{n1},$$

$$L_1 \frac{di_{1c}}{dt} = -R_1 i_{1c} + V_{sc} - V_{n1}, \tag{9.16}$$

Equation (9.16) is equivalent to

$$L_1 \frac{d\vec{i_1}}{dt} = -R_1 \vec{i_1} + \vec{V_s}. \tag{9.17}$$

---

[2]The exception is an ideal, independent, current-sourced load for which i_{Ld} and i_{Lq} are independent of V_{sd}, V_{sq}, and ω.

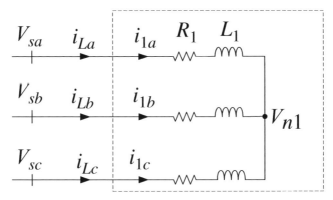

FIGURE 9.4   Series RL load of Example 9.1.

Substituting for $\overrightarrow{f} = (f_d + jf_q)e^{j\rho(t)}$ in (9.17) and letting $d\rho/dt = \omega(t)$, we obtain

$$\frac{di_{1d}}{dt} = -\frac{R_1}{L_1}i_{1d} + \omega i_{1q} + \frac{1}{L_1}V_{sd} = f_1(i_{1d}, i_{1q}, V_{sd}, V_{sq}),$$

$$\frac{di_{1q}}{dt} = -\omega i_{1d} - \frac{R_1}{L_1}i_{1q} + \frac{1}{L_1}V_{sq} = f_2(i_{1d}, i_{1q}, V_{sd}, V_{sq}). \tag{9.18}$$

In addition,

$$i_{Ld} = i_{1d} = g_1(i_{1d}, i_{1q}, V_{sd}, V_{sq}),$$

$$i_{Lq} = i_{1q} = g_2(i_{1d}, i_{1q}, V_{sd}, V_{sq}). \tag{9.19}$$

Thus, in general, the dynamic system representing the load of Figure 9.4 has two state variables, three inputs, and two outputs. It is interesting to note that even if $\omega(t)$ is a variable, (9.18) and (9.19) represent a nonlinear dynamic system.

EXAMPLE 9.2   Dynamic Model of a Composite Load

Consider the load system of Figure 9.5, which is composed of parallel connection of the load of Figure 9.4 and a series RLC branch. The RLC branch of the load is described by

$$L_2\frac{di_{2a}}{dt} = -R_2i_{2a} + V_{sa} - V_a - V_{n2},$$

$$L_2\frac{di_{2b}}{dt} = -R_2i_{2b} + V_{sb} - V_b - V_{n2},$$

$$L_2\frac{di_{2c}}{dt} = -R_2i_{2c} + V_{sc} - V_c - V_{n2}, \tag{9.20}$$

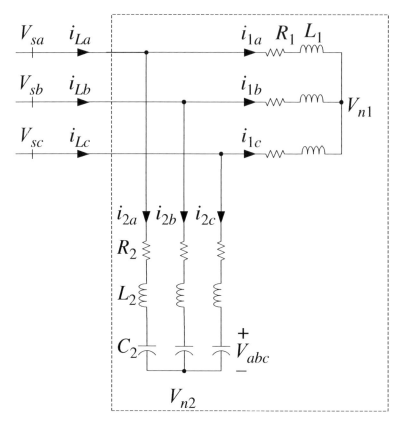

FIGURE 9.5   Composite load of Example 9.2.

and

$$C_2 \frac{dV_a}{dt} = i_{2a},$$

$$C_2 \frac{dV_b}{dt} = i_{2b},$$

$$C_2 \frac{dV_c}{dt} = i_{2c}. \tag{9.21}$$

Following the same procedures as those employed for derivation of (9.18) from (9.16), based on (9.20) and (9.21) we derive

$$\frac{di_{2d}}{dt} = -\frac{R_2}{L_2} i_{2d} + \omega i_{2q} + \frac{1}{L_2} V_{sd} - \frac{1}{L_2} V_d = f_1(\cdot),$$

$$\frac{di_{2q}}{dt} = -\omega i_{2d} - \frac{R_2}{L_2} i_{2q} + \frac{1}{L_2} V_{sq} - \frac{1}{L_2} V_q = f_2(\cdot), \tag{9.22}$$

and

$$\frac{dV_d}{dt} = \omega V_q + \frac{1}{C_2} i_{2d} = f_3(\cdot),$$

$$\frac{dV_q}{dt} = -\omega V_d + \frac{1}{C_2} i_{2q} = f_4(\cdot). \tag{9.23}$$

Dynamics of the *RL* branch of the load are described by (9.16) and (9.18), Example 9.1. Thus,

$$\frac{di_{1d}}{dt} = -\frac{R_1}{L_1} i_{1d} + \omega i_{1q} + \frac{1}{L_1} V_{sd} = f_5(\cdot),$$

$$\frac{di_{1q}}{dt} = -\omega i_{1d} - \frac{R_1}{L_1} i_{1q} + \frac{1}{L_1} V_{sq} = f_6(\cdot). \tag{9.24}$$

Then, since $i_{Labc} = i_{1abc} + i_{2abc}$, we obtain

$$i_{Ld} = i_{1d} + i_{2d} = g_1(\cdot),$$

$$i_{Lq} = i_{1q} + i_{2q} = g_2(\cdot). \tag{9.25}$$

Equations (9.22)–(9.25) describe dynamics of the load of Figure 9.5 and include six state variables, three inputs, and two outputs. These equations also represent a nonlinear system if ω is a variable.

9.4   VOLTAGE CONTROL

The control objective for the controlled-frequency VSC system of Figure 9.1 is to regulate the amplitude and the frequency of the load voltage, V_{sabc}. Figure 9.1 illustrates that the control is exercised in a *dq*-frame that is defined by the angle ρ with respect to the $\alpha\beta$-frame and rotates with the angular velocity $d\rho/dt = \omega$. Therefore, the frequency of the *abc*-frame variables, including V_{sabc}, is also ω and can be readily imposed. On the other hand, regulation of the amplitude of V_{sabc}, that is, $\widehat{V}_s = \sqrt{V_{sd}^2 + V_{sq}^2}$, is equivalent to ensuring that the tip of $\overrightarrow{V}_s$ resides on a circle, as shown in Figure 9.6. However, $\overrightarrow{V}_s$ can assume different angles with respect to the *d*-axis, depending on the values of V_{sd} and V_{sq}. One possible combination, which we adopt in this chapter, is that of $(V_{sd}, V_{sq}) = (\widehat{V}_{sn}, 0)$, where $\widehat{V}_{sn}$ is the nominal value of $\widehat{V}_s$. This combination is analogous to the case of the grid-imposed frequency VSC system where V_{sq} is forced to zero by a PLL[3].

---

[3] In Ref. [84], a slightly different control strategy has been adopted where the frequency, ω, is dynamically controlled and regulated by the control of V_{sq}.

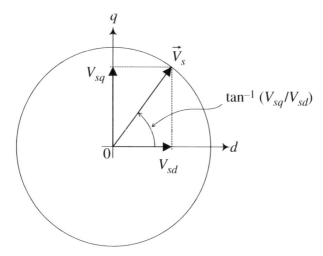

FIGURE 9.6   Loci of the load voltage space phasor if the amplitude is constant.

The control process for the controlled-frequency VSC system of Figure 9.1 can be based on the dynamic model illustrated in Figure 9.3. The block diagram of Figure 9.3 suggests that V_{sd} and V_{sq} can be controlled by i_{dref} and i_{qref}. Equations (9.11) and (9.12) indicate that V_{sd} and V_{sq} are coupled and thus the system of Figure 9.3 is a multi-input-multi-output (MIMO) system. Moreover, the system of Figure 9.3 includes the impacts of i_{Ld} and i_{Lq}. As shown in Examples 9.1 and 9.2, even for simple linear configurations, the load dynamic model (generically described by (9.14) and (9.15)) is typically of a high order, strongly intercoupled, time varying, and even nonlinear. Consequently, the design of a control scheme for the system of Figure 9.3 that guarantees the closed-loop stability or fulfills a prespecified dynamic performance is not a straightforward task.

Figure 9.7 illustrates a possible control structure for the controlled-frequency VSC system (whose open-loop model is illustrated in Fig. 9.3) that largely overcomes the aforementioned difficulties. Figure 9.7 shows that the coupling between V_{sd} and V_{sq} is eliminated by a decoupling feed-forward compensation [83]. The feed-forward compensation is similar to the one utilized to decouple i_d and i_q in a current-controlled VSC (see Fig. 9.2 or Section 8.4.1) and makes it possible to control V_{sd} by i_{dref} and V_{sq} by i_{qref}. Figure 9.7 also shows that i_{Ld} and i_{Lq} are compensated by means of another feed-forward compensation strategy. Thus, measures of i_{Ld} and i_{Lq} are added to i_{dref} and i_{qref}, respectively. Therefore, the compensated system performs under all load conditions almost the same way as the system without the load-compensating feed-forward would perform under a no-load condition.

With reference to Figure 9.7, i_{dref} and i_{qref} are determined as

$$i_{dref} = u_d - C_f(\omega V_{sq}) + i_{Ld}, \tag{9.26}$$

$$i_{qref} = u_q + C_f(\omega V_{sd}) + i_{Lq}. \tag{9.27}$$

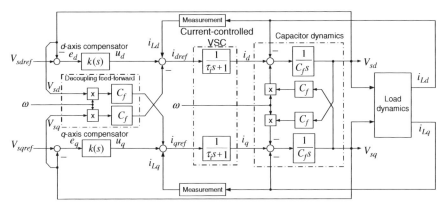

FIGURE 9.7   Control block diagram of the controlled-frequency VSC system of Figure 9.1.

where u_d and u_q are two new control inputs. Then, based on (9.4) and (9.5), i_d and i_q respond to i_{dref} and i_{qref} as

$$I_d(s) = G_i(s)U_d(s) - C_f G_i(s)\, \pounds\, \{\omega V_{sq}\} + G_i(s)I_{Ld}(s), \qquad (9.28)$$

$$I_q(s) = G_i(s)U_q(s) + C_f G_i(s)\, \pounds\, \{\omega V_{sd}\} + G_i(s)I_{Lq}(s), \qquad (9.29)$$

where $\pounds\,\{\cdot\}$ denotes the Laplace transform operator. Substitution of $i_d(s)$ and $i_q(s)$ from (9.28) and (9.29) in the Laplace transforms of (9.11) and (9.12) yields

$$C_f s V_{sd}(s) = G_i(s)U_d(s) + C_f[1 - G_i(s)]\, \pounds\, \{\omega V_{sq}\} - [1 - G_i(s)]I_{Ld}(s), \quad (9.30)$$

$$C_f s V_{sq}(s) = G_i(s)U_q(s) - C_f[1 - G_i(s)]\, \pounds\, \{\omega V_{sd}\} - [1 - G_i(s)]I_{Lq}(s). \quad (9.31)$$

The transfer function $G_i(s) = 1/(\tau_i s + 1)$ has a unity DC gain, and thus $[1 - G_i(s)] = \tau_i s/(\tau_i s + 1)$ has a zero DC gain. Therefore, if τ_i is small, $[1 - G_i(s)]$ is insignificant over a relatively wide range of frequencies and can be approximated by zero. Thus, (9.30) and (9.31) can be simplified and rewritten as

$$\frac{V_{sd}(s)}{U_d(s)} \approx G_i(s)\frac{1}{C_f s}, \qquad (9.32)$$

$$\frac{V_{sq}(s)}{U_q(s)} \approx G_i(s)\frac{1}{C_f s}, \qquad (9.33)$$

which represent two, linear, decoupled systems with u_d and u_q as inputs and V_{sd} and V_{sq} as respective outputs. Hence, V_{sd} and V_{sq} can be independently controlled, respectively, by u_d and u_q.

Figure 9.7 shows that u_d and u_q are the outputs of two independent compensators. The first compensator processes $e_d = V_{sdref} - V_{sd}$ and generates u_d. The other compensator processes $e_q = V_{sqref} - V_{sq}$ and generates u_q. Then, i_{dref} and i_{qref} are

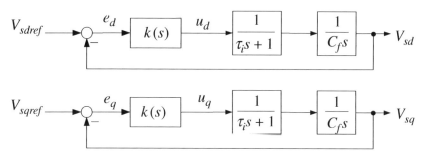

FIGURE 9.8 Simplified control block diagram equivalent to Figure 9.7.

constructed based on (9.26) and (9.27) and delivered to the corresponding d- and q-axis current-control loops. Based on (9.32) and (9.33), one can simplify the closed-loop block diagram of Figure 9.7 to that of Figure 9.8, which is more insightful for the compensator design, since it represents two decoupled single-input-single-output (SISO) control loops.

Each of the d- and q-axis control loops shown in Figure 9.8 includes one integral term, that is, one pole at $s = 0$, in addition to the real pole at $s = -1/\tau_i$. For a system of this type, the simplest compensator to fulfill fast regulation and zero steady-state error is a PI compensator. Let the PI compensator be

$$k(s) = k\frac{s + z}{s}. \tag{9.34}$$

Then, the loop gain is

$$\ell(s) = \frac{k}{\tau_i C_f}\left(\frac{s + z}{s + \tau_i^{-1}}\right)\frac{1}{s^2}. \tag{9.35}$$

At low frequencies, $\angle\ell(j\omega) \approx -180°$ due to the double pole at $s = 0$. If $z < \tau_i^{-1}$, $\angle\ell(j\omega)$ first increases until it reaches a maximum, δ_m, at a certain frequency, ω_m. Then, for $\omega > \omega_m$, $\angle\ell(j\omega)$ drops and asymptotically approaches $-180°$. δ_m and ω_m are given by

$$\delta_m = \sin^{-1}\left(\frac{1 - \tau_i z}{1 + \tau_i z}\right) \tag{9.36}$$

and

$$\omega_m = \sqrt{z\tau_i^{-1}}. \tag{9.37}$$

If the gain crossover frequency ω_c is chosen as ω_m, then δ_m becomes the phase margin. In order for this to hold, the compensator proportional gain, k, must satisfy the condition $|\ell(j\omega_c)| = |\ell(j\omega_m)| = 1$. This yields

$$k = C_f \omega_c. \tag{9.38}$$

The foregoing method of compensator design is referred to as the method of *symmetrical optimum* [43] and is suitable for a loop gain that has two poles at $s = 0$ (including the PI compensator pole) and one real pole. Based on the symmetrical optimum method, the resultant closed-loop system is of the third order. It can be shown that the closed-loop system always has one real pole at $s = -\omega_c$ in addition to two other (complex-conjugate) poles that are located on a circle with a radius of ω_c. The exact locations of the two poles on the circle depend on the selected phase margin that, typically, is selected between $30°$ and $75°$. Two particular choices of interest are (i) a phase margin of $\delta_m = 45°$, which renders the two poles complex conjugate and with a damping ratio of $\zeta = 0.707$, and (ii) a phase margin of $\delta_m = 53°$, which makes the two poles coincide with $s = -\omega_c$. With the second choice, the closed-loop system will have a triple pole at $s = -\omega_c$.

Example 9.3 demonstrates an application of the symmetrical optimum method and the performance of the closed-loop system of Figure 9.7.

EXAMPLE 9.3   Compensator Design for Voltage Regulation Loop

Consider the controlled-frequency VSC system of Figure 9.1 for which $C_f = 2500\,\mu\text{F}$, $L = 100\,\mu\text{H}$, $R = 1.19\,\text{m}\Omega$, $r_{on} = 0.88\,\text{m}\Omega$, $V_{DC} = 1500\,\text{V}$, and $f_s = 1800\,\text{Hz}$. The VSC is a three-level NPC that is switched based on the third-harmonic injected PWM and is current-controlled with $\tau_i = 0.5\,\text{ms}$ ($\tau_i^{-1} = 2000\,\text{rad/s}$). The load voltage is to be regulated at $(V_{sdref}, V_{sqref}) = (400, 0)$ V and a frequency of 377 rad/s. Thus, the VCO input is set to $\omega_0 = 377$ rad/s (Fig. 9.1). Figure 9.9 shows a schematic diagram of the load that can be re-configured through Switch #1 and Switch #2. The load parameters are given in Table 9.1.

Assuming that a phase margin of $53°$ is desired, based on (9.36) one finds $z = 224$ rad/s. It then follows from (9.37) and (9.38) that $\omega_c = 669$ rad/s and

TABLE 9.1   Load Parameters: Example 9.3

Quantity	Value
R_1	83 mΩ
L_1	137 μH
R_2	50 mΩ
L_2	68 μH
C_2	13.55 mF

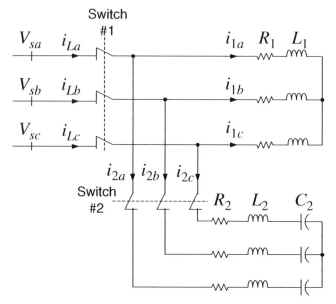

FIGURE 9.9   Load circuit of Example 9.3.

$k = 1.673 \ \Omega^{-1}$. Thus,

$$k(s) = 1.673 \left(\frac{s + 224}{s} \right) \quad [\Omega^{-1}]. \tag{9.39}$$

Figure 9.10 illustrates the loop-gain magnitude and phase plots of the load voltage regulator using the compensator of (9.39). Figure 9.10 shows that the phase margin is about $53°$ at $\omega_c = 669$ rad/s.

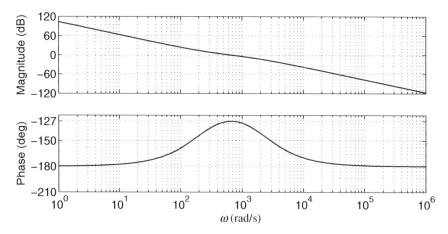

FIGURE 9.10   Open-loop frequency response; Example 9.3.

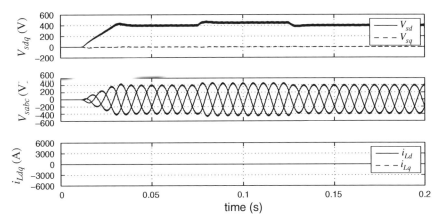

FIGURE 9.11 Start-up transient and step response of the controlled-frequency VSC system of Example 9.3 under the no-load condition.

Figures 9.11–9.13 illustrate the dynamic performance of the controlled-frequency VSC system of Figure 9.1, in response to the start-up process and stepwise changes in V_{sdref} ($V_{sqref} = 0$), under the following three load conditions:

- The no-load condition, that is, when Switch #1 and Switch #2 are open. The system response for this case is shown in Figure 9.11.
- A partially loaded condition, that is, when Switch #1 is closed but Switch #2 is open. Under this configuration, the load dynamic equations are the same as those in Example 9.1. The system response for this case is shown in Figure 9.12.

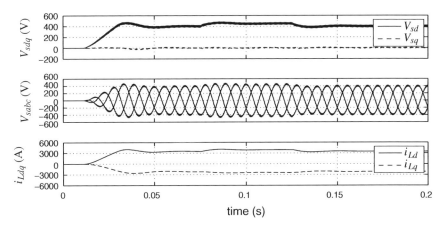

FIGURE 9.12 Start-up transient and step response of the controlled-frequency VSC system of Example 9.3 under a partially loaded condition.

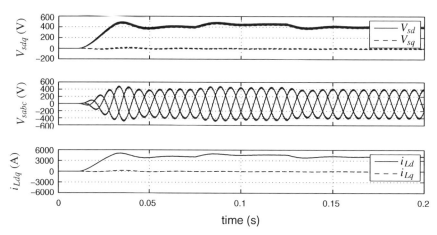

FIGURE 9.13   Start-up transient and step response of the controlled-frequency VSC system of Example 9.3 under the fully loaded condition.

- The fully loaded condition, that is, when Switch #1 and Switch #2 are both closed. For this configuration, the load dynamic equations are the same as those in Example 9.2 and the response is illustrated in Figure 9.13.

For all three cases, initially the VSC gating signals are blocked and all controllers are inactive. At $t = 0.01$ s, gating signals are unblocked, controllers are enabled, and V_{sdref} is ramped up from zero to 400 V. At $t = 0.075$ s, V_{sdref} is subjected to a step change from 400 to 450 V. At $t = 0.125$ s, V_{sdref} is subjected to another step change from 450 to 400 V.

Figures 9.11–9.13 show that, as expected, in all three cases V_{sq} remains relatively regulated at zero, despite the changes in V_{sd}. This is a result of the adopted decoupling feed-forward control. Moreover, since we have adopted a load-compensating feed-forward control in the closed-loop system (see (9.26) and (9.27), and Fig. 9.7), we expect the system response to be invariant to the load conditions. A comparison of Figures 9.12 and 9.13 with Figure 9.11 reveals that the system responses under the loaded conditions are quite similar in pattern to that of the no-load condition. The reason for discrepancies in responses is that the bandwidths of d- and q-axis current controllers are limited. Consequently, the load dynamics are not perfectly masked and manifest themselves in closed-loop responses, as (9.30) and (9.31) indicate. Theoretically, if τ_i in (9.4) and (9.5) approaches zero, the responses shown in Figures 9.12 and 9.13 become identical to that of the no-load condition shown in Figure 9.11.

Figure 9.14 illustrates the performance of the controlled-frequency VSC system in response to sudden changes in the load. Initially, the system is under a steady-state condition while both Switch #1 and Switch #2 are open and thus the load current is zero as Figure 9.14(a) illustrates. At $t = 0.2$ s, Switch #1 is closed and the RL branch of the load is switched on. Based on parameters of Table 9.1, the load rating is 2.1 MW at 0.85 lagging power factor. Following

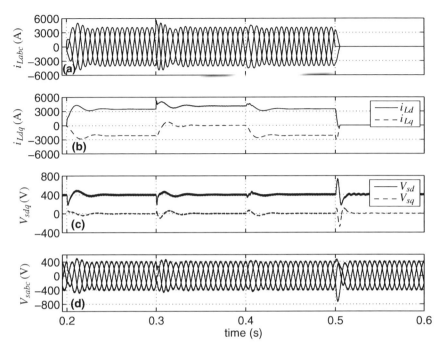

FIGURE 9.14 Dynamic response of the controlled-frequency VSC system of Example 9.3 to sudden changes in the load configuration.

the load energization, the load current increases (Fig. 9.14(a)), and i_{Ld} and i_{Lq} assume nonzero values (Fig. 9.14(b)). Since i_{Labc} lags V_{sabc}, that is, the load is inductive, i_{Lq} is negative (Fig. 9.14(b)). Figure 9.14(c) shows that V_{sd} and V_{sq} undergo transients due to the disturbance, but rapidly revert to their corresponding predisturbance values. Figure 9.14(d) shows that following the disturbance V_{sabc} undergoes fluctuations that are damped in less than three 60-Hz cycles.

At $t = 0.3$ s, Switch #2 is closed and the RLC branch of the load is also energized. Thus, the resultant load assumes the same configuration as the one introduced in Example 9.2. Based on parameters of Table 9.1, the composite load has a rating of 2.5 MW at unity power factor. Therefore, as Figure 9.14(b) illustrates, i_{Lq} becomes zero in steady state. Figure 9.14(b) also illustrates that i_{Ld} exhibits a large, sharp overshoot due to the charging current of the capacitor C_2, although R_2 and L_2 limit the amplitude and the rate of change of the charging current. Figure 9.14(c) shows that the transients of V_{sd} and V_{sq} are damped in less than three 60-Hz cycles. The maximum deviation of the amplitude of V_{abc} is about 20% of the nominal voltage of 400 V (Fig. 9.14(d)).

At $t = 0.4$ s, Switch #2 opens and the RLC branch is deenergized. Therefore, the load configuration and the operating condition become identical to those during the time interval between 0.2 and 0.3 s. Figure 9.14(a) shows that the steady-state amplitude of i_{Labc} does not significantly change compared to that under the predisturbance condition. The reason is that although the opening of Switch #2 reduces the load real power from 2.5 to 2.1 MW, it also results in a reduction of the load power factor from unity to 0.85. Figure 9.14(d) illustrates that the impact of the disturbance on V_{sbc} is negligible.

At $t = 0.5$ s, Switch #1 is opened and the rest of the load is also deenergized. Figure 9.14(a) shows that the current of one phase becomes zero, and then the currents of the other two phases simultaneously become zero. The reason is that the switch current interruptions occur at the current zero crossings. Figure 9.14(b) shows that i_{Ld} and i_{Lq} become zero when the switch opens. Figure 9.14(c) illustrates that the disturbance causes large transients in V_{sd} and V_{sq}. Consequently, V_{sc} reaches about -720 V, at $t = 0.503$ s, as Figure 9.14(d) shows.

9.4.1 Autonomous Operation

A VSC system may need to be required to operate in the grid-imposed frequency mode, as well as in the controlled-frequency mode. Consider the VSC system of Figure 9.15(a) in which the controlled-frequency VSC system of Figure 9.1 can be connected to an AC system in addition to the load. The connection to the AC system is possible through a *main switch*, and the point of connection is referred to as PCC. Thus, the entity including the VSC system and the load may operate in either the grid-connected mode or the autonomous (islanded) mode.

First assume that the main switch is closed and, therefore, the load voltage V_{sabc} is the same as the PCC voltage V_{gabc} and imposed by the AC system. Under this condition, the VSC system is synchronized to V_{gabc} and acts as a real-/reactive-power controller. Thus, the power delivered to the AC system is equal to the power delivered by the VSC system minus the load power. The real and reactive power delivered by the VSC system are, in turn, determined by the commands i_{dref} and i_{qref}, as discussed in Chapter 8.

Now assume that a disturbance occurs in the AC system and results in considerable changes in the voltage and inductance of the (Thevenin equivalent circuit of the) AC system, that is, $V_{\infty abc}$ and L_g. Except under special circumstances, the disturbance entails considerable deviations of V_{sabc}, V_{gabc}, and the frequency, ω, from their respective nominal values; dynamics of the voltage(s) and frequency depend primarily on the set points i_{dref} and i_{qref}, the load configuration and dynamic properties, and the PLL dynamic characteristics. The disturbance may be intentionally initiated, for example, to allow maintenance operations within the AC system, or it may be a consequence of a fault within the AC system.

If the disturbance is preplanned or it can be rapidly detected [85], the circuit composed of the VSC system and the load can be isolated from the AC system by opening the main switch. Thus, the operational mode of the VSC system can be switched from the grid-imposed frequency mode to the controlled-frequency mode, such that the load voltage and frequency remain regulated at about their respective nominal values. The latter mode of operation is also known as the *autonomous mode of operation*. For the system of Figure 9.15, the autonomous mode is realized if (i) the VCO input is

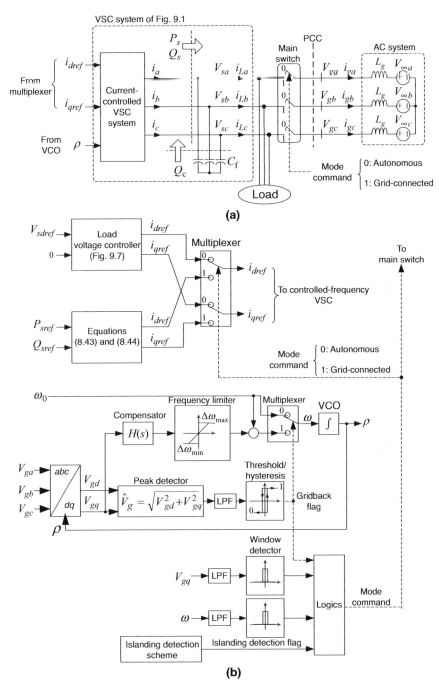

FIGURE 9.15 Schematic diagram of a VSC system and its supervisory control scheme for operation in both grid-connected and islanded modes.

set to ω_0, (ii) i_{dref} and i_{qref} are obtained from the d- and q-axis compensators of a voltage regulation scheme, as discussed in Section 9.4, and (iii) the main switch is commanded to open and thus i_{gabc} is forced to become zero.

When the AC system resumes its normal condition and permits the VSC system to operate in the grid-imposed frequency mode, the PCC voltage retrieves its nominal qualities. Then, the VSC system must be resynchronized to V_{gabc} before reclosure of the main switch (Fig. 9.15(a)). Although in the controlled-frequency mode the frequency is regulated at the nominal value of the AC system frequency, a phase error is likely to exist between V_{sabc} and V_{gabc} when the AC system is restored. As Figure 9.15(b) shows, the AC system restoration is detected by comparing the amplitude of V_{gabc} with a threshold, for example, 90% of its corresponding nominal value. Thus, once $\widehat{V}_g$ is larger than the threshold, the AC system is considered as restored and the VCO is switched from the free-running mode, in which $\omega(t) = \omega_0$, to the PLL mode in which $\omega(t) = \omega_0 + \Delta\omega(t)$. $\Delta\omega(t)$ is dynamically controlled by the compensator $H(s)$ that attempts to regulate V_{gq} at zero and is constrained by a saturation block to upper and lower limits. To prevent multiple transitions between the free-running mode and the PLL mode, $\widehat{V}_g$ is passed through a low-pass filter (LPF) and a hysteresis block. Figure 9.15(b) also shows that when V_{gq} and $\omega(t)$ are adequately close to their steady-state values of, respectively, zero and ω_0, the synchronization scheme changes the *mode command* from 0 to 1. This, in turn, results in (i) the closure of the main switch and (ii) the issuance of i_{dref} and i_{qref} in proportion to the required real and reactive power. Thereafter, the VSC system continues its operation in the grid-imposed frequency mode.

The following examples demonstrate the operation of the system of Figure 9.15 in the autonomous and grid-connected modes.

EXAMPLE 9.4   Operation Under Accidental Islanding Condition

Consider the system of Figure 9.15 in which the parameters and the controllers of the VSC system are the same as those of the VSC system of Example 9.3. The load configuration is illustrated in Figure 9.16 in which $R_1 = 83$ mΩ, $R_2 = R_3 = 25$ Ω, $L_1 = 137$ μH, $L_2 = L_3 = 45$ μH, $C_2 = 163$ μF, and $C_3 = 186$ μF. The Thevenin equivalent voltage and inductance of the AC system are 480 Vrms line-to-line and $L_g = 2.0$ μH, respectively.

Initially, the main switch is closed and the VSC system operates in the grid-connected (grid-imposed frequency) mode, for which $i_{dref} = 2000$ A corresponding to a real power of 1.17 MW delivered by the VSC system. Even though i_{qref} is set to zero, the reactive power delivered by the VSC system is equal to 217 kVAr due to the filter capacitor C_f. The load absorbs the real and reactive power of 2.0 MW and 1.21 MVAr. Thus, the AC system supplies the balance of 830 kW and 993 KVAr. At $t = 0.1$ s, the AC system line is disconnected (by a circuit breaker that is not shown in the schematic diagram), that is, i_{gabc} is forced to zero, while the main switch is still closed. Consequently, the VSC system and the load form an island. However, the VSC system remains in the grid-imposed frequency mode of operation until the islanding condition is

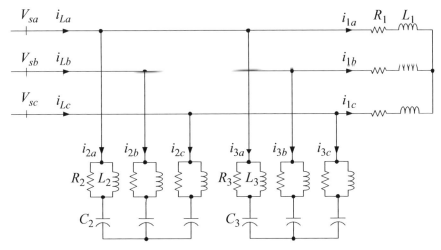

FIGURE 9.16   Load circuit of Example 9.4.

detected and the operational mode is switched to the autonomous (controlled-frequency) mode of operation.

Figure 9.17 illustrates the system response under the condition that the is-landing condition is detected in 50 ms and the operational mode is changed. It

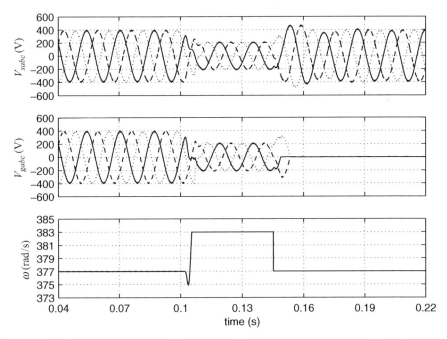

FIGURE 9.17   Response of the VSC system of Figure 9.15 to an unplanned islanding con-dition with an islanding detection delay of 50 ms; Example 9.4.

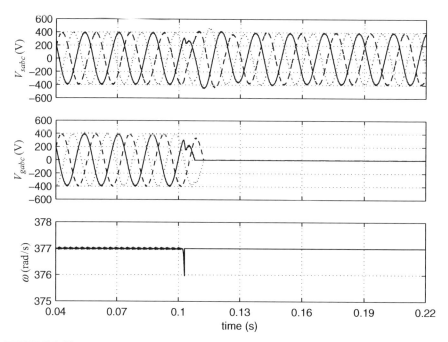

FIGURE 9.18   Response of the VSC system of Figure 9.15 to an unplanned islanding condition with an islanding detection delay of 3.0 ms; Example 9.4.

can be observed that subsequent to the formation of the island and until the operational mode is modified, V_{sabc} and V_{gabc} reduce to half whereas ω increases to its upper limit of 383 rad/s (the upper limit is imposed by the saturation block of the PLL, Fig. 9.15(b)). When the formation of the island is detected, the operational mode is switched to the controlled-frequency mode, ω is set to $\omega_0 = 377$ rad/s, and the main switch is opened. Thus, V_{gabc} becomes zero whereas V_{sabc} is regulated at its nominal value by the controlled-frequency system.

Figure 9.18 shows the response of the VSC system under the condition that the formation of the island is detected after 3.0 ms. Figure 9.18 illustrates that, in contrast to the previous case study, ω and V_{sabc} remain relatively regulated. Ideally, if the islanding condition is instantaneously detected, V_{sabc} and ω do not experience any excursions with respect to their predisturbance condition.

EXAMPLE 9.5   Resynchronization to the AC System Voltage

Consider the system of Example 9.4 in the context of Figure 9.15, under the condition that the main switch is open and the VSC system operates in the controlled-frequency mode. Thus, the load voltage is regulated at its nominal value corresponding to $V_{sd} = 400$ V and $V_{sq} = 0$. Figures 9.19–9.22

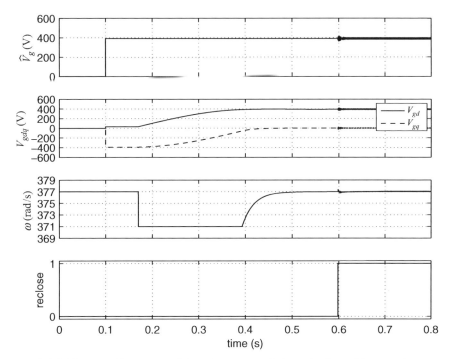

FIGURE 9.19   Overall response of the VSC system of Figure 9.15 to grid restoration, synchronization, and reconnection; Example 9.5.

demonstrate the resynchronization of the load voltage to the PCC voltage, followed by reclosure of the main switch.

As Figure 9.19 illustrates, at $t = 0.1$ s the AC system is restored and the amplitude of V_{gabc} increases. Consequently, V_{gq} assumes a negative value due to the phase shift between V_{sabc} and V_{gabc}. The restoration of the AC system is detected after a delay of about 0.08 s, corresponding to the response time of the LPF. Therefore, the operational mode of the VCO is switched from the free-running mode to the PLL mode, at $t = 0.18$ s, and $H(s)$ reduces ω to regulate V_{gq} at zero. Initially, V_{gq} is large (in absolute value) and ω is saturated at its lower limit of 371 rad/s. However, as V_{gq} tends to be zero, ω approaches 377 rad/s. Figure 9.19 illustrates that at about $t = 0.6$ s, both V_{gq} and ω are within their respective permissible limits, and the main switch is commanded to close. Figure 9.20 provides a closer look at V_{sa}, V_{ga}, ω, V_{gd}, and V_{gq}, from $t = 0$ to $t = 0.35$ s. Figure 9.20 illustrates that subsequent to AC system restoration, V_{ga} lags V_{sa} by about 90°. However, when ω is decreased by the PLL compensator to 371 rad/s, both V_{gq} and the phase shift approach zero.

Figure 9.21 illustrates the waveforms of V_{sabc} and V_{gabc} about the instant when *mode command* is changed from 0 to 1 and the main switch is closed. As shown in Figure 9.21, since the two voltages are synchronized, the reclosure

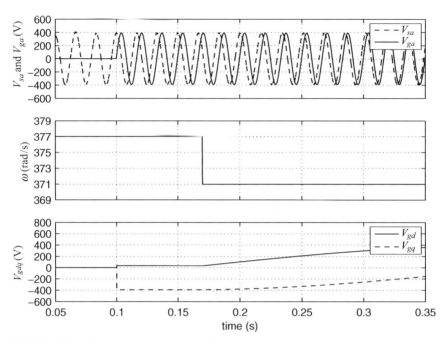

FIGURE 9.20   System response during the synchronization process subsequent to grid restoration; Example 9.5.

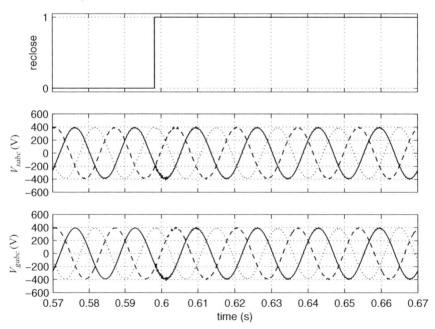

FIGURE 9.21   Responses of the load and PCC voltages to the reconnection disturbance; Example 9.5.

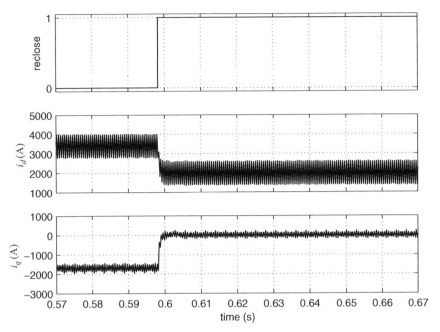

FIGURE 9.22   Responses of the VSC d- and q-axis current components to the switching in the operational mode; Example 9.5.

does not result in significant transients. When mode command is changed, the commands i_{dref} and i_{qref} are withdrawn from the outputs of the load voltage controller and get set at two independent values corresponding to real- and reactive-power requirements. Figure 9.22 shows that, before the mode command change at $t = 0.6$ s, i_d and i_q are equal to 3500 and -1655 A, respectively, as commanded by the load voltage controller of Figure 9.7. Once mode command is changed from 0 to 1, i_d and i_q track their respective reference values of $i_{dref} = 2000$ A and $i_{qref} = 0$, corresponding to $P_s = 1.17$ MW and $Q_s = 0$, respectively. Under the nominal voltage, the load demands real and reactive power of 2.0 MW and 1.21 MVAr, and C_f supplies the reactive power of $Q_c = 0.217$ MVAr. Hence, in the grid-connected mode, the AC system supplies real and reactive power of 0.830 MW and 0.993 MVAr.

10 Variable-Frequency VSC System

10.1 INTRODUCTION

This chapter introduces a class of VSC systems that we refer to as *variable-frequency VSC system*. A variable-frequency VSC system constitutes the kernel of an electromechanical energy conversion system in which the VSC AC side is interfaced with the stator terminals of an electric machine[1]. The VSC system controls the machine voltage and frequency such that the torque is controlled and the flux is regulated at its nominal value. The torque control in turn permits control of the machine power. In a variable-frequency VSC system, the machine speed is usually variable and thus the VSC system operates in a variable-frequency environment. Typical applications of a variable-frequency VSC system include variable-frequency wind-power units, gas microturbine units, and industrial regenerative drive systems. The first of the aforementioned applications is discussed in Chapter 13.

In this chapter, we first adopt the model developed in Appendix A for a symmetrical three-phase AC machine. Then, based on the model, we present the methodology of *vector control* (or *vectorial control*) for fast torque control and robust flux regulation. We introduce proper control schemes for the asynchronous machine, the doubly-fed asynchronous machine, and the permanent-magnet synchronous machine (PMSM).

10.2 STRUCTURE OF VARIABLE-FREQUENCY VSC SYSTEM

Figure 10.1 shows a schematic diagram of a variable-frequency VSC system to control the asynchronous machine or the PMSM. Methodologies of this chapter can also be applied to the conventional (wound-rotor) synchronous machine. However, since dynamics and control of the synchronous machine are very similar to those of the PMSM, the methodologies are not repeated for the synchronous machine; the main difference is that the rotor flux in the synchronous machine is generated by the rotor winding, whereas in the PMSM a set of permanent magnets generates the rotor flux.

---

[1]If the machine is of the doubly-fed asynchronous type, the VSC is interfaced with the machine rotor terminals, whereas the machine stator is directly interfaced with the AC system.

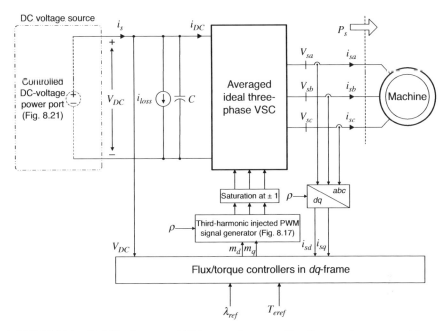

FIGURE 10.1   Variable-frequency VSC system for controlling the asynchronous machine or the PMSM.

The variable-frequency VSC system of Figure 10.1 employs a three-phase VSC that is directly interfaced with the machine stator. Commonly, no inductors are installed between the VSC AC-side terminals and the machine terminals since, due to the machine relatively large inductance, the VSC switching voltage harmonics result in negligible current harmonics. The VSC can be a two-level VSC or a three-level NPC, supplied from a DC voltage source or a rectified AC voltage source. The VSC can also be supplied by a controlled DC-voltage power port (see Section 8.6 for details) and establishes a bidirectional power flow path between the AC system interfaced with the controlled DC-voltage power port and the mechanical load/mover interfaced with the machine shaft. If the VSC is a three-level NPC, its net DC-bus voltage must be equally divided between the two DC-side capacitors. Thus, a partial DC-side voltage equalizing scheme is required to guarantee proper operation of the VSC system (see Sections 6.7, 7.4, and 8.5 for details). Irrespective of its type, the VSC can be modeled by an ideal (lossless) VSC, a current source connected in parallel with the VSC DC side, and three resistors each in series with one of the VSC AC-side terminals (Fig. 5.2). As discussed in Chapter 5, the current source mainly represents the VSC switching power loss, whereas the resistors represent the on-state resistance of each VSC switch cell. The on-state resistance can be taken into account as part of the machine stator resistance.

Figure 10.2 shows a variable-frequency VSC system that is used to control the doubly-fed asynchronous machine. For the doubly-fed asynchronous machine, the

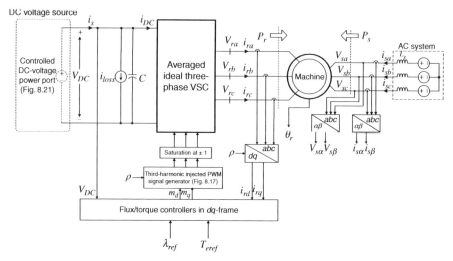

FIGURE 10.2   Variable-frequency VSC system for controlling the doubly-fed asynchronous machine.

VSC is interfaced with the machine rotor windings, whereas the stator windings are directly connected to a constant-frequency AC system. The variable-frequency VSC system of Figure 10.2 is employed when the variable speed capability is required over a fairly limited range of speed [43]; the main advantage is that the VSC handles a fraction of the machine total power and is therefore of a smaller size and power rating. The variable-frequency VSC system of Figure 10.2 has been proposed for MW-range wind-power units [86, 87].

Figures 10.1 and 10.2 illustrate that the variable-frequency VSC system is controlled in a *dq*-frame. The $\alpha\beta$-frame control is not an attractive methodology for the variable-frequency VSC system since the variable-frequency nature of the waveforms complicates the design of constant-parameter compensators. In case of the grid-imposed frequency VSC system, compensators in $\alpha\beta$-frame include complex-conjugate poles at the operating frequency, to provide zero steady-state error (see Example 7.1, Section 7.3.3). Such an implementation is not straightforward for the variable-frequency VSC system since the operating frequency is not constant. In this chapter, we identify *dq* frames suitable to control the asynchronous machine, the doubly-fed asynchronous machine, and the PMSM. The control inputs of the variable-frequency VSC system are the torque command, T_{eref}, and the reference flux, λ_{ref}. We assume that the reference flux is set to the machine rated flux and is normally not changed. Depending on the application, the torque command is usually the output of another control loop. For example, in a wind-power unit, T_{eref} is forced to change proportional to the square of the machine rotor speed, to maximize the output power [88].

10.3   CONTROL OF VARIABLE-FREQUENCY VSC SYSTEM

The equations governing dynamics of a symmetrical three-phase AC machine, as presented in Appendix A, are

$$\frac{d\vec{\lambda_s}}{dt} = \vec{V_s} - R_s \vec{i_s}, \tag{10.1}$$

$$\frac{d\vec{\lambda_r}}{dt} = \vec{V_r} - R_r \vec{i_r}, \tag{10.2}$$

$$\vec{\lambda_s} = L_m \left[(1 + \sigma_s) \vec{i_s} + e^{j\theta_r} \vec{i_r} \right], \tag{10.3}$$

$$\vec{\lambda_r} = L_m \left[(1 + \sigma_r) \vec{i_r} + e^{-j\theta_r} \vec{i_s} \right], \tag{10.4}$$

where the stator and rotor leakage factors σ_s and σ_r are defined as

$$\sigma_s = \frac{L_s}{L_m} - 1, \tag{10.5}$$

$$\sigma_r = \frac{L_r}{L_m} - 1. \tag{10.6}$$

L_s and L_r are the stator and rotor inductances, respectively, and L_m is the magnetizing inductance. Substituting for $\vec{\lambda_s}$ and $\vec{\lambda_r}$ in (10.1) and (10.2), from (10.3) and (10.4), we obtain

$$L_m \frac{d}{dt} \left[(1 + \sigma_s) \vec{i_s} + e^{j\theta_r} \vec{i_r} \right] = \vec{V_s} - R_s \vec{i_s}, \tag{10.7}$$

$$L_m \frac{d}{dt} \left[(1 + \sigma_r) \vec{i_r} + e^{-j\theta_r} \vec{i_s} \right] = \vec{V_r} - R_r \vec{i_r}. \tag{10.8}$$

The machine electrical torque is

$$T_e = \left(\frac{3}{2} L_m \right) Im \left\{ e^{-j\theta_r} \vec{i_s} \vec{i_r}^* \right\}. \tag{10.9}$$

The machine rotor speed, ω_r, is related to the machine torque by

$$J \frac{d\omega_r}{dt} = T_e - T_{ext}, \tag{10.10}$$

where T_{ext} is the torque corresponding to the mechanical load/source that also embeds friction and windage losses, and

$$\frac{d\theta_r}{dt} = \omega_r. \tag{10.11}$$

In some applications, T_{ext} does not depend (or depends weakly) on ω_r or θ_r and therefore can be regarded as a disturbance input to the control system. However, in most energy conversion applications, T_{ext} is a function of ω_r, θ_r, and other exogenous variables. For example, in a wind turbine T_{ext} is a nonlinear function of the machine rotor speed, wind speed, and turbine pitch angle. If the pulsating torque of a horizontal-axis wind turbine is also considered, then T_{ext} is also a nonlinear function of the rotor position, θ_r. In general,

$$T_{ext} = f(\omega_r, \theta_r, u_1, ..., u_n), \tag{10.12}$$

where $u_1, ..., u_n$ are exogenous variables. Equations (10.7)–(10.12) describe an electromechanical system consisting of the machine and the mechanical system. Equations (10.7)–(10.9) correspond to the machine electrical dynamics whereas (10.10)–(10.12) describe the mechanical system. The electrical equations can be made decoupled from those of the mechanical system if participation of θ_r in (10.7)–(10.9) is compensated. As such, the electromechanical system can be divided into two subsystems: the electrical subsystem for which $\overrightarrow{V_s}$ and $\overrightarrow{V_r}$ are the control inputs and T_e is the output, and the mechanical subsystem for which T_e is the control input. Depending on the application, ω_r, θ_r, or the mechanical power $P_m = T_{ext}\omega_r$ can be defined as the output. In this chapter, we concentrate on the control of T_e by the VSC system.

10.3.1 Asynchronous Machine

10.3.1.1 Machine Model in Rotor-Field Coordinates   In this section, we present the controls for the VSC system of Figure 10.1 that is interfaced with an asynchronous machine. We consider a squirrel-cage asynchronous machine or, equivalently, a wound-rotor asynchronous machine whose rotor terminals are short circuited and rotor current is not measurable. Thus, in (10.7)–(10.9), $\overrightarrow{V_r} = 0$ and $\overrightarrow{i_r}$ is not measurable.

Let us introduce the fictitious space phasor current $\overrightarrow{i_{mr}} = \hat{i}_{mr}e^{j\rho}$ and the change of variable $(1 + \sigma_r)e^{j\theta_r}\overrightarrow{i_r} + \overrightarrow{i_s} = \overrightarrow{i_{mr}}$, [43]. Then, we have

$$(1 + \sigma_r)\overrightarrow{i_r} + e^{-j\theta_r}\overrightarrow{i_s} = \hat{i}_{mr}e^{j(\rho - \theta_r)}. \tag{10.13}$$

Solving for $\overrightarrow{i_r}$ in (10.13), we deduce

$$\overrightarrow{i_r} = \frac{\hat{i}_{mr}e^{j\rho} - \overrightarrow{i_s}}{1 + \sigma_r}e^{-j\theta_r}. \tag{10.14}$$

Substituting for $\vec{i_r}$ from (10.14) in (10.9), we obtain

$$
\begin{aligned}
T_e &= \frac{3}{2}\left(\frac{L_m}{1+\sigma_r}\right) Im\left\{\vec{i_s}\left(\widehat{i}_{mr}e^{-j\rho}-\vec{i_s}^*\right)\right\} \\
&= \frac{3}{2}\left(\frac{L_m}{1+\sigma_r}\right)\widehat{i}_{mr} Im\left\{\vec{i_s}\,e^{-j\rho}\right\}.
\end{aligned}
\tag{10.15}
$$

Based on (4.65), the term $\vec{i_s}\,e^{-j\rho}$ corresponds to an $\alpha\beta$- to dq-frame transformation for which the angle is ρ, as illustrated in Figure 10.1. Thus, $Im\left\{\vec{i_s}\,e^{-j\rho}\right\}$ is the q-axis component of $\vec{i_s}$, denoted by i_{sq}, in that dq-frame, and

$$
T_e = \frac{3}{2}\left(\frac{L_m}{1+\sigma_r}\right)\widehat{i}_{mr}i_{sq}.
\tag{10.16}
$$

The advantage of the change of variable (10.13) now becomes apparent; based on (10.16), if $\widehat{i}_{mr}$ is maintained at a constant value, the machine torque becomes a linear function of the q-axis component of the stator current, i_{sq}.

Considering $\vec{V_r} = 0$, dynamics of $\widehat{i}_{mr}$ and ρ can be derived by substituting for $(1+\sigma_r)\vec{i_r} + e^{-j\theta_r}\vec{i_s}$ and $\vec{i_r}$ in (10.8), from (10.13) and (10.14). The result is

$$
\tau_r\frac{d}{dt}\left[\widehat{i}_{mr}e^{j(\rho-\theta_r)}\right] = -\widehat{i}_{mr}e^{j(\rho-\theta_r)} + \vec{i_s}\,e^{-j\theta_r},
\tag{10.17}
$$

where the rotor time constant τ_r is defined as

$$
\tau_r = \frac{(1+\sigma_r)L_m}{R_r}.
\tag{10.18}
$$

Calculating the derivative in (10.17) and simplifying the resultant, we deduce

$$
\tau_r\frac{d\widehat{i}_{mr}}{dt} + j\left(\omega - \omega_r\right)\tau_r\widehat{i}_{mr} = -\widehat{i}_{mr} + \vec{i_s}\,e^{-j\rho},
\tag{10.19}
$$

where

$$
\omega = \frac{d\rho}{dt},
\tag{10.20}
$$

and ω_r, the rotor speed, is defined by (10.11). Decomposing (10.19) into its real and imaginary components, we obtain

$$\tau_r \frac{d\hat{i}_{mr}}{dt} = -\hat{i}_{mr} + Re\left\{ \vec{i}_s\, e^{-j\rho} \right\} = -\hat{i}_{mr} + i_{sd},$$
(10.21)

$$\omega = \omega_r + \frac{Im\left\{ \vec{i}_s\, e^{-j\rho} \right\}}{\tau_r \hat{i}_{mr}} = \omega_r + \frac{i_{sq}}{\tau_r \hat{i}_{mr}}.$$
(10.22)

Equations (10.20)–(10.22) describe the machine dynamics in the *rotor-field co-ordinates* [43]. Equation (10.21) represents a first-order, linear system with a unity DC gain, for which i_{sd}, $\hat{i}_{mr}$, and τ_r are, respectively, the input, the output, and the time constant. Based on (10.21), to regulate $\hat{i}_{mr}$ (which is required by (10.16) for a linear torque control by i_{sq}), it is sufficient to regulate i_{sd} at a constant value, and this ensures $\hat{i}_{mr} = i_{sd}$ in about $5\tau_r$. However if a faster time response is required, and if an estimate of $\hat{i}_{mr}$ is available for feedback, a closed-loop scheme can be employed to regulate $\hat{i}_{mr}$ through dynamic control of i_{sd}. Regulation of $\hat{i}_{mr}$, whether open loop or closed loop, is normally exercised (i) during the VSC system start-up process, (ii) when the machine is at the standstill condition, and (iii) while i_{sq} is kept at zero. These conditions ensure that the machine flux is established before any torque demand. Later in this section we present the schemes for controlling i_{sd} and i_{sq}.

Equations (10.22) and (10.20) describe a nonlinear system that determines ρ based on the values of ω_r, i_{sq}, and $\hat{i}_{mr}$. In turn, $\hat{i}_{mr}$ is the output of the first-order system of (10.21). This combination is illustrated in Figure 10.3 and is called a *flux observer* or a *flux model* [43]. The flux observer of Figure 10.3 receives i_{sd}, i_{sq}, and ω_r as inputs, and delivers $\hat{i}_{mr}$, ρ, and ω. Current components i_{sd} and i_{sq} are the outputs of an *abc-* to *dq*-frame transformer (Fig. 10.1), for which the angle, ρ, is provided by the flux observer; ρ is also required for the pulse-width modulation (PWM) signal generator (see Fig. 8.17 for details). The flux observer also delivers $\hat{i}_{mr}$ and ω. The former output can be used as a feedback signal for flux regulation, while the latter output is utilized in a feed-forward compensation strategy to decouple the dynamics of i_{sd} and i_{sq}; these will be detailed in the next subsection.

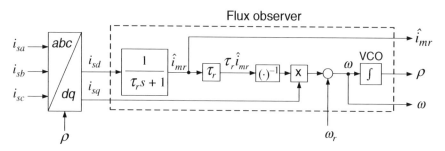

FIGURE 10.3   Schematic diagram of rotor-flux observer for the squirrel-cage asynchronous machine.

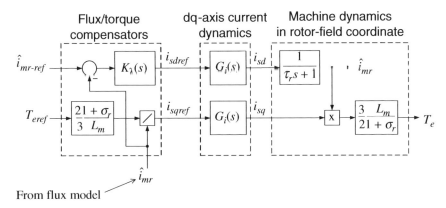

FIGURE 10.4   Control block diagram of the vector-controlled squirrel-cage asynchronous machine in rotor-field coordinates.

It is noted that the flux observer of Figure 10.3 requires ω_r as an input, obtained from a speed sensor. The flux observer can, however, be constructed based on an alternative approach in which ω_r is not required, but V_{sdq} and i_{sdq} are the inputs. This type of flux observer delivers $\widehat{i}_{mr}$, ρ, and ω_r as the outputs;[2] each type of flux observer has its own merits and demerits [43, 89, 90], and further details are outside the scope of this text.

10.3.1.2   *Machine Vector Control in Rotor-Field Coordinates*   Figure 10.4 depicts

a block diagram of a scheme for vectorial control of the squirrel-cage asynchronous machine, in the rotor-field coordinates. Figure 10.4 illustrates that the flux compensator $K_\lambda(s)$ processes the error $\widehat{i}_{mr-ref} - \widehat{i}_{mr}$ and issues the reference command for the d-axis current control loop, that is, i_{sdref}. $\widehat{i}_{mr-ref}$ is a constant set point, and $\widehat{i}_{mr}$ is obtained from the flux observer of Figure 10.3. $K_\lambda(s)$, in its simplest form, is a proportional-integral (PI) compensator whose parameters can be tuned for a fast closed-loop response. Due to the large rotor time constant, i_{sdref} and i_{sd} assume large transient values if the flux regulation loop is fast. This, however, does not overload the machine or the VSC, since flux regulation normally takes place during the start-up process of the VSC system, when $i_{sq} = 0$. If a fast closed-loop response is not required, the flux compensator is omitted and i_{sdref} is assigned a constant value, that is, the desired steady-state value of $\widehat{i}_{mr}$; then, i_{sd} will settle at this value in about $5\tau_r$.

Figure 10.4 also illustrates the machine torque control loop. Based on (10.16), the machine torque is $\frac{3}{2}(\frac{L_m}{1+\sigma_r})\widehat{i}_{mr}$ times i_{sq}. Therefore, as Figure 10.4 shows, the reference value for the q-axis current controller, i_{sqref}, is determined by dividing T_{eref} by $\frac{3}{2}(\frac{L_m}{1+\sigma_r})\widehat{i}_{mr}$, where $\widehat{i}_{mr}$ can be either obtained from a flux observer or assigned a constant value. It can be argued that the torque controller part of Figure 10.4 is an open-loop scheme and therefore not adequately robust to model uncertainties. However,

---

[2]The flux observer of Figure 10.3 is also known as *current model*, whereas the other type is referred to as *voltage model*. The names are, however, not descriptive.

it should be noted that T_{eref} is often either the output of another compensator or the product of an inherent feedback loop. An example of the former is a variable-speed motor drive where T_{eref} is determined by a speed (or position) control loop that regulates ω_r (or θ_r). An example of the latter is a variable-speed wind-power unit where T_{eref} is set to be proportional to ω_r^2. In both cases, ω_r is related to T_e and thus the loop is closed.

The flux regulator and torque controller parts of Figure 10.4 issue the commands i_{sdref} and i_{sqref} which, in turn, control i_{sd} and i_{sq} based on an inherently, nonlinear, two-input-two-output system. The next subsection shows that by proper control and feed-forward compensation techniques, one can transform the system into two decoupled, single-input–single-output (SISO), linear time-invariant subsystems, each characterized by a first-order transfer function, $G_i(s)$ (Fig. 10.4); one subsystem relates i_{sd} to i_{sdref} whereas the other subsystem relates i_{sq} to i_{sqref}. We will also show that by proper selection of parameters of (d- and q-axis) compensators, one can make the time constant of $G_i(s)$ arbitrarily small.

10.3.1.3   Machine Current Control by VSC System   As discussed in the previous subsection, the machine flux and torque are controlled, respectively, by i_{sd} and i_{sq}. However, the VSC can only control the stator voltage.[3] Therefore, one must first develop mathematical expressions to relate i_{sd} and i_{sq} to V_{sd} and V_{sq}.

The stator terminal voltage and current are related based on (10.1) and (10.3). Substituting for $\overrightarrow{i_r}$ from (10.14) in (10.3), and then for $\overrightarrow{\lambda_s}$ from the resultant in (10.1), we obtain

$$L_m \frac{d}{dt} \left[\frac{(1+\sigma_s)(1+\sigma_r)-1}{1+\sigma_r} \overrightarrow{i_s} + \frac{1}{1+\sigma_r} \hat{i}_{mr} e^{j\rho} \right] = \overrightarrow{V_s} - R_s \overrightarrow{i_s} . \quad (10.23)$$

Defining the machine total leakage factor, σ, as [43]

$$\sigma = 1 - \frac{1}{(1+\sigma_r)(1+\sigma_s)}, \quad (10.24)$$

we can rewrite (10.23) as

$$L_m \sigma (1+\sigma_s) \frac{d\overrightarrow{i_s}}{dt} + L_m (1-\sigma)(1+\sigma_s) \frac{d}{dt} \left(\hat{i}_{mr} e^{j\rho} \right) = \overrightarrow{V_s} - R_s \overrightarrow{i_s} . \quad (10.25)$$

Dividing both sides of (10.25) by R_s, we deduce

$$\sigma \tau_s \frac{d\overrightarrow{i_s}}{dt} + (1-\sigma)\tau_s \frac{d}{dt} \left(\hat{i}_{mr} e^{j\rho} \right) = \frac{1}{R_s} \overrightarrow{V_s} - \overrightarrow{i_s} , \quad (10.26)$$

---

[3]The VSC can directly control the stator current if a hysteresis-band current-control strategy is employed rather than the PWM strategy. The main disadvantage of the method is the variable switching frequency.

where τ_s is the stator time constant and defined as

$$\tau_s = \frac{L_m(1+\sigma_s)}{R_s}. \tag{10.27}$$

Substituting for $\overrightarrow{i_s} = (i_{sd} + i_{sq})e^{j\rho}$ and $\overrightarrow{V_s} = (V_{sd} + V_{sq})e^{j\rho}$ in (10.26), calculating the derivatives, and multiplying both sides of the resultant by $e^{-j\rho}$, we obtain

$$\sigma\tau_s\frac{di_{sdq}}{dt} + i_{sdq} = -j\sigma\tau_s\omega i_{sdq} - j(1-\sigma)\tau_s\omega\widehat{i}_{mr} - (1-\sigma)\tau_s\frac{d\widehat{i}_{mr}}{dt} + \frac{1}{R_s}V_{sdq}, \tag{10.28}$$

where $\omega = d\rho/dt$ and we have employed f_{dq} as a compact representation for $f_d + jf_q$. Decomposing (10.28) into real and imaginary components, we deduce

$$\left(\sigma\tau_s\frac{di_{sd}}{dt} + i_{sd}\right) = \sigma\tau_s\omega i_{sq} - (1-\sigma)\tau_s\frac{d\widehat{i}_{mr}}{dt} + \frac{1}{R_s}V_{sd}, \tag{10.29}$$

$$\left(\sigma\tau_s\frac{di_{sq}}{dt} + i_{sq}\right) = -\sigma\tau_s\omega i_{sd} - (1-\sigma)\tau_s\omega\widehat{i}_{mr} + \frac{1}{R_s}V_{sq}. \tag{10.30}$$

Equations (10.29) and (10.30) represent a nonlinear system for which V_{sd} and V_{sq} are inputs, and i_{sd} and i_{sq} are outputs (and also the state variables). Based on (10.29) and (10.30), dynamics of i_{sd} and i_{sq} are coupled. Moreover, the system is nonlinear due to the terms ωi_{sd}, ωi_{sq}, and $\omega\widehat{i}_{mr}$, in view of the fact that ω and $\widehat{i}_{mr}$ are both functions of i_{sd} and i_{sq}. This nonlinearity can be avoided in the control if two new control inputs, u_d and u_q, are defined as

$$u_d = \sigma\tau_s\omega i_{sq} - (1-\sigma)\tau_s\frac{d\widehat{i}_{mr}}{dt} + \frac{1}{R_s}V_{sd}, \tag{10.31}$$

$$u_q = -\sigma\tau_s\omega i_{sd} - (1-\sigma)\tau_s\omega\widehat{i}_{mr} + \frac{1}{R_s}V_{sq}. \tag{10.32}$$

Substituting for V_{sd} and V_{sq} in (10.29) and (10.30), from (10.31) and (10.32), we obtain

$$\left(\sigma\tau_s\frac{di_{sd}}{dt} + i_{sd}\right) = u_d, \tag{10.33}$$

$$\left(\sigma\tau_s\frac{di_{sq}}{dt} + i_{sq}\right) = u_q. \tag{10.34}$$

Equations (10.33) and (10.34) represent two decoupled first-order subsystems with unity DC gains. The first subsystem controls i_{sd} by u_d whereas the second subsystem

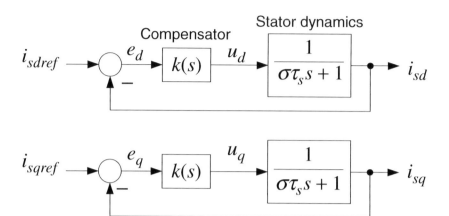

FIGURE 10.5   Control block diagram for stator current control, based on (10.33) and (10.34).

controls i_{sq} via u_q. In turn, u_d and u_q are obtained from two corresponding PI compensators, as illustrated by Figure 10.5. One compensator processes $i_{sdref} - i_{sd}$ and provides u_d, whereas the other one processes $i_{sqref} - i_{sq}$ and provides u_q. Parameters of the compensators can be determined based on the block diagrams of Figure 10.5. Let us assume

$$k(s) = \frac{k_p s + k_i}{s}. \tag{10.35}$$

If parameters k_p and k_i are

$$k_p = \frac{\sigma \tau_s}{\tau_i}, \tag{10.36}$$

$$k_i = \frac{1}{\tau_i}, \tag{10.37}$$

then d- and q-axis current-control loops assume the following, first-order, transfer functions:

$$\frac{I_{sd}(s)}{I_{sdref}(s)} = G_i(s) = \frac{1}{\tau_i s + 1}, \tag{10.38}$$

$$\frac{I_{sq}(s)}{I_{sqref}(s)} = G_i(s) = \frac{1}{\tau_i s + 1}, \tag{10.39}$$

where the time constant τ_i is a design choice. Equations (10.38) and (10.39) correspond to the middle outlined part of the block diagram of Figure 10.4. The actual control signals V_{sd} and V_{sq} are calculated based on (10.31) and (10.32), as

$$V_{sd} = R_s \left[u_d - \sigma \tau_s \omega i_{sq} + (1 - \sigma)\tau_s \frac{d\widehat{i}_{mr}}{dt} \right], \tag{10.40}$$

$$V_{sq} = R_s \left[u_q + \sigma \tau_s \omega i_{sd} + (1 - \sigma)\tau_s \omega \widehat{i}_{mr} \right], \tag{10.41}$$

where the signals $\widehat{i}_{mr}$ and ω are obtained from the flux observer of Figure 10.3.

Selection of compensator(s) parameters based on (10.36) and (10.37) results in cancellation of the plant pole $p = -1/\sigma\tau_s$ by the PI compensator zero. L_m and σ are obtained from measurements, subject to errors. Furthermore, R_s significantly changes with the machine temperature. Hence, $\sigma\tau_s$ is not precisely known and the pole cancellation will not be perfect. A similar concern arises with respect to (10.40) and (10.41) where not only $\sigma\tau_s$ is uncertain but R_s is also subject to uncertainties and variations due to the temperature changes. However, compensator design based on the machine nominal parameters often renders the compensations effective, and the resultant closed-loop subsystems can still be characterized by the first-order transfer functions of (10.38) and (10.39).

Figure 10.6 illustrates implementation of the dq-frame current-control scheme that receives i_{sdref} and i_{sqref} from the flux and torque compensators (Fig. 10.4), and $\widehat{i}_{mr}$ and ω from the flux observer (Fig. 10.3). Then, V_{sd} and V_{sq}, which are to be reproduced by the VSC, are calculated based on (10.40) and (10.41). To compensate for the VSC voltage conversion gain, m_d and m_q are calculated by dividing V_{sd} and V_{sq} by $V_{DC}/2$, and delivered to the PWM signal generator (Fig. 8.17) to generate the gating pulses. Since, the machine is interfaced with the VSC via a three-wire connection, the third-harmonic injected PWM can be adopted to permit a lower DC-bus voltage.

Figure 10.6 shows that the controller implementation is slightly modified with respect to (10.40); the term $(1 - \sigma)\tau_s d\widehat{i}_{mr}/dt$ is omitted from the expression for V_{sd}. The reason is to avoid differentiation and its associated noise-amplifying characteristic. This, however, poses no issue since, except during the system start-up process, $\widehat{i}_{mr}$ is regulated at a constant value and thus (the average of) $d\widehat{i}_{mr}/dt$ is zero.

10.3.1.4 *Machine Magnetizing Current*   In previous sections, we presented the rotor-field oriented control of the asynchronous machine by introducing the fictitious current component $\widehat{i}_{mr}$, (10.13). Based on (10.16) we also showed that the machine electrical torque is proportional to the product of $\widehat{i}_{mr}$ and i_{sq}. Thus, if $\widehat{i}_{mr}$ is kept constant, the torque can be linearly controlled by i_{sq}. In this section, we show that $\widehat{i}_{mr}$ is equivalent to the machine magnetizing current [43], and thus its reference value $\widehat{i}_{mr-ref}$ must also be set at the machine nominal magnetizing current.

The magnetizing current of a squirrel-cage machine can be calculated based on the machine steady-state equivalent circuit of Figure A.3, which, for ease of reference, is repeated as Figure 10.7. Figure 10.7 shows that if $\omega_r = \omega$, that is, if the machine rotor speed is equal to the stator frequency, then $i'_r = 0$ and the stator current is equal to the magnetizing current, that is, $i_s = i_m$. Assuming a negligible voltage drop across

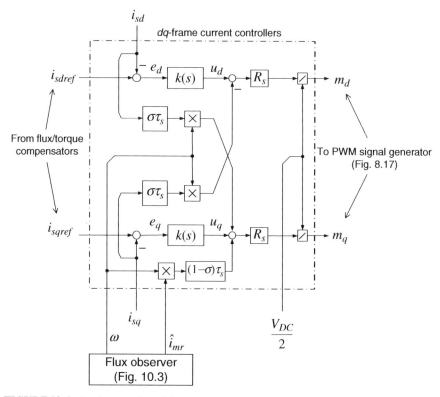

FIGURE 10.6   Implementation of d- and q-axis current controllers for asynchronous machine.

R_s, based on Figure 10.7, we deduce

$$|\underline{i}_s| = |\underline{i}_m| \simeq \frac{|V_s|}{(1+\sigma_s)L_m\omega}, \quad \text{if } \omega = \omega_r, \tag{10.42}$$

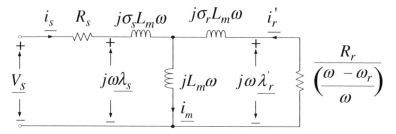

FIGURE 10.7   Steady-state equivalent circuit of the asynchronous machine with short-circuited rotor (see Appendix A for details).

where $|i_s|$ is the stator peak current and $|V_s|$ is the stator line-to-neutral peak voltage. The stator peak current can be expressed as $|i_s| = \hat{i}_s = \sqrt{i_{sd}^2 + i_{sq}^2}$. Thus, (10.42) can be rewritten as

$$\sqrt{i_{sd}^2 + i_{sq}^2} \simeq \frac{\sqrt{2}\left(\frac{V_{s-ll}}{\sqrt{3}}\right)}{(1+\sigma_s)L_m\omega} = \sqrt{\frac{2}{3}}\frac{V_{s-ll}}{(1+\sigma_s)L_m\omega}, \quad \text{if } \omega = \omega_r, \quad (10.43)$$

where V_{s-ll} is the stator, line-to-line, rms voltage. Based on (10.21), in a steady state $\hat{i}_{mr} = i_{sd}$. Moreover, based on (10.22), $\omega_r = \omega$ corresponds to $i_{sq} = 0$. Thus,

$$\sqrt{i_{sd}^2 + i_{sq}^2} = i_{sd} = \hat{i}_{mr}, \quad \text{if } \omega = \omega_r. \quad (10.44)$$

Comparing (10.44) and (10.43), we conclude

$$\hat{i}_{mr} \simeq \sqrt{\frac{2}{3}}\frac{V_{s-ll}}{(1+\sigma_s)L_m\omega_r}. \quad (10.45)$$

That is, $\hat{i}_{mr}$ is equivalent to the machine magnetizing current (peak value). Therefore, $\hat{i}_{mr-ref}$ must be set at the machine nominal magnetizing current, that is, corresponding to the rated voltage and $\omega_r = \omega_0$, as

$$\hat{i}_{mr-ref} = \sqrt{\frac{2}{3}}\frac{V_{sn}}{(1+\sigma_s)L_m\omega_0}, \quad (10.46)$$

where V_{sn} is the nominal stator line-to-line rms voltage and ω_0 is the machine nominal frequency.

The following example illustrates operation of the variable-frequency VSC system of Figure 10.1 controlling a squirrel-cage asynchronous machine.

EXAMPLE 10.1   Vector Control of Asynchronous Machine

Consider the variable-frequency VSC system of Figure 10.1. The machine is a 1.68 MW, wound-rotor, asynchronous machine whose rotor terminals are short circuited. The machine parameters are given in Table 10.1 [53]. The VSC is a three-level NPC with a capacitive voltage divider (Fig. 6.18). The third-harmonic injected PWM strategy with a switching frequency of $f_s = 1680$ Hz is adopted for the VSC. The capacitance of each DC-side capacitor is $2C = 19250 \ \mu\text{F}$. The voltages of the capacitors are equalized by the partial DC-side voltage balancer discussed in Section 6.7. The net DC-bus voltage is $V_{DC} = 3500$ V. The on-state resistance of each switch cell is about 1.0 mΩ. Thus, the effective internal resistance of each phase of the VSC is approximately $r_{on} = 2 \times 1.0 = 2$ mΩ, which is considerably smaller than R_s and ignored in the control design process. If a time constant of $\tau_i = 3.0$ ms is desired

TABLE 10.1   Asynchronous Machine Parameters: Example 10.1

Quantity	Value	Comment
Nominal power	2250 hp (1.678 MW)	
Nominal voltage	2300 V (line-to-line rms)	V_{sn}
Nominal frequency	377 rad/s	ω_0
R_s	29 mΩ	
R_r	22 mΩ	
L_m	34.6 mH	
L_s	35.2 mH	
L_r	35.2 mH	
σ_s	0.0173	
σ_r	0.0173	
σ	0.0337	
τ_s	1.213 s	
τ_r	1.6 s	

for the current controllers, one obtains $k_p = 13.6$ and $k_i = 333$ s^{-1} based on (10.36) and (10.37). The transfer function of the flux compensator is $K_\lambda(s) = 180(s + 0.625)/s$, and $\widehat{i}_{mr-ref}$ is set to 141 A, based on (10.46). Figures 10.8–10.11 illustrate the response of the variable-frequency VSC system to a constant-torque free acceleration, that is, no external mechanical torque is exerted.

Initially, the machine is stalled, all the controllers are disabled, and the PWM gating signals are blocked. At $t = 0.1$ s, the controllers are enabled and the gating signals are released. Thus, to regulate $\widehat{i}_{mr}$, the flux compensator steps up i_{sdref} (Fig. 10.8(a)), which is rapidly tracked by i_{sd}. Figure 10.8(a) also shows that i_{sdref} is limited to 1200 A to prevent exposure of the machine and the VSC to an over-current condition. While $i_{sd} = 1200$ A, $\widehat{i}_{mr}$ rises toward its reference value (Fig. 10.8(b)) at a fairly low speed due to the large rotor time constant. At $t = 0.305$ s, $\widehat{i}_{mr}$ reaches the set point of 141 A (Fig. 10.8(b)), and the flux controller reduces i_{sdref} (and therefore i_{sd}) to about 141 A (Fig. 10.8(a)); $\widehat{i}_{mr}$, i_{sdref}, and i_{sd} become equal in the steady state. Figure 10.8(c) illustrates that i_{sqref} is set at zero while the machine flux is being regulated. Consequently, the machine torque and rotor also remain at zero (Fig. 10.8(d) and (e)).

At $t = 0.4$ s, i_{sqref} is stepped to 1200 A. Thus, i_{sq} and T_e quickly track their respective reference values, as Figures 10.8(c) and (d) illustrate. Once T_e becomes nonzero, the machine starts to move and ω_r increases linearly with time (Fig. 10.8(e)). The reason is that, except for the first few milliseconds when i_{sq} changes exponentially, T_e assumes a constant value.

Figures 10.9(a) and (b) provide a closer look at i_{sq} and T_e around $t = 0.4$ s. Figure 10.9(a) shows that the step response of i_{sq} is a first-order exponential time function that reaches its steady-state value in about 15 ms, that is, $5\tau_i$. Figure 10.9(b) shows that T_e is proportional to i_{sq} and thus follows the same pattern of variation. It should be noted that the waveforms of i_{sq} and T_e are identical in their per-unit terms.

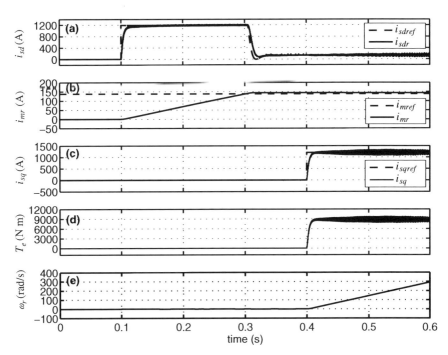

FIGURE 10.8   Response of the variable-frequency VSC system of Example 10.1 to free acceleration.

Figures 10.10(a)–(d) illustrate the frequency of the VSC system, the angle for *abc-* to *dq*-frame transformation, the PWM modulating signal, and the stator current. Figures 10.10(a) and (b) illustrate that from $t = 0.1$ to 0.4 s when ω_r and i_{sq} are both zero, ω is zero and ρ remains unchanged. Therefore, the modulating signal and the stator current are DC waveforms, as illustrated by

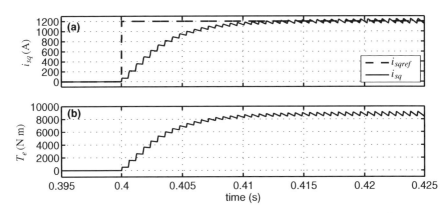

FIGURE 10.9   Zoomed waveforms of the *q*-axis stator current and machine torque; Example 10.1.

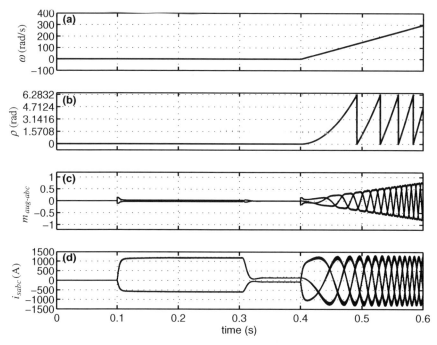

FIGURE 10.10 Waveforms of the stator frequency, dq-frame synchronization angle, modulating signals, and stator current; Example 10.1.

Figures 10.10(c) and (d). Figure 10.10(d) also shows that from $t = 0.1$ to 0.4 s, $i_b = i_c = -i_a/2$; this is consistent with $i_{sq} = 0$, in view of $i_{sd} + ji_{sq} = \overrightarrow{i_s}\, e^{-\rho}$, $\overrightarrow{i_s} = (2/3)[i_a + e^{j(2\pi/3)}i_b + e^{j(4\pi/3)}i_c]$, and $\rho = 0$. It is interesting to note that, from $t = 0.1$ to 0.4 s, the VSC system maintains DC currents in the machine stator windings to regulate the flux.

From $t = 0.4$ s onward when i_{sq} and ω_r rise, ω increases accordingly, as Figure 10.10(a) shows. Since $i_{sq} > 0$, ω is slightly larger than ω_r, as understood from (10.22). Figure 10.10(b) illustrates that ρ and its slope increase as ω increases, so does the frequency of the modulating signal and that of the stator current, as Figures 10.10(c) and (d) show. Figure 10.10(d) illustrates that the amplitude of i_{sabc} is constant after $t = 0.4$ s. The reason is that i_{sd} and i_{sq} are both constants during the foregoing time interval. However, the amplitude of $m_{aug-abc}$ increases with ω. This is expected due to the dependence of V_{sd} and V_{sq} on ω, based on (10.40) and (10.41).

Figures 10.11(a) and (b), respectively, show the machine electrical power, $P_e = T_e\omega_r$, and (the average of) the VSC DC-side power, $P_{DC} = i_s V_{DC}$ (see Fig. 10.1). From $t = 0.1$ to 0.4 s, $P_e = 0$ (Fig. 10.11(a)), while P_{DC} assumes a small positive value (Fig. 10.11(b)). P_e is zero since $T_e = 0$ over that period of time. P_{DC} is nonzero since $i_{sd} \neq 0$ from $t = 0.1$ to 0.4 s; thus, a small amount

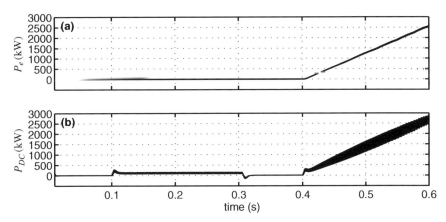

FIGURE 10.11   Waveforms of the machine electric power and the VSC DC-side power during free acceleration; Example 10.1.

of power is drawn from the DC-side voltage source to compensate for the stator resistive losses. After $t = 0.4$ s, when ω_r increases linearly and T_e is constant, P_e and P_{DC} increase linearly, as Figures 10.11(a) and (b) illustrate. Due to the stator and rotor resistive losses, P_{DC} is slightly larger than P_e.

Rapid torque control of an AC machine can be utilized to control the power exchange between the mechanical load/source interfaced with the machine and the entity interfaced with the VSC DC side. Figure 10.12 illustrates the performance of the variable-frequency VSC system of Example 10.1 when the machine torque is suddenly switched, at $t = 0.7$ s, from zero to a profile characterized by $T_{eref} = P_{eref}(t)/\omega_r(t)$ where $P_{eref} = -1350$ kW. Consequently, T_e becomes negative and ω_r starts to drop from its predisturbance value of 340 rad/s (Fig. 10.12(a) and (b)). As ω_r decreases, (the absolute value of) T_{eref} and T_e become larger. Thus, the stator current increases (Fig. 10.12(c)). Figure 10.12(d) illustrates that P_e settles at -1350 kW and remains constant thereafter. Figure 10.12(e) shows that P_{DC} exhibits a similar pattern of variations but, due to losses, is slightly smaller (in absolute value) than P_e. Since P_{DC} is negative, about 1350 kW is drawn from the machine inertia and transferred to the DC voltage source.

A variable-frequency VSC system that is controlled with the foregoing operational strategy can be used as a *flywheel energy storage system* to level out power fluctuations caused by another energy conversion system. For example, the flywheel energy storage system has been proposed for mitigation of power fluctuations caused by wind-power units [91]. A flywheel system can achieve this goal by rapid conversion of the machine kinetic energy to the electric power. If the DC-side voltage source in Figure 10.1 is realized by a controlled DC-voltage power port (Fig. 8.21, Section 8.6), the machine kinetic energy is transferred to the AC system that is interfaced with the controlled DC-voltage power port. If a higher moment of inertia is provided, the VSC system can

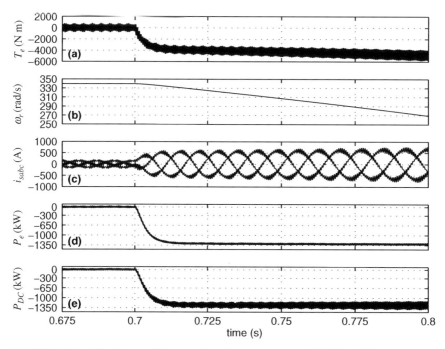

FIGURE 10.12 Waveforms of the machine electric power and VSC DC-side power in the flywheel mode; Example 10.1.

maintain the power flow for a longer period of time. It should be noted that a flywheel energy storage system can maintain the power only until the machine torque and current limits are reached; thereafter, the power linearly drops as ω_r decreases to zero. To bring ω_r back to its rated value, T_{eref} must be maintained at a positive value, for example, via a speed regulation loop.

10.3.2 Doubly-Fed Asynchronous Machine

10.3.2.1 Machine Model in Stator-Field Coordinates In this section, we present the controls for the variable-frequency VSC system of Figure 10.2 to control a doubly-fed asynchronous machine. Figure 10.2 illustrates that the machine stator is directly connected to an AC system with the nominal frequency ω_0, whereas the rotor is interfaced with the variable-frequency VSC system. The main advantage of the scheme manifests itself when the rotor speed varies only a few percentages about the AC system frequency, that is, $\omega_r = \omega_0 \pm \Delta\omega$. Under this condition, the ratio of the real power flowing through the rotor circuit to the total real power of the machine is proportional to $\Delta\omega$ [43]. Thus, the VSC can be of a smaller power rating compared to the machine, and most of the power is directly exchanged between the machine stator circuit and the AC system. A practical case for the application of the system of Figure 10.2 is the case of a variable-speed wind-power unit where the extracted wind power can be significantly varied if the turbine speed is controlled only over a narrow range [86, 87]. The wind power can then be controlled (and maximized) by changing ω_r through the variable-frequency VSC system of Figure 10.2, while the VSC handles only a small fraction of the total machine power. This application is extensively discussed in Chapter 13.

The model of a symmetrical three-phase AC machine, represented by (10.1)–(10.4) and (10.9), is equally applicable to the doubly-fed asynchronous machine. Let us express $\overrightarrow{\lambda_s}$, based on (4.6), as

$$\overrightarrow{\lambda_s} = \widehat{\lambda}_s e^{j\theta(t)}, \tag{10.47}$$

where θ is the angle of the space phasor and a function of time, in general. Defining the change of variable $\theta(t) = \rho(t) + \theta_r(t)$, we can rewrite (10.47) as

$$\overrightarrow{\lambda_s} = \widehat{\lambda}_s e^{j(\rho+\theta_r)}, \tag{10.48}$$

where θ_r is the rotor angle. Substituting for $\overrightarrow{\lambda_s}$ in (10.3), from (10.48), and solving for $\overrightarrow{i_s}$, we obtain

$$\overrightarrow{i_s} = \frac{\widehat{\lambda}_s e^{j\rho} - L_m \overrightarrow{i_r}}{(1+\sigma_s)L_m} e^{j\theta_r}. \tag{10.49}$$

Substituting for $\overrightarrow{i_s}$ in (10.9) from (10.49), and considering that $Im\left\{\overrightarrow{i_r}\,\overrightarrow{i_r}^{\,*}\right\} = 0$, we deduce

$$T_e = \left(\frac{3}{2}\right)\frac{1}{1+\sigma_s}\widehat{\lambda}_s Im\left\{\left(\overrightarrow{i_r}\,e^{-j\rho}\right)^*\right\}. \tag{10.50}$$

The term $\overrightarrow{i_r}\,e^{-j\rho}$ represents a transformation from the $\alpha\beta$-frame to the dq-frame, with ρ being the angle for transformation as indicated in Figure 10.2. Substituting for $\overrightarrow{i_r}\,e^{-j\rho} = i_{rd} + ji_{rq}$ in (10.50), we deduce

$$T_e = -\left(\frac{3}{2}\right)\frac{1}{1+\sigma_s}\widehat{\lambda}_s i_{rq}. \tag{10.51}$$

Equation (10.51) indicates that the machine torque can be linearly controlled by i_{rq}, provided that $\widehat{\lambda}_s$ is regulated at a constant value.

The mathematical equations describing the dynamics of $\widehat{\lambda}_s$ and ρ are developed by substituting in (10.1) for $\overrightarrow{\lambda_s} = \widehat{\lambda}_s e^{j(\rho+\theta_r)}$, and for $\overrightarrow{i_s}$ from (10.49):

$$\frac{d}{dt}\left[\widehat{\lambda}_s e^{j(\rho+\theta_r)}\right] = \overrightarrow{V_s} - R_s\frac{\widehat{\lambda}_s e^{j\rho} - L_m \overrightarrow{i_r}}{(1+\sigma_s)L_m} e^{j\theta_r}. \tag{10.52}$$

Calculating the derivative, multiplying both sides of the resultant by $e^{-j(\rho+\theta_r)}$, and substituting for $\overrightarrow{i_r}\,e^{-j\rho} = i_{rd} + ji_{rq}$, we deduce

$$\left(\tau_s \frac{d\widehat{\lambda}_s}{dt} + \widehat{\lambda}_s\right) + j\tau_s\,(\omega + \omega_r)\,\widehat{\lambda}_s = \tau_s\,\overrightarrow{V_s}e^{-j(\rho+\theta_r)} + L_m i_{rdq}, \qquad (10.53)$$

where $\omega = d\rho/dt$, ω_r is the rotor speed, and τ_s is the stator time constant as defined by (10.27). Assume that the stator terminal voltage is sinusoidal and balanced as

$$V_{sa}(t) = \widehat{V}_s \cos(\omega_0 t + \theta_0),$$

$$V_{sb}(t) = \widehat{V}_s \cos\left(\omega_0 t + \theta_0 - \frac{2\pi}{3}\right),$$

$$V_{sc}(t) = \widehat{V}_s \cos\left(\omega_0 t + \theta_0 - \frac{4\pi}{3}\right), \qquad (10.54)$$

where $\widehat{V}_s$ is the peak value of the line-to-neutral voltage, ω_0 is the AC system angular frequency, and θ_0 is the initial phase angle of the stator terminal voltage. Then, based on (4.42), we have

$$\overrightarrow{V_s} = V_{s\alpha} + jV_{s\beta} = \widehat{V}_s e^{j(\omega_0 t + \theta_0)}. \qquad (10.55)$$

Substituting for $\overrightarrow{V_s}$ from (10.55) in (10.53), we obtain

$$\left(\tau_s \frac{d\widehat{\lambda}_s}{dt} + \widehat{\lambda}_s\right) + j\tau_s\,(\omega + \omega_r)\,\widehat{\lambda}_s = \tau_s\,\widehat{V}_s e^{j(\omega_0 t + \theta_0 - \rho - \theta_r)} + L_m i_{rdq}, \qquad (10.56)$$

where f_{dq} denotes $f_d + jf_q$. Decomposing (10.56) into real and imaginary components, we deduce

$$\left(\tau_s \frac{d\widehat{\lambda}_s}{dt} + \widehat{\lambda}_s\right) = \overbrace{\tau_s\,\widehat{V}_s \cos(\omega_0 t + \theta_0 - \rho - \theta_r)}^{V'_{sd}} + L_m i_{rd}, \qquad (10.57)$$

$$\omega = -\omega_r + \frac{\overbrace{\tau_s\,\widehat{V}_s \sin(\omega_0 t + \theta_0 - \rho - \theta_r)}^{V'_{sq}} + L_m i_{rq}}{\tau_s \widehat{\lambda}_s}. \qquad (10.58)$$

Except for the presence of the terms $V'_{sd} = \widehat{V}_s \cos(\omega_0 t + \theta_0 - \rho - \theta_r)$ and $V'_{sq} = \widehat{V}_s \sin(\omega_0 t + \theta_0 - \rho - \theta_r)$, (10.57) and (10.58) are analogous to (10.21) and (10.22), which were developed for the squirrel-cage asynchronous machine. Equation (10.57) describes the dynamics of the stator flux, whereas (10.58) formulates the rotating speed of the dq-frame, ω, as a function of i_{rq}. Based on (10.51), i_{rq} is proportional to the electrical torque and thus ω is also a function of the torque. Equation (10.57) suggests that the stator flux can be controlled through i_{rd}. However, in the doubly-fed asynchronous machine, $\widehat{\lambda}_s$ is naturally regulated by the AC system and is fairly invariant with respect to i_{rd}. This can be demonstrated by linearizing (10.57) and (10.58) around the steady-state operating point corresponding to $i_{rd} = i_{rq} = 0$, as follows.

Let us first introduce the following change of variable:

$$\gamma = \omega_0 t + \theta_0 - \rho - \theta_r. \tag{10.59}$$

Based on (10.59), we can rewrite (10.57) and (10.58) as

$$\left(\tau_s \frac{d\widehat{\lambda}_s}{dt} + \widehat{\lambda}_s\right) = \tau_s \widehat{V}_s \cos\gamma + L_m i_{rd}, \tag{10.60}$$

$$\frac{d\gamma}{dt} = \omega_0 - \frac{\tau_s \widehat{V}_s \sin\gamma + L_m i_{rq}}{\tau_s \widehat{\lambda}_s}. \tag{10.61}$$

The perturbed variables are defined as

$$\widehat{\lambda}_s = \widehat{\lambda}_{s0} + \Delta\widehat{\lambda}_s,$$

$$\gamma = \gamma_0 + \Delta\gamma,$$

$$i_{rd} = \Delta i_{rd},$$

$$i_{rq} = \Delta i_{rq}. \tag{10.62}$$

Substituting for the perturbed variables in (10.60) and (10.61), and considering only the first-order terms, we deduce

$$\left(\tau_s \frac{d\Delta\widehat{\lambda}_s}{dt} + \Delta\widehat{\lambda}_s\right) = -\left(\tau_s \widehat{V}_s \sin\gamma_0\right)\Delta\gamma + L_m \Delta i_{rd}, \tag{10.63}$$

$$\frac{d\Delta\gamma}{dt} = -\left(\frac{\widehat{V}_s}{\widehat{\lambda}_{s0}}\cos\gamma_0\right)\Delta\gamma + \left(\frac{\widehat{V}_s}{\widehat{\lambda}_{s0}^2}\sin\gamma_0\right)\Delta\widehat{\lambda}_s - \left(\frac{L_m}{\tau_s \widehat{\lambda}_{s0}}\right)\Delta i_{rq}, \tag{10.64}$$

where

$$\sin\gamma_0 = \frac{\omega_0 \widehat{\lambda}_{s0}}{\widehat{V}_s}, \tag{10.65}$$

$$\cos\gamma_0 = \frac{\widehat{\lambda}_{s0}}{\tau_s \widehat{V}_s}. \tag{10.66}$$

It then follows from (10.65), (10.66), and identity $\sin^2(\cdot) + \cos^2(\cdot) = 1$ that

$$\widehat{\lambda}_{s0} = \frac{\tau_s \widehat{V}_s}{\sqrt{(\tau_s \omega_0)^2 + 1}} \approx \frac{\widehat{V}_s}{\omega_0}, \tag{10.67}$$

$$\sin \gamma_0 = \frac{\tau_s \omega_0}{\sqrt{(\tau_s \omega_0)^2 + 1}} \approx 1 \implies \gamma_0 \approx \frac{\pi}{2}, \tag{10.68}$$

$$\cos \gamma_0 = \frac{1}{\sqrt{(\tau_s \omega_0)^2 + 1}} \approx \frac{1}{\tau_s \omega_0}, \tag{10.69}$$

where the approximation is based on $\tau_s \omega_0 \gg 1$. Based on (10.67)–(10.69), (10.63) and (10.64) are rewritten as

$$\frac{d\Delta\widehat{\lambda}_s}{dt} = -\left(\frac{1}{\tau_s}\right)\Delta\widehat{\lambda}_s - \left(\widehat{V}_s\right)\Delta\gamma - \left(\frac{L_m}{\tau_s}\right)\Delta i_{rd}, \tag{10.70}$$

$$\frac{d\Delta\gamma}{dt} = \left(\frac{\omega_0^2}{\widehat{V}_s}\right)\Delta\widehat{\lambda}_s - \left(\frac{1}{\tau_s}\right)\Delta\gamma + \left(\frac{L_m\omega_0}{\tau_s\widehat{V}_s}\right)\Delta i_{rq}. \tag{10.71}$$

Equations (10.70) and (10.71), expressed in the classical state-space form, represent a two-input–two-output linear system for which Δi_{rd} and Δi_{rq} are inputs, and $\Delta\widehat{\lambda}_s$ and $\Delta\gamma$ are outputs. Taking the Laplace transform from both sides of (10.70) and (10.71), and solving for $\Delta\widehat{\lambda}_s(s)$ and $\Delta\gamma(s)$, we obtain the following transfer functions:

$$\Delta\widehat{\lambda}_s(s) = -\frac{\left(\frac{L_m}{\tau_s}\right)\left(s + \frac{1}{\tau_s}\right)}{\left(s + \frac{1}{\tau_s}\right)^2 + \omega_0^2}\Delta I_{rd}(s) - \frac{\left(\frac{L_m\omega_0}{\tau_s}\right)}{\left(s + \frac{1}{\tau_s}\right)^2 + \omega_0^2}\Delta I_{rq}(s), \tag{10.72}$$

$$\Delta\gamma(s) = -\frac{\left(\frac{L_m\omega_0^2}{\tau_s\widehat{V}_s}\right)}{\left(s + \frac{1}{\tau_s}\right)^2 + \omega_0^2}\Delta I_{rd}(s) + \frac{\left(\frac{L_m\omega_0}{\tau_s\widehat{V}_s}\right)\left(s + \frac{1}{\tau_s}\right)}{\left(s + \frac{1}{\tau_s}\right)^2 + \omega_0^2}\Delta I_{rq}(s), \tag{10.73}$$

where s denotes the complex frequency. Inspection of (10.72) and (10.73) reveals that the transfer functions $\Delta\widehat{\lambda}_s(s)/\Delta I_{rd}(s)$, $\Delta\widehat{\lambda}_s(s)/\Delta I_{rq}(s)$, $\Delta\gamma(s)/\Delta I_{rd}(s)$, and $\Delta\gamma(s)/\Delta I_{rq}(s)$ have small DC gains. For example, based on the parameters of Table 10.1, $\Delta\widehat{\lambda}_s(0)/\Delta I_{rd}(0)$ and $\Delta\widehat{\lambda}_s(0)/\Delta I_{rq}(0)$ are, respectively, equal to -1.65×10^{-4} Wb/kA and -7.56×10^{-2} Wb/kA for the machine of Example 10.1. Therefore, the transient and steady-state impacts of Δi_{rd} and Δi_{rq}, on $\Delta\widehat{\lambda}_s$ and $\Delta\gamma$, are insignificant and can be ignored in our subsequent developments. This, in turn, indicates that $\widehat{\lambda}_s$ and γ can be approximated by their corresponding steady-state values of

$$\widehat{\lambda}_s \approx \widehat{\lambda}_{s0} = \frac{\widehat{V}_s}{\omega_0}, \tag{10.74}$$

$$\gamma \approx \gamma_0 = \frac{\pi}{2}. \tag{10.75}$$

The conclusions of the foregoing mathematical analysis could have been drawn also through the inspection of the equivalent circuit of Figure A.2. In the doubly-fed asynchronous machine, if the stator is directly supplied from a stiff voltage source, the voltage across the machine magnetizing inductance and thus the flux are tightly regulated since the stator resistance is relatively small. Consequently, to change the voltage of the magnetizing branch, a large rotor current would be required. Thus, although $\widehat{\lambda}_s$ is perturbed as i_{rq} is changed for torque control, there is practically no need to regulate $\widehat{\lambda}_s$ via i_{rd} as deviations are insignificant. Hence, i_{rd} can be set to zero to keep the VSC current at minimum.

Substituting for $\widehat{\lambda}_s \approx \widehat{V}_s/\omega_0$ in (10.51), we can express the machine torque as

$$T_e = -\left(\frac{3}{2}\right) \frac{\widehat{V}_s}{(1+\sigma_s)\omega_0} i_{rq}. \tag{10.76}$$

Substituting for $\gamma \approx \pi/2$ in (10.59), taking derivative from both sides of the resultant, and solving for $\omega = d\rho/dt$, we deduce

$$\omega = \frac{d\rho}{dt} \approx \omega_0 - \omega_r. \tag{10.77}$$

Equation (10.77) indicates that the rotational speed of the dq-frame, ω, is approximately equal to the difference between the AC system frequency and the rotor speed. ω is used for decoupling and feed-forward compensations in the d- and q-axis current-control loops of the VSC system.

10.3.2.2 *Machine Vector Control in Stator-Field Coordinates*   As explained in the previous section, the machine torque in the variable-frequency VSC system of Figure 10.2 can be linearly controlled through i_{rq}, (10.76). It was also discussed that i_{rd} has an insignificant impact on the machine flux and can be treated as a free control variable. The aforementioned control strategy is illustrated by the block diagram of Figure 10.13. Figure 10.13 shows that the torque command T_{eref} is divided by $-\frac{3\widehat{V}_s}{2(1+\sigma_s)\omega_0}$ to provide i_{rqref}. Although i_{rdref} can be assigned any value within the range of the VSC current rating, it preferably should be set to zero, as Figure 10.13 suggests, to keep the VSC AC-side current at minimum. This, in turn, reduces the VSC and rotor power losses and enhances the overall system efficiency. The dq-frame current-control scheme ensures that i_{rd} and i_{rq} rapidly track their corresponding reference commands. Thus, i_{rd} is regulated at zero, whereas i_{rq} is varied according to the torque demand.

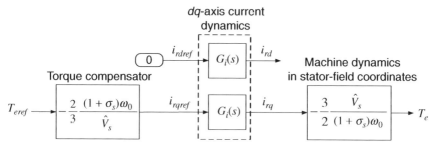

FIGURE 10.13 Control block diagram of the vector-controlled doubly-fed asynchronous machine in stator-field coordinates.

Figure 10.2 illustrates that the VSC system is controlled in a dq-frame that is synchronized to angle ρ. We require ρ to (i) obtain i_{rd} and i_{rq} as feedback signals for the dq-frame current controllers and (ii) transform m_d and m_q to $m_{abc}(t)$ for the PWM signal generator. ρ is not directly available and must be obtained from a flux observer. The block diagram of a flux observer that is based on (10.57) and (10.58) is shown in Figure 10.14. The flux observer of Figure 10.14 is similar to its counterpart for the squirrel-cage asynchronous machine (Fig. 10.3). The flux observer of Figure 10.14 is, however, more complicated than that of Figure 10.3 since calculations of $V'_{sd} = \hat{V}_s \cos(\omega_0 t + \theta_0 - \rho - \theta_r)$ and $V'_{sq} = \hat{V}_s \sin(\omega_0 t + \theta_0 - \rho - \theta_r)$ require abc- to dq-frame transformations, in addition to the one that is required for the i_{rabc} to i_{rdq} transformation. Moreover, as discussed in the previous section, dynamics of $\omega = d\rho/dt$ and $\hat{\lambda}_s$ are fairly insensitive to i_{rd} and i_{rq}. Consequently, at the system start-up when both ρ and $\omega = d\rho/dt$ are in zero-state conditions, if ω_r is considerably different from ω_0, the flux observer may enter a limit cycle that takes a long time to fade away. Depending on initial values, the flux observer may never reach a steady state. The observer start-up issue is similar to that of the phase-locked loop (PLL) discussed in Section 8.3.4.

An alternative flux observer that does not encounter the limit-cycle issue of the observer of Figure 10.14 can be developed based on (10.1) and (10.48). Considering that $\vec{f} = f_\alpha + jf_\beta$, (10.1) can be decomposed into

$$\frac{d\lambda_{s\alpha}}{dt} = V_{s\alpha} - R_s i_{s\alpha}, \tag{10.78}$$

$$\frac{d\lambda_{s\beta}}{dt} = V_{s\beta} - R_s i_{s\beta}. \tag{10.79}$$

Integrating both sides of (10.78) and (10.79), we deduce

$$\lambda_{s\alpha} = \int_0^t (V_{s\alpha} - R_s i_{s\alpha}) d\tau, \tag{10.80}$$

$$\lambda_{s\beta} = \int_0^t (V_{s\beta} - R_s i_{s\beta}) d\tau. \tag{10.81}$$

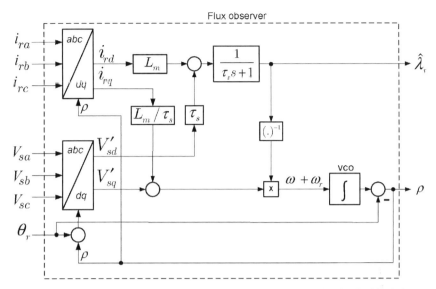

FIGURE 10.14 Block diagram of stator-flux model (flux observer) for the doubly-fed asynchronous machine.

Then, based on (10.48), we have

$$\widehat{\lambda}_s e^{j\rho} = \overrightarrow{\lambda}_s e^{-j\theta_r}$$
$$= \underbrace{(\lambda_{s\alpha} + j\lambda_{s\beta}) \, e^{-j\theta_r}}_{\lambda'_{sd} + j\lambda'_{sq}}, \tag{10.82}$$

where $\lambda_{s\alpha}$ and $\lambda_{s\beta}$ are the outcomes of (10.80) and (10.81), respectively. Equation (10.82) represents an $\alpha\beta$- to dq-frame transformation for which the angle is θ_r. Therefore,

$$\widehat{\lambda}_s e^{j\rho} = \lambda'_{sd} + j\lambda'_{sq}, \tag{10.83}$$

where λ'_{sd} and λ'_{sq} are the outputs of the $\alpha\beta$- to dq-frame transformation. Decomposing (10.83) into real and imaginary components, we conclude

$$\widehat{\lambda}_s = \sqrt{\lambda'^2_{sd} + \lambda'^2_{sq}}, \tag{10.84}$$

$$\sin \rho = \frac{\lambda'_{sq}}{\sqrt{\lambda'^2_{sd} + \lambda'^2_{sq}}}, \tag{10.85}$$

$$\cos \rho = \frac{\lambda'_{sd}}{\sqrt{\lambda'^2_{sd} + \lambda'^2_{sq}}}. \tag{10.86}$$

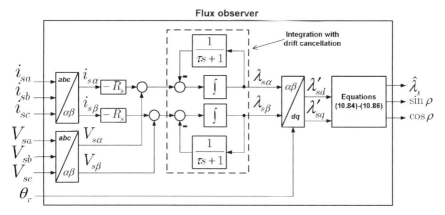

FIGURE 10.15   Block diagram of an alternative stator-flux model (flux observer) for the doubly-fed asynchronous machine.

Figure 10.15 illustrates a block diagram of the alternative flux observer. The flux observer of Figure 10.15 is based on (10.80)–(10.82), and (10.84)–(10.86). The outlined part of the flux observer of Figure 10.15 illustrates α- and β-axis integrators. To prevent drifts at the outputs of the integrators, which are mainly due to offsets and/or numerical errors, each integrator is augmented with an internal feedback loop. The internal feedback is closed through a first-order low-pass filter with the unity gain and a large time constant, τ. The low-pass filter feeds any DC component of the integrator output back to its input. Thus, an offset component at the input cannot result in a drift at the output. On the other hand, since τ is large, the feedback path is effectively open at the AC system frequency and the loop gain is equivalent to an integrator.

The flux observer of Figure 10.15 is an appropriate choice for the doubly-fed asynchronous machine, as it is simple, robust, and not subject to complicated start-up transients. Moreover, it deals with the relatively clean signals V_{sabc}, i_{sabc}, and θ_r. The flux observer of Figure 10.15, with minor modifications, may also be employed for the squirrel-cage asynchronous machine and is referred to as the voltage model in the technical literature [43]. However, for the squirrel-cage machine, the proper operation of the flux observer requires that the machine speed be adequately higher than zero. The reason is that at low speeds V_{sabc}, which is a switched waveform in case of the squirrel-cage machine, has a small fundamental component resulting in inaccurate integration. This issue does not exist for the doubly-fed asynchronous machine since V_{sabc} is a relatively clean sinusoidal voltage supported by the AC system.

10.3.2.3   Machine Current Control by VSC   As discussed in the previous section, the torque of the doubly-fed asynchronous machine is controlled by i_{rq}, while i_{rd} can be regulated at any value within the VSC rating (Fig. 10.13). In this section, we present a control scheme to regulate i_{rd} and i_{rq}, respectively at i_{rdref} and i_{rqref}. Since the VSC controls the rotor terminal voltage of the doubly-fed machine, we must relate i_{rd} and i_{rq} to the control inputs V_{rd} and V_{rq}.

The rotor terminal voltages and currents are related based on (10.2) and (10.4). Substituting for $\overrightarrow{i_s}$ from (10.49) in (10.4), we obtain

$$\overrightarrow{\lambda_r} - \upsilon(1 + \upsilon_r)L_m \overrightarrow{i_r} + \frac{1}{1 + \sigma_s}\widehat{\lambda_s}e^{j\rho}, \tag{10.87}$$

where σ_s, σ_r, and σ are defined by (10.5), (10.6), and (10.24), respectively. Substituting for $\overrightarrow{\lambda_r}$ in (10.2) from (10.87), and substituting for $\overrightarrow{i_r} = i_{rdq}e^{j\rho}$ and $\overrightarrow{V_r} = V_{rdq}e^{j\rho}$ in the resultant, we find

$$\sigma(1 + \sigma_r)L_m\frac{d}{dt}\left(i_{rdq}e^{j\rho}\right) + \left(\frac{1}{1 + \sigma_s}\right)\frac{d}{dt}\left(\widehat{\lambda_s}e^{j\rho}\right) = V_{rdq}e^{j\rho} - R_r i_{rdq}e^{j\rho}, \tag{10.88}$$

where f_{dq} is a compact representation of $f_d + jf_q$. Calculating the derivatives in (10.88) and multiplying both sides of the resultant by $e^{-j\rho}/R_r$, we obtain

$$\left(\sigma\tau_r\frac{di_{rdq}}{dt} + i_{rdq}\right) = -j\sigma\tau_r\omega i_{rdq} - \frac{(1-\sigma)\tau_r}{L_m}\frac{d\widehat{\lambda_s}}{dt} - j\frac{(1-\sigma)\tau_r}{L_m}\omega\widehat{\lambda_s} + \frac{V_{rdq}}{R_r}, \tag{10.89}$$

where τ_r is defined by (10.18). Decomposing (10.89) into real and imaginary components, we deduce

$$\left(\sigma\tau_r\frac{di_{rd}}{dt} + i_{rd}\right) = \sigma\tau_r\omega i_{rq} - \frac{(1-\sigma)\tau_r}{L_m}\frac{d\widehat{\lambda_s}}{dt} + \frac{V_{rd}}{R_r}, \tag{10.90}$$

$$\left(\sigma\tau_r\frac{di_{rq}}{dt} + i_{rq}\right) = -\sigma\tau_r\omega i_{rd} - \frac{(1-\sigma)\tau_r}{L_m}\omega\widehat{\lambda_s} + \frac{V_{rq}}{R_r}. \tag{10.91}$$

Equations (10.90) and (10.91) represent a two-input–two-output system for which V_{rd} and V_{rq} are inputs and i_{rd} and i_{rq} are outputs (and also the state variables). Based on (10.90) and (10.91), dynamics of i_{rd} and i_{rq} are coupled. However, as opposed to the case of the squirrel-cage induction machine, that is, (10.29) and (10.30), the system can be regarded as a linear time-varying system. The reason is that $\widehat{\lambda_s}$ and ω are almost independent of i_{rd} and i_{rq}, as indicated by (10.74) and (10.77). Dynamics of i_{rd} and i_{rq} can be made decoupled by introducing the two new control inputs

$$u_d = \sigma\tau_r\omega i_{rq} - \frac{(1-\sigma)\tau_r}{L_m}\frac{d\widehat{\lambda_s}}{dt} + \frac{1}{R_r}V_{rd}, \tag{10.92}$$

$$u_q = -\sigma\tau_r\omega i_{rd} - \frac{(1-\sigma)\tau_r}{L_m}\omega\widehat{\lambda_s} + \frac{1}{R_r}V_{rq}. \tag{10.93}$$

Then, (10.90) and (10.91) can be expressed as

$$\left(\sigma\tau_r\frac{di_{rd}}{dt} + i_{rd}\right) = u_d, \tag{10.94}$$

$$\left(\sigma\tau_r\frac{di_{rq}}{dt} + i_{rq}\right) = u_q. \tag{10.95}$$

Equations (10.94) and (10.95) represent two decoupled, first-order subsystems with unity DC gains. The first subsystem controls i_{rd} by u_d whereas the second subsystem controls i_{rq} via u_q. In turn, u_d and u_q are delivered by two corresponding PI compensators, as illustrated by Figure 10.16. One compensator processes $i_{rdref} - i_{rd}$ and provides u_d; the other compensator processes $i_{rqref} - i_{rq}$ and provides u_q. The parameters of the compensators can be calculated based on the block diagram of Figure 10.16, similar to the case of the squirrel-cage asynchronous machine. Let us assume

$$k(s) = \frac{k_p s + k_i}{s}, \tag{10.96}$$

where k_p and k_i are

$$k_p = \frac{\sigma\tau_r}{\tau_i} \tag{10.97}$$

$$k_i = \frac{1}{\tau_i}. \tag{10.98}$$

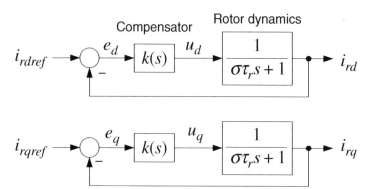

FIGURE 10.16 Block diagrams of the rotor current-control loops, based on (10.94) and (10.95).

Then, the d- and q-axis current-control loops are described by the following, first-order, transfer functions:

$$\frac{I_{rd}(s)}{I_{rdref}(s)} = G_i(s) = \frac{1}{\tau_i s + 1}, \tag{10.99}$$

$$\frac{I_{rq}(s)}{I_{qref}(s)} = G_i(s) = \frac{1}{\tau_i s + 1}, \tag{10.100}$$

where the time constant τ_i is a design choice. Equations (10.99) and (10.100) correspond to the middle outlined part of the block diagram of Figure 10.13. V_{rd} and V_{rq} are computed based on (10.92) and (10.93) as

$$V_{rd} = R_r \left[u_d - \sigma \tau_r \omega i_{rq} + \frac{(1-\sigma)\tau_r}{L_m} \frac{d\widehat{\lambda}_s}{dt} \right], \tag{10.101}$$

$$V_{rq} = R_r \left[u_q + \sigma \tau_r \omega i_{rd} + \frac{(1-\sigma)\tau_r}{L_m} \omega \widehat{\lambda}_s \right]. \tag{10.102}$$

Figure 10.17 illustrates a block diagram of the dq-frame current-control scheme that receives i_{rdref} (usually set to zero) and i_{rqref} (from the torque compensator) (Fig. 10.13). V_{rd} and V_{rq}, which are to be reproduced by the VSC, are calculated based on (10.101) and (10.102) for which $\widehat{\lambda}_s$ is assumed to be constant based on (10.74) and ω is determined based on (10.77). Since, based on (5.22) and (5.23), the VSC has a gain of $V_{DC}/2$, the calculated signals V_{rd} and V_{rq} are divided by $V_{DC}/2$ to yield m_d and m_q, respectively. The PWM signal generator (Fig. 8.17), receives m_d and m_q and generates the gating pulses. It should be pointed out that $\cos\rho$ and $\sin\rho$, required for the PWM signal generator in addition to transforming i_{rabc} to i_{rdq}, are obtained from the flux observer of Figure 10.15. Since, the machine is interfaced with the VSC via a three-wire connection (Fig. 10.2), one is permitted to employ the third-harmonic injected PWM strategy to allow a lower DC-bus voltage. Figure 10.17 also reveals that the term $R_r[\tau_r(1-\sigma)/L_m]d\widehat{\lambda}_s/dt$ in expression (10.101) for V_{rd} is omitted in implementation. The justification is that $\widehat{\lambda}_s$ is fairly constant based on (10.74), and its derivative has a negligible average value.

10.3.2.4 VSC Power Rating

In this section, we show that in the VSC system of Figure 10.2, if the doubly-fed asynchronous machine rotates at an angular speed close to the AC system frequency ω_0, the real power that flows through the rotor terminals constitutes a small fraction of the machine total electric power and, thus, the VSC can be of a smaller power rating compared to the machine power rating.

Based on (4.40), the real power delivered by the VSC system to the rotor circuit is

$$P_r = \frac{3}{2} Re \left\{ \left(\overrightarrow{V}_r - R_r \overrightarrow{i_r} \right) \overrightarrow{i_r}^* \right\} = \frac{3}{2} Re \left\{ \frac{d\overrightarrow{\lambda}_r}{dt} \overrightarrow{i_r}^* \right\}. \tag{10.103}$$

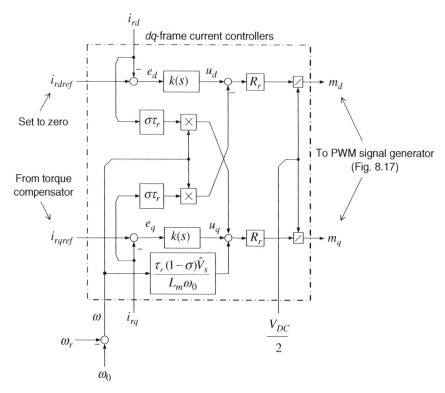

FIGURE 10.17 Implementation of d- and q-axis current-control loops for the doubly-fed asynchronous machine.

Substituting for $\vec{i_r} = i_{rdq}e^{j\rho}$ in (10.87), and for $\vec{\lambda_r}$ from the resultant in (10.103), we have

$$P_r = \frac{3}{2}Re\left\{\frac{d}{dt}\left[\sigma(1+\sigma_r)L_m i_{rdq}e^{j\rho} + \frac{1}{1+\sigma_s}\widehat{\lambda}_s e^{j\rho}\right]i^*_{rdq}e^{-j\rho}\right\}. \qquad (10.104)$$

Under the steady-state condition, i_{rdq} and $\widehat{\lambda}_s$ are constant and (10.104) can be simplified to

$$P_r = \frac{3}{2}Re\left\{j\left[\sigma(1+\sigma_r)L_m i_{rdq}\omega e^{j\rho} + \frac{1}{1+\sigma_s}\widehat{\lambda}_s \omega e^{j\rho}\right]i^*_{rdq}e^{-j\rho}\right\}$$

$$= \left(\frac{3}{2}\right)\frac{1}{1+\sigma_s}\widehat{\lambda}_s \omega i_{rq}. \qquad (10.105)$$

Based on (4.40), the apparent power exchanged with the stator circuit is

$$S_s = \frac{3}{2} \left(\overrightarrow{V}_s - R_s \overrightarrow{i}_s \right) \overrightarrow{i}_s^* = \frac{3}{2} \frac{d \overrightarrow{\lambda}_s}{dt} \overrightarrow{i}_s^*. \tag{10.106}$$

Substituting for $\overrightarrow{\lambda}_s = \widehat{\lambda}_s e^{j(\rho + \theta_r)}$ and $\overrightarrow{i}_s$ in (10.106), respectively, from (10.48) and (10.49), we obtain

$$S_s = \frac{3}{2} \frac{d}{dt} \left[\widehat{\lambda}_s e^{j(\rho + \theta_r)} \right] \frac{\widehat{\lambda}_s e^{-j(\rho + \theta_r)} - L_m i_{rdq}^* e^{-j(\rho + \theta_r)}}{(1 + \sigma_s) L_m}$$

$$= j \left(\frac{3}{2} \right) (\omega + \omega_r) \widehat{\lambda}_s \frac{\widehat{\lambda}_s - L_m i_{rdq}^*}{(1 + \sigma_s) L_m}. \tag{10.107}$$

The real power delivered by the AC system to the stator is

$$P_s = Re\left\{ S_s \right\} = - \left(\frac{3}{2} \right) \frac{1}{(1 + \sigma_s)} \widehat{\lambda}_s (\omega + \omega_r) i_{rq}. \tag{10.108}$$

Based on $P_e = T_e \omega_r$ and (10.51), the machine electrical power can be expressed as

$$P_e = - \left(\frac{3}{2} \right) \frac{1}{(1 + \sigma_s)} \widehat{\lambda}_s \omega_r i_{rq}. \tag{10.109}$$

Comparing (10.105) and (10.108) with (10.109), and considering that $\omega = \omega_0 - \omega_r$ based on (10.77), we conclude

$$P_r = \left(1 - \frac{\omega_0}{\omega_r} \right) P_e, \tag{10.110}$$

$$P_s = \left(\frac{\omega_0}{\omega_r} \right) P_e. \tag{10.111}$$

Based on (10.110) and (10.111), if ω_r is close to ω_0, the VSC system exchanges only a small fraction of the machine net power with the rotor; the rest of the machine power is directly exchanged between the stator and the AC system. Thus, the VSC can be of a fairly small power rating if the rotor speed is close to the AC system frequency. Let us assume

$$\omega_r = \omega_0 + \Delta \omega_r, \qquad |\Delta \omega_r| \ll \omega_0, \tag{10.112}$$

where $\Delta\omega_r$ is the speed deviation from ω_0. Substituting for ω_r from (10.112) in (10.110), we deduce

$$P_r = \left[1 - \frac{\omega_0}{\omega_0\left(1 + \frac{\Delta\omega_r}{\omega_0}\right)}\right] P_e$$

$$= \left(1 - \frac{1}{1 + \frac{\Delta\omega_r}{\omega_0}}\right) P_e. \tag{10.113}$$

If $\Delta\omega_r \ll \omega_0$, then $1/(1 + \Delta\omega_r/\omega_0) \approx (1 - \Delta\omega_r/\omega_0)$. Thus, (10.113) can be rewritten as

$$P_r = \left(\frac{\Delta\omega_r}{\omega_0}\right) P_e. \tag{10.114}$$

Equation (10.114) indicates that P_r is proportional to $\Delta\omega_r$, and decreases as $\Delta\omega_r$ becomes smaller. It should be noted that the real power exchanged between the VSC system and the rotor is slightly different from P_r due to the rotor resistive losses.

The exchange of reactive power between the AC system and the stator is characterized by

$$Q_s = Im\{S_s\}$$
$$= \left(\frac{3}{2}\right) \frac{1}{(1 + \sigma_s)L_m} \widehat{\lambda}_s^2(\omega + \omega_r) - \left(\frac{3}{2}\right) \frac{1}{(1 + \sigma_s)} \widehat{\lambda}_s(\omega + \omega_r)i_{rd}. \tag{10.115}$$

Substituting for $\widehat{\lambda}_s$ and ω in (10.115), from (10.74) and (10.77), we deduce

$$Q_s = \left(\frac{3}{2}\right) \frac{\widehat{V}_s^2}{(1 + \sigma_s)L_m\omega_0} - \left(\frac{3}{2}\right) \frac{1}{(1 + \sigma_s)} \widehat{V}_s i_{rd}. \tag{10.116}$$

Equation (10.116) indicates that the reactive power drawn from the AC system has two components. The first component corresponds to the machine magnetizing current and is constant. The second component is proportional to i_{rd}. Due to the machine large magnetizing current, the constant component of the reactive power is significant. As (10.116) suggests, the constant component of the reactive power can be compensated if $i_{rd} = \widehat{V}_s/(L_m\omega_0)$. This, of course, results in an increase in the AC-side current of the VSC. With $i_{rd} = 0$, the AC-side current remains minimal at the expense of a stator lagging power-factor.

The following example illustrates the operation of the variable-frequency VSC system of Figure 10.2.

EXAMPLE 10.2   Vector Control of Doubly-Fed Asynchronous Machine

Consider the variable-frequency VSC system of Figure 10.2 to control a 1.68 MW doubly-fed asynchronous machine with the parameters given in Table 10.1. The machine stator is directly interfaced with an AC system whose frequency is $\omega_0 = 377$ rad/s. The AC system is represented by a Thevenin equivalent circuit with a 2300-V (line-to-line, rms) voltage source and a series inductance of $L_g = 750$ µH. The inductance can be regarded as the leakage inductance of a 1.6 MVA interface transformer, corresponding to a leakage reactance of 0.09 per-unit (pu).

In the VSC system of Figure 10.2, the machine rotor is controlled by a two-level VSC. Figure 10.18 illustrates that a three-phase switch connects the rotor terminals to the corresponding AC-side terminals of the VSC. At the start-up, the switch is open and will not be closed until the machine accelerates (by an external mechanical torque) and ω_r lies within a narrow range about ω_0. The need for this mechanism can be explained as follows. Since the stator is supplied from the AC system, $\widehat{\lambda}_s$ is fairly constant at its nominal value and $d\widehat{\lambda}_s/dt = 0$. Now assume that during the start-up i_{rd} and i_{rq} are both zero and ω_r is small (zero under a standstill condition). Then, based on (10.77), $\omega = \omega_0 - \omega_r$ has a large value (equal to ω_0 for the standstill condition). Thus, based on (10.101) and (10.102), $V_{rd} = 0$ whereas V_{rq} has a large value. Consequently, V_{rabc} has a large amplitude, and a large DC-bus voltage is required for the VSC to properly operate.[4] However, if ω_r is close to ω_0 the amplitude of V_{rabc} is considerably smaller and a significantly lower DC-bus voltage is allowed.

In the system of Figure 10.2, the DC-bus voltage of the VSC is 1500 V. The VSC adopts the third-harmonic injected PWM strategy with a switching frequency of $f_s = 1620$ Hz. The on-state resistance of each switch cell is about 0.9 mΩ, which is considerably smaller than R_r and can be ignored in the control design process. Assuming that a time constant of $\tau_i = 3.0$ ms is desired for the current-control loops, based on (10.97) and (10.98) we find $k_p = 17.97$ and $k_i = 333$ s^{-1}.

Figure 10.18 also shows that a three-phase star-connected series RLC filter is connected in parallel with the machine stator terminals. This filter is to mitigate harmonic distortion of V_{sabc}, which is measured for flux observation (Fig. 10.15). Without the filter, V_{sabc} includes switching voltage notches that may adversely impact the flux observation process. The reason is that, once the switch is closed and the VSC is activated, the rotor terminal voltage becomes a switched waveform. On the other hand, the rotor and stator windings behave analogous to the windings of a transformer. Consequently, the rotor voltage is transferred to the stator circuit and, due to the AC system inductance, switching notches will appear at the stator terminals. The filter parameters are $R_f = 20$ mΩ, $L_f = 90$ µH, and $C_f = 100$ µF.

---

[4]For this example, we find $V_{rd} = 0$ and $V_{rq} = 1845$ V, for $\omega_r = 0$. Therefore, the amplitude of V_{rabc} is 1845 V in standstill. Consequently, even using the third-harmonic injected PWM, the VSC DC-bus voltage must be at least 3210 V to avoid overmodulation.

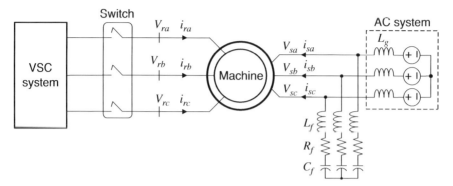

FIGURE 10.18   Variable-frequency VSC system of Example 10.2.

Figure 10.19 illustrates the response of the variable-frequency VSC system of Figure 10.2 to a forced acceleration followed by a flywheel mode of operation. Initially, the machine is in standstill condition, the switch is open, all controllers are inactive, and the VSC gating pulses are blocked. Thus, the stator carries the magnetizing current (Fig. 10.19(a)), and the flux rises (Fig. 10.19(b)).

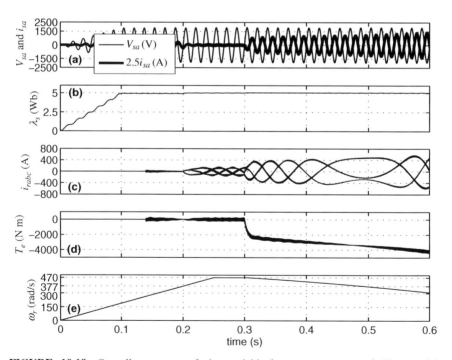

FIGURE 10.19   Overall response of the variable-frequency system of Figure 10.2; Example 10.2.

Since the magnetizing branch is dominantly inductive, the stator current lags the corresponding phase voltage by almost $90°$. Since the rotor current is zero (Fig. 10.19(c)), the machine generates no torque (Fig. 10.19(d)). However, a mechanical energy source exerts a constant torque on the machine, and consequently the rotor speed linearly increases (Fig. 10.19(e)).

At $t = 0.14$ s, the switch is closed, the controllers are activated, and the VSC gating pulses are unblocked. However, since T_{eref} (and thus i_{rqref}) and i_{rdref} are set to zero, the rotor current and the machine torque remain zero (Figs. 10.19(c) and (d)). Substituting for $Q_s = 0$ and solving for i_{rd} based on (10.116), we deduce that the stator reactive power can be compensated if i_{rd} is regulated at 144 A. Thus, at $t = 0.2$ s, i_{rdref} is changed from zero to 144 A and the rotor current increases as Figure 10.19(c) shows. This results in elimination of the magnetizing current component from the stator circuit and, therefore, the stator current becomes zero, as Figure 10.19(a) illustrates. However, λ_s remains regulated as Figure 10.19(b) illustrates. At $t = 0.25$ s, the mechanical torque is removed, and the rotor speed remains constant at $\omega_r = 470$ rad/s (Fig. 10.19(e)).

At $t = 0.3$ s, T_{eref} is switched from zero to a dynamic value determined by $T_{eref} = P_{eref}/\omega_r$ where $P_{eref} = -1340$ kW; since the torque control is fast, that is, $T_e \approx T_{eref}$, this, essentially, corresponds to a constant-power operation. Thus, the stator and rotor currents increase (Figs. 10.19(a) and (c)), the machine generates a negative torque (Fig. 10.19(d)), and ω_r decreases (Fig. 10.19(e)). Furthermore, as Figure 10.19(a) shows, the stator current is $-180°$ phase shifted with respect to the corresponding stator phase voltage since Q_s is nullified. Based on $T_{eref} = P_{eref}/\omega_r$, the absolute value of T_e increases with time (Fig. 10.19(d)), as ω_r decreases. Thus, the stator and rotor currents increase, as Figures 10.19(a) and (c) show.

Figure 10.19(c) shows that in addition to the amplitude, the frequency of the rotor current changes with ω_r, such that at $t = 0.5$ s, that is, when $\omega_0 \approx \omega_r$, the rotor current freezes and its phase sequence reverses thereafter. To better demonstrate this phenomenon, Figure 10.20 illustrates $\omega = \omega_0 - \omega_r$, $\sin \rho$ and $\cos \rho$, the modulating signal $m_{\text{aug-abc}}$, and the rotor current i_{rabc}. As discussed earlier in this chapter and also illustrated by Figure 10.2, the VSC PWM generator and the rotor terminal voltage are synchronized to the angle ρ and, therefore, they change with the frequency $\omega = d\rho/dt$. Based on (10.77), this frequency is equal to the difference between the rotor angular speed and the AC system frequency. Hence, at the standstill when $\omega_r = 0$, ω is equal to ω_0 whereas it is zero at the synchronous speed, that is, when $\omega_r = \omega_0$. It is also noted that the phase sequences of the rotor voltage/current are reversed when ω_r surpasses ω_0, that is, when ω becomes negative (Example 4.3 and Fig. 4.12).

Figure 10.21 provides a closer look at the rotor q-axis current component and the machine torque. Figure 10.21(b) illustrates that $-T_e$ follows the same pattern of variation as i_{rq} (Fig. 10.21(a)), as predicted by (10.51).

Figure 10.22 illustrates waveforms of the machine electrical power, P_e, the VSC DC-side power, P_{DC}, and the stator (AC system) real and reactive power,

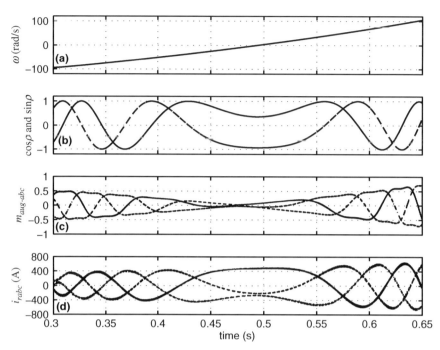

FIGURE 10.20   Response of the VSC system around the synchronous speed; Example 10.2.

P_s and Q_s. As Figure 10.22 shows, until $t = 0.2$ s, P_e, P_{DC}, and P_s are zero, whereas $Q_s = 400$ kVAr due to the machine magnetizing current supplied by the AC system. After $t = 0.2$ s, Q_s reduces to zero (Fig. 10.22(d)), since i_{rdref} is changed from zero to 144 A, as described in Figure 10.19. From $t = 0.3$ s

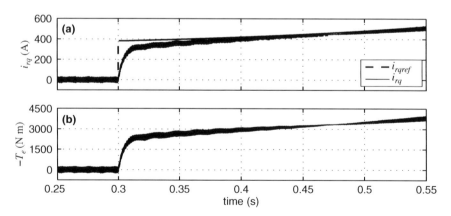

FIGURE 10.21   Responses of the rotor q-axis current component and the machine torque; Example 10.2.

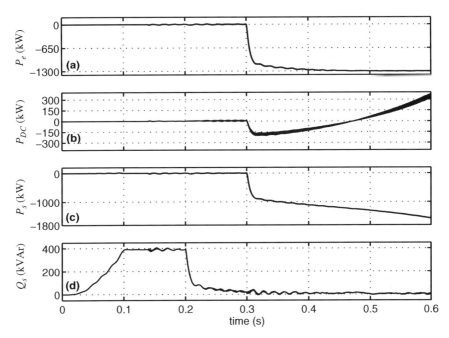

FIGURE 10.22   Waveforms of various power components in the variable-frequency VSC system of Figure 10.2; Example 10.2.

on, P_e is regulated at about -1300 kW as Figure 10.22(a) shows. Thus, the kinetic energy of the machine is absorbed and delivered to the AC system. Figure 10.22(b) illustrates that the VSC DC-side power varies between -200 and 320 kW, that is, within less than $\pm 25\%$ of P_e. Depending on whether ω_r is larger or smaller than ω_0, P_{DC} can assume different polarities, according to (10.114). Figure 10.22(c) illustrates that most of the machine power is directly exchanged between the stator and the AC system, and $P_e = P_s + P_{DC}$ holds.

10.3.3   Permanent-Magnet Synchronous Machine

The variable-frequency VSC system of Figure 10.1 can also be used to control a PMSM. In the *abc* and $\alpha\beta$ frames, the PMSM model is time varying. As demonstrated in Example 4.10, the model becomes time invariant if it is expressed in a *dq*-frame that is synchronized to the rotor angle. Hence, for the VSC system of Figure 10.1 that controls a PMSM, the angle ρ is equal to the rotor angle, θ_r. The rotor angle may be either measured using a shaft encoder or observed based on an estimation technique [50–52].

10.3.3.1 PMSM Model in Rotor-Field Coordinates
Let us adopt the model of a salient-pole PMSM, derived in Example 4.10 and represented by equations[5]

$$\begin{bmatrix} \lambda_{sd} \\ \lambda_{sq} \end{bmatrix} = \begin{bmatrix} L_d & 0 \\ 0 & L_q \end{bmatrix} \begin{bmatrix} i_{sd} \\ i_{sq} \end{bmatrix} + \begin{bmatrix} \lambda_m \\ 0 \end{bmatrix},$$ (10.117)

$$\frac{d}{dt} \begin{bmatrix} \lambda_{sd} \\ \lambda_{sq} \end{bmatrix} = \begin{bmatrix} 0 & \omega_r \\ -\omega_r & 0 \end{bmatrix} \begin{bmatrix} \lambda_{sd} \\ \lambda_{sq} \end{bmatrix} + \begin{bmatrix} -R_s & 0 \\ 0 & -R_s \end{bmatrix} \begin{bmatrix} i_{sd} \\ i_{sq} \end{bmatrix} + \begin{bmatrix} V_{sd} \\ V_{sq} \end{bmatrix},$$ (10.118)

and

$$T_e = \frac{3}{2} (L_d - L_q) i_{sd} i_{sq} + \frac{3}{2} \lambda_m i_{sq},$$ (10.119)

where λ_{sdq} represents the stator flux components, i_{sdq} denotes the stator current components, V_{sdq} represents the stator voltage components, and ω_r signifies the rotor speed. L_d and L_q are the d- and q-axis stator inductances that depend on the machine geometry and rotor saliency; in a round (nonsalient) rotor machine $L_d = L_q$. The parameter λ_m represents the maximum amount of flux generated by the rotor magnets and linked by the stator windings.

10.3.3.2 PMSM Control in Rotor-Field Coordinates
Equation (10.119) indicates that the machine torque is composed of two components. One component is linearly proportional to only i_{sq}, whereas the other component is proportional to the product of i_{sd} and i_{sq}. The proportionality constant of the latter torque component, $L_d - L_q$, represents the rotor saliency; $L_d - L_q$ becomes larger as the rotor becomes more salient. In a current-control scheme where i_{sd} and i_{sq} can be independently controlled, T_e can be controlled based on different combinations of trajectories for i_{sd} and i_{sq}. The desired combination is usually selected based on a performance criterion, for example, the machine efficiency, torque per current ratio, etc., [90, 92–94]. However, if the rotor saliency is not significant, $L_d - L_q$ is small and thus the torque component proportional to $i_{sd} i_{sq}$ does not have a considerable contribution to T_e. Therefore, in this case i_{sd} can be regulated at zero to minimize the line current and ohmic losses.

To control i_{sd} and i_{sq}, (10.117) and (10.118) are transformed to the standard state-space form by eliminating λ_{sd} and λ_{sq} between the two equations. The result is

$$L_d \frac{di_{sd}}{dt} = -R_s i_{sd} + L_q \omega_r i_{sq} + V_{sd},$$ (10.120)

$$L_q \frac{di_{sq}}{dt} = -R_s i_{sq} - L_d \omega_r i_{sd} - \lambda_m \omega_r + V_{sq}.$$ (10.121)

---

[5]Equations of a simplified, nonsalient-rotor PMSM are derived in Section A.5.

Introducing the two new control variables

$$u_d = L_q \omega_r i_{sq} + V_{sd}, \tag{10.122}$$

$$u_q - L_d \omega_r i_{sd} - \lambda_m \omega_r + V_{sq}, \tag{10.123}$$

one can simplify (10.120) and (10.121) as

$$L_d \frac{di_{sd}}{dt} + R_s i_{sd} = u_d, \tag{10.124}$$

$$L_q \frac{di_{sq}}{dt} + R_s i_{sq} = u_q. \tag{10.125}$$

Equations (10.124) and (10.125) represent two decoupled, first-order, single-input-single-output (SISO) subsystems. Therefore, two independent feedback loops can be employed to regulate i_{sd} and i_{sq} at their respective reference commands of i_{sdref} and i_{sqref}, as shown in Figure 10.23.

Figure 10.23 shows that to control i_{sd}, the d-axis compensator $k_d(s)$ processes the error $e_d = i_{sdref} - i_{sd}$ and commands u_d. Similarly, the q-axis compensator $k_q(s)$ processes the error $e_q = i_{sqref} - i_{sq}$ and delivers u_q. Assuming that the closed-loop transfer functions $I_{sd}(s)/I_{sdref}(s)$ and $I_{sq}(s)/I_{sqref}(s)$ are of the first order with a time constant of τ_i, we have

$$k_d(s) = \frac{L_d s + R_s}{\tau_i s}, \tag{10.126}$$

$$k_q(s) = \frac{L_q s + R_s}{\tau_i s}. \tag{10.127}$$

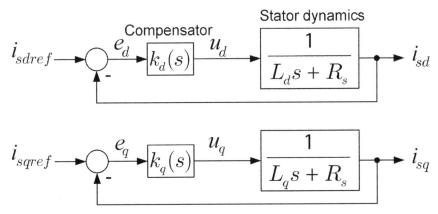

FIGURE 10.23   Closed-loop d- and q-axis current controllers for PMSM, based on (10.124) and (10.125).

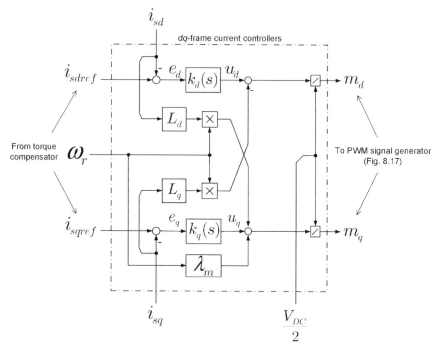

FIGURE 10.24 Implementation of d- and q-axis current controllers for PMSM.

To implement the control, V_{sd} and V_{sq} must be determined, respectively, from u_d and u_q, based on (10.122) and (10.123). The result is

$$V_{sd} = u_d - L_q \omega_r i_{sq}, \tag{10.128}$$

$$V_{sq} = u_q + L_d \omega_r i_{sd} + \lambda_m \omega_r. \tag{10.129}$$

V_{sd} and V_{sq}, calculated based on (10.128) and (10.129), are then divided by $V_{DC}/2$ to yield m_d and m_q. The process is illustrated in the block diagram of Figure 10.24.

PART II
Applications

11 Static Compensator (STATCOM)

11.1 INTRODUCTION

In this chapter, we analyze the *static compensator (STATCOM)*.[1] The STATCOM is a VSC system whose prime function is to exchange reactive power with the host AC system. In an electric power transmission system, the STATCOM can be used to increase the line power transmission capacity [1], to enhance the voltage/angle stability [95], or to damp the system oscillatory modes [96]. In a distribution system, the STATCOM is mainly used for voltage regulation [97]; however, it can also supply real power to the loads in the case of a blackout if it is augmented with an energy storage device, for example, a battery storage system. Moreover, the STATCOM may also be employed to balance a distribution network by compensating for load imbalances.

In this chapter, we demonstrate that the STATCOM is a special case of the *controlled DC-voltage power port* that was introduced in Section 8.6 (Section 7.5 for $\alpha\beta$-frame control). Although this model enables one to analyze different applications of the STATCOM, in this chapter we merely focus on the application of the STATCOM for AC voltage regulation. Even though the STATCOM can also be controlled in $\alpha\beta$-frame, in this chapter we focus on the dq-frame control due to its widespread acceptance in the technical literature. Furthermore, based on a dq-frame platform we are able to formulate and analyze dynamics of the phase-locked loop (PLL). The presented methodology is applicable to most cases where a VSC system is interfaced with a weak AC system.

11.2 CONTROLLED DC-VOLTAGE POWER PORT

The kernel of the STATCOM is a controlled DC-voltage power port. As we discussed in Section 8.6 (Section 7.5 for $\alpha\beta$-frame control), a controlled DC-voltage power port is a VSC system whose DC bus is connected in parallel with an exogenous apparatus (Figs. 8.21 and 7.21). The exogenous apparatus/system, which we refer to as a power source, exchanges power with the DC side of the VSC; the VSC DC-bus voltage is regulated via a closed-loop mechanism and, thus, the power

---

[1] In the technical literature, the STATCOM is also referred to as *static condenser (STATCON)*.

Voltage-Sourced Converters in Power Systems, by Amirnaser Yazdani and Reza Iravani
Copyright © 2010 John Wiley & Sons, Inc.

313

imposed by the power source is transferred to an AC system. The controlled DC-voltage power port can also exchange a prespecified reactive power with the AC system. In most VSC systems, a controlled DC-voltage power port is employed primarily for real-power exchange. In the STATCOM, however, reactive-power exchange is the main control objective.

11.3   STATCOM STRUCTURE

Figure 11.1 illustrates the schematic diagram of a STATCOM interfaced with an AC system. The AC system is represented by an ideal three-phase voltage source, V_{gabc}, connected in series with a transmission line. The combined inductance of the line and any interface transformer (not shown in the figure) is represented by L_g. For a less cluttered formulation, the resistances of the line and transformer are assumed to be negligible in this chapter. A comparison between Figures 11.1 and 8.21 reveals that the STATCOM is a special case of the controlled DC-voltage power port (Section 8.6), where the exogenous power source is eliminated, that is, $i_{ext} = 0$. The STATCOM DC-bus voltage, V_{DC}, is regulated by controlling the real power exchanged with the rest of the system, that is, P_s, as detailed in Section 8.6. In the steady state, P_s is small as it only compensates for the VSC power loss, associated with i_{loss}. The STATCOM of Figure 11.1 is different from the controlled DC-voltage power port of Figure 8.21 since (i) the AC system has a considerable internal inductance, L_g, and

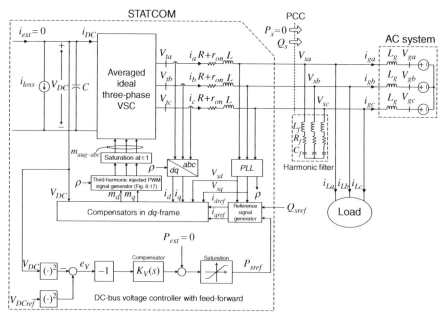

FIGURE 11.1   Schematic diagram of the STATCOM.

(ii) the reactive-power reference Q_{sref}, which is commonly set to zero in the controlled DC-voltage power port, is controlled via a closed-loop mechanism, as explained later in this chapter.

With reference to Figure 11.1, the electrical nodes where the STATCOM three phases are connected to the corresponding phases of the AC system constitute the point of common coupling (PCC). The PCC voltage is labeled as V_{sabc}. Figure 11.1 shows that the VSC synchronization signals, that is, the PLL inputs, are obtained from the PCC. A three-phase load is also supplied from the PCC. Since the VSC AC-side terminal voltages V_{tabc} are modulated waveforms and L_g is relatively large, V_{sabc} includes large voltage notches that distort both the load voltage and the feedback signals V_{sd} and V_{sq}. Therefore, a three-phase series RLC filter is connected in parallel with the STATCOM at the PCC (Fig. 11.1). Each RLC branch is usually tuned to the dominant pulse-width modulation (PWM) side-band harmonic, but exhibits a large impedance at the grid frequency. Thus, the VSC current harmonics flow through the RLC filter and do not penetrate the grid.

Due to the grid inductance, the magnitude and the phase angle of V_{sabc} can be different from those of V_{gabc}, depending on load conditions. Moreover, load switching incidents result in abrupt and large deviations in V_{sabc}. In this chapter, we demonstrate that V_{sabc} can be regulated via a closed-loop system that adjusts the reactive-power component Q_s of the STATCOM.

11.4 DYNAMIC MODEL FOR PCC VOLTAGE CONTROL

11.4.1 Large-Signal Model of PCC Voltage Dynamics

The function of the STATCOM of Figure 11.1 is to regulate the PCC voltage, V_{sabc}, in the presence of i_{Labc}, by controlling i_{abc}. These variables are related by

$$V_{sa} = L_g \frac{di_{ga}}{dt} + V_{ga} + V_{null}, \tag{11.1}$$

$$V_{sb} = L_g \frac{di_{gb}}{dt} + V_{gb} + V_{null}, \tag{11.2}$$

$$V_{sc} = L_g \frac{di_{gc}}{dt} + V_{gc} + V_{null}, \tag{11.3}$$

and

$$i_{ga} = i_a - i_{La}, \tag{11.4}$$

$$i_{gb} = i_b - i_{Lb}, \tag{11.5}$$

$$i_{gc} = i_c - i_{Lc}, \tag{11.6}$$

where V_{null} is the voltage of the AC system neutral point with respect to the VSC (virtual or actual) DC-bus midpoint. Multiplying both sides of (11.1)–(11.3), respectively, by $(2/3)e^{j0}$, $(2/3)e^{j2\pi/3}$, and $(2/3)e^{j4\pi/3}$, and adding the corresponding sides, based on (4.2), we obtain

$$\vec{V_s} = L_g \frac{d\vec{i_g}}{dt} + \vec{V_g}. \tag{11.7}$$

Since $e^{j0} + e^{j2\pi/3} + e^{j4\pi/3} \equiv 0$, V_{null} does not appear in the space-phasor expression (11.7). Similarly, (11.4)–(11.6) can be combined and expressed as

$$\vec{i_g} = \vec{i} - \vec{i_L}. \tag{11.8}$$

Let the AC system Thevnin voltage be

$$V_{ga} = \widehat{V_g} \cos(\omega_0 t + \theta_0),$$

$$V_{gb} = \widehat{V_g} \cos\left(\omega_0 t + \theta_0 - \frac{2\pi}{3}\right),$$

$$V_{gc} = \widehat{V_g} \cos\left(\omega_0 t + \theta_0 - \frac{4\pi}{3}\right), \tag{11.9}$$

where $\widehat{V_g}$ is the amplitude of the line-to-neutral voltage, ω_0 is the AC system frequency, and θ_0 is the initial phase angle of V_{gabc}. Then based on (4.2), V_{gabc} is equivalent to

$$\vec{V_g} = \widehat{V_g} e^{j(\omega_0 t + \theta_0)}. \tag{11.10}$$

Figure 11.1 illustrates that the STATCOM is controlled in a dq-frame, synchronized to angle ρ. Thus, substituting in (11.7) for $\vec{V_s} = V_{sdq}e^{j\rho}$, $\vec{i_g} = i_{gdq}e^{j\rho}$, and $\vec{V_g} = \widehat{V_g}e^{j(\omega_0 t + \theta_0)}$, we have

$$V_{sdq}e^{j\rho} = L_g \frac{d}{dt}(i_{gdq}e^{j\rho}) + \widehat{V_g}e^{j(\omega_0 t + \theta_0)}. \tag{11.11}$$

Similarly, it follows from substituting in (11.8) for $\vec{i_g} = i_{gdq}e^{j\rho}$, $\vec{i} = i_{dq}e^{j\rho}$, and $\vec{i_L} = i_{Ldq}e^{j\rho}$ that

$$i_{gdq} = i_{dq} - i_{Ldq}, \tag{11.12}$$

which can be decomposed into

$$i_{gd} = i_d - i_{Ld}, \tag{11.13}$$

$$i_{gq} = i_q - i_{Lq}. \tag{11.14}$$

Calculating the derivative in (11.11), multiplying both sides by $e^{j\rho}$, and decomposing the resultant into real and imaginary components, we obtain

$$V_{sd} = L_g \frac{di_{gd}}{dt} - L_g \omega i_{gq} + \widehat{V}_g \cos(\omega_0 t + \theta_0 - \rho), \tag{11.15}$$

$$V_{sq} = L_g \frac{di_{gq}}{dt} + L_g \omega i_{gd} + \widehat{V}_g \sin(\omega_0 t + \theta_0 - \rho), \tag{11.16}$$

where $\omega = d\rho/dt$. As detailed in Section 8.3.4, ω is controlled by the PLL (Fig. 8.5), based on the control law

$$\frac{d\rho}{dt} = \omega(t) = H(p)V_{sq}(t), \tag{11.17}$$

where $p = d(\cdot)/dt$ is the differentiation operator and $H(s)$ is the transfer function of the PLL compensator. Thus, $H(p)f(t)$ ($f(t)$ is an arbitrary function of time) represents the zero-state response of $H(s)$ to the input $f(t)$. As explained in Section 8.3.4, the PLL compensator includes one integral term and thus $\omega(t)$ assumes a nonzero steady-state value when V_{sq} settles at zero. Equations (11.13)–(11.17) represent a dynamic system for which V_{sd} is the output, i_d and i_q are the control inputs, and i_{Ld} and i_{Lq} are the disturbance inputs. The system is nonlinear due to the presence of the terms $\widehat{V}_g \cos(\omega_0 t + \theta_0 - \rho)$ and $\widehat{V}_g \sin(\omega_0 t + \theta_0 - \rho)$. Moreover, the frequency of the VSC system, ω, is a dynamic variable that depends on the operating point. To further clarify this point, let us substitute for V_{sq}, from (11.16), in (11.17):

$$\frac{d\rho}{dt} = L_g H(p) \left(\frac{di_{gq}}{dt} + \omega i_{gd} \right) + \widehat{V}_g H(p) \sin(\omega_0 t + \theta_0 - \rho). \tag{11.18}$$

Equation (11.18) indicates that dynamic responses of ρ and ω, in addition to their natural transient components corresponding to $i_{gd} = i_{gq} = 0$, include forced components that are functions of i_{gd} and i_{gq}. This is in contrast to the case of a stiff grid described by (8.24). Based on (8.24), if the VSC is interfaced with a stiff AC system, the responses of ρ and ω merely include natural transient components; the PLL dynamics are decoupled from those of the rest of the system and the operating point and, therefore, once the PLL reaches the steady state, $\rho = \omega_0 t + \theta_0$ and $\omega = \omega_0$.

11.4.2 Small-Signal Model of PCC Voltage Dynamics

Small-signal dynamics of the PCC voltage are derived by linearizing (11.15)–(11.17) around a steady-state operating point. Let us define the following perturbed variables:

$$V_{sd} = V_{sd0} + \tilde{V}_{sd},$$
$$V_{sq} = 0 + \tilde{V}_{sq},$$
$$i_{gd} = i_{gd0} + \tilde{i}_{gd},$$
$$i_{gq} = i_{gq0} + \tilde{i}_{gq},$$
$$\omega_0 t + \theta_0 - \rho = -(\rho_0 + \tilde{\rho}) \Longrightarrow \underbrace{d\rho/dt}_{\omega} = \omega_0 + \underbrace{d\tilde{\rho}/dt}_{\tilde{\omega}}. \qquad (11.19)$$

We also note that if $\tilde{\rho}/\rho_0 \ll 1$, then

$$\cos(\rho_0 + \tilde{\rho}) \approx \cos \rho_0 - (\sin \rho_0)\tilde{\rho},$$
$$\sin(\rho_0 + \tilde{\rho}) \approx \sin \rho_0 + (\cos \rho_0)\tilde{\rho}. \qquad (11.20)$$

It then follows from (11.20) and substitution of the perturbed variables of (11.19) in (11.15) and (11.16) that

$$V_{sd0} = -L_g\omega_0 i_{gq0} + \widehat{V}_g \cos \rho_0, \qquad (11.21)$$
$$0 = L_g\omega_0 i_{gd0} - \widehat{V}_g \sin \rho_0, \qquad (11.22)$$

and

$$\tilde{V}_{sd} = L_g\frac{d\tilde{i}_{gd}}{dt} - L_g\omega_0\tilde{i}_{gq} - L_g i_{gq0}\tilde{\omega} - (\widehat{V}_g \sin \rho_0)\tilde{\rho}, \qquad (11.23)$$

$$\tilde{V}_{sq} = L_g\frac{d\tilde{i}_{gq}}{dt} + L_g\omega_0\tilde{i}_{gd} + L_g i_{gd0}\tilde{\omega} - (\widehat{V}_g \cos \rho_0)\tilde{\rho}. \qquad (11.24)$$

Substituting for $\widehat{V}_g \cos \rho_0$ and $\widehat{V}_g \sin \rho_0$ in (11.23) and (11.24), respectively, from (11.21) and (11.22), we deduce

$$\tilde{V}_{sd} = L_g\frac{d\tilde{i}_{gd}}{dt} - L_g\omega_0\tilde{i}_{gq} - L_g i_{gq0}\frac{d\tilde{\rho}}{dt} - L_g\omega_0 i_{gd0}\tilde{\rho}, \qquad (11.25)$$

$$\tilde{V}_{sq} = L_g\frac{d\tilde{i}_{gq}}{dt} + L_g\omega_0\tilde{i}_{gd} + L_g i_{gd0}\frac{d\tilde{\rho}}{dt} - \left(V_{sd0} + L_g\omega_0 i_{gq0}\right)\tilde{\rho}. \qquad (11.26)$$

Similarly, substituting for the perturbed variables of (11.19) in (11.17), we deduce

$$\frac{d\widetilde{\rho}}{dt} = \widetilde{\omega} = H(p)\widetilde{V}_{sq}. \tag{11.27}$$

Equations (11.25)–(11.27) can be expressed in the Laplace domain as

$$\widetilde{V}_{sd}(s) = L_g s \widetilde{I}_{gd}(s) - L_g \omega_0 \widetilde{I}_{gq}(s) - L_g \left(i_{gq0}s + \omega_0 i_{gd0} \right) \widetilde{\rho}(s), \tag{11.28}$$

$$\begin{aligned} \widetilde{V}_{sq}(s) = L_g s \widetilde{I}_{gq}(s) + L_g \omega_0 \widetilde{I}_{gd}(s) \\ + \left[L_g i_{gd0}s - \left(V_{sd0} + L_g \omega_0 i_{gq0} \right) \right] \widetilde{\rho}(s), \end{aligned} \tag{11.29}$$

$$\widetilde{\rho}(s) = \frac{H(s)}{s} \widetilde{V}_{sq}(s). \tag{11.30}$$

Equations (11.25)–(11.27), or their Laplace domain counterparts (11.28)–(11.30), describe a linear system that is the small-signal equivalent of the system described by (11.15)–(11.17). To express the dynamics of $\widetilde{V}_{sd}(s)$ in terms of $\widetilde{I}_{gd}(s)$ and $\widetilde{I}_{gq}(s)$, first $\widetilde{V}_{sq}$ can be eliminated between (11.29) and (11.30), and then from the resultant equation $\widetilde{\rho}$ can be substituted in (11.28). The result is

$$\widetilde{V}_{sd}(s) = G_d(s)\widetilde{I}_{gd}(s) + G_q(s)\widetilde{I}_{gq}(s), \tag{11.31}$$

where $G_d(s)$ and $G_q(s)$ are two linear transfer functions whose parameters are functions of i_{gd0} and i_{gq0}. In the system of Figure 11.1, $i_d \approx 0$ and therefore $i_{d0} = \widetilde{i}_d \approx 0$, since the STATCOM exchanges a small amount of real power with the PCC, that is, $P_s \approx 0$, corresponding to the DC-side power of $P_{loss} = V_{DC}i_{loss}$ (Fig. 11.1). Thus, based on (11.13) and (11.14) we have

$$i_{gd0} \approx -i_{Ld0}, \tag{11.32}$$

$$i_{gq0} = i_{q0} - i_{Lq0}, \tag{11.33}$$

and

$$\widetilde{i}_{gd} \approx -\widetilde{i}_{Ld}, \tag{11.34}$$

$$\widetilde{i}_{gq} = \widetilde{i}_q - \widetilde{i}_{Lq}. \tag{11.35}$$

Substituting for $\widetilde{i}_{gd}$ and $\widetilde{i}_{gq}$, from (11.34) and (11.35), in (11.31), we conclude

$$\widetilde{V}_{sd}(s) = \underbrace{-G_d(s)\widetilde{I}_{Ld}(s) - G_q(s)\widetilde{I}_{Lq}(s)}_{\text{load effect}} + \underbrace{G_q(s)\widetilde{I}_q(s)}_{\text{control effect}}. \tag{11.36}$$

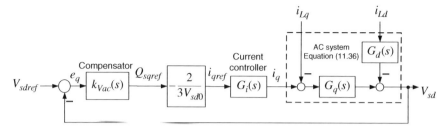

FIGURE 11.2   Control block diagram of the STATCOM PCC voltage regulator.

Equation (11.36) describes a dynamic system with $\widetilde{V}_{sd}$ as the output and $\widetilde{i}_q$ as the control input. The inputs $\widetilde{i}_{Ld}$ and $\widetilde{i}_{Lq}$ are, in general, functions of $\widetilde{V}_{sd}$ and $\widetilde{V}_{sq}$ (see Examples 9.1 and 9.2) and therefore cannot be called disturbances. They also embed dynamics of the harmonic filters. If i_{Labc} can be measured, or a model is identified for the load, the impact of $\widetilde{i}_{Ld}$ and $\widetilde{i}_{Lq}$ on $\widetilde{V}_{sd}$ can be mitigated by means of appropriate feed-forward compensation techniques. However, these conditions are seldom satisfied in practice. Therefore, load dynamics are commonly ignored and the stability of the control system is delegated to a robust design of the compensator(s).

Figure 11.2 shows a control block diagram of the STATCOM PCC voltage regulator and illustrates that a compensator, $k_{Vac}(s)$, processes $V_{sdref} - V_{sd}$ and provides Q_{sref}. Assuming $V_{sd} \approx V_{sd0}$, based on (8.44), Q_{sref} is divided by $-2/(3V_{sd0})$ to provide i_{qref}. Then, i_q tracks i_{qref}, based on a closed-loop transfer function, $G_i(s)$. As discussed in Section 8.4.1, parameters of the dq-frame current controllers can be selected such that $G_i(s)$ is a first-order transfer function with an arbitrarily small time constant. For most nested control structures, the closed-loop bandwidth of the voltage control loop should be adequately lower than that of $G_i(s)$, such that $G_i(s)$ can be approximated by a unity gain.

11.4.3   Steady-State Operating Point

The steady-state operating point of the STATCOM of Figure 11.1 can be derived by eliminating $\cos \rho_0$ and $\sin \rho_0$ between (11.21) and (11.22), and solving for i_{gq0} in view of (11.32). The process yields

$$i_{gq0} = \frac{-V_{sd0} + \sqrt{\widehat{V}_g^2 - \left(L_g\omega_0 i_{Ld0}\right)^2}}{L_g\omega_0} \tag{11.37}$$

or

$$i_{gq0} = \frac{-V_{sd0} - \sqrt{\widehat{V}_g^2 - \left(L_g\omega_0 i_{Ld0}\right)^2}}{L_g\omega_0}. \tag{11.38}$$

Based on (11.37) and (11.38), for a given i_{Ld0} there are two possible values for i_{gq0} that result in the same V_{sd0}. The two possibilities are illustrated in the phasor diagrams

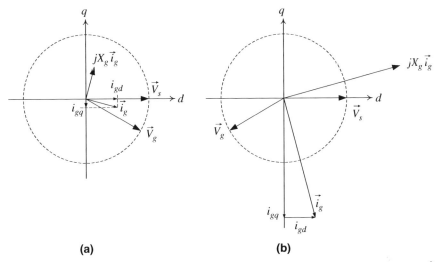

FIGURE 11.3 Phasor diagram of the two operating scenarios corresponding to $V_{sd0} = \widehat{V}_g$ and a given i_{Ld0}.

of Figures 11.3(a) and (b), which correspond to (11.37) and (11.38), respectively. Figures 11.3(a) and (b) show that i_{Ld0} and V_{sd0} are the same in both operating scenarios, indicating that the same amount of real power is exchanged between the load and the AC system in both cases. However, in case of Figure 11.3(b), i_{gq0} is considerably larger than that of Figure 11.3(a). This, based on (11.33), implies that the STATCOM has to provide a much larger reactive power in case of Figure 11.3(b), which would be significantly larger than its MVA rating. In contrast, (11.37) and Figure 11.3(a) represent a realistic operating scenario, and

$$i_{q0} = i_{Lq0} + \frac{-V_{sd0} + \sqrt{\widehat{V}_g^2 - \left(L_g\omega_0 i_{Ld0}\right)^2}}{L_g\omega_0}. \qquad (11.39)$$

Usually, V_{sd0} is regulated at $\widehat{V}_g$. Therefore, if $\widehat{V}_g \gg |L_g\omega_0 i_{Ld0}|$ then, based on (11.39), $i_{q0} \approx i_{Lq0}$.

11.5 APPROXIMATE MODEL OF PCC VOLTAGE DYNAMICS

If transient excursions of ρ and ω are ignored in (11.25), that is, $\widetilde{\rho} = d\widetilde{\rho}/dt = 0$, then a simplified dynamic model can be derived as

$$\widetilde{V}_{sd} \approx L_g\frac{d\widetilde{i}_{gd}}{dt} - L_g\omega_0\widetilde{i}_{gq}. \qquad (11.40)$$

Since $\widetilde{\rho}$ and $\widetilde{\omega} = d\widetilde{\rho}/dt$ are functions of $\widetilde{i}_{gd}$ and $\widetilde{i}_{gq}$, (11.40) provides an adequately accurate description of the PCC voltage dynamics provided that $\widetilde{i}_{gd}$ and $\widetilde{i}_{gq}$ change

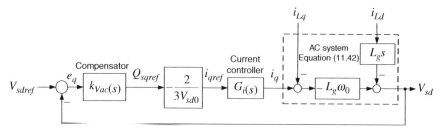

FIGURE 11.4   Control block diagram of the PCC voltage regulator, based on the approximate model of (11.42).

slowly. Formally, this requires that (i) the PCC voltage control loop is adequately slower than the d- and q-axis closed-loop current controllers and (ii) i_{Ld} and i_{Lq} have reasonably low rates of change. Substituting for $\tilde{i}_{gd}$ and $\tilde{i}_{gq}$ from (11.34) and (11.35) in (11.40), we obtain

$$\tilde{V}_{sd} \approx -L_g \frac{d\tilde{i}_{Ld}}{dt} + L_g \omega_0 \tilde{i}_{Lq} - L_g \omega_0 \tilde{i}_q. \tag{11.41}$$

Equation (11.41) can be expressed in the Laplace domain as

$$\tilde{V}_{sd}(s) \approx -L_g s \tilde{I}_{Ld}(s) + L_g \omega_0 \tilde{I}_{Lq}(s) - L_g \omega_0 \tilde{I}_q(s). \tag{11.42}$$

Comparing (11.42) with (11.36) we conclude that based on the approximate model $G_d(s)$ and $G_q(s)$ are

$$G_d(s) \approx L_g s, \tag{11.43}$$

$$G_q(s) \approx -L_g \omega_0. \tag{11.44}$$

If the simplified model of (11.42) is considered as a design basis for $k_{Vac}(s)$, then the control block diagram of Figure 11.2 can be altered to that of Figure 11.4. Thus, an inspection of the control loop suggests that $k_{Vac}(s)$ in its simplest form can be a proportional-integral (PI) compensator.

11.6   STATCOM CONTROL

Comparing Figure 11.1 with Figure 8.21, one notes that the STATCOM and the controlled DC-voltage power port are conceptually similar. The difference is that in the STATCOM no power source is interfaced with the VSC DC-side terminals and the DC bus is merely terminated to the DC-bus capacitor(s). Therefore, the STATCOM can be regarded as a special case of the controlled DC-voltage power port. Thus, the STATCOM DC-bus voltage is regulated by controlling P_s, based on the same

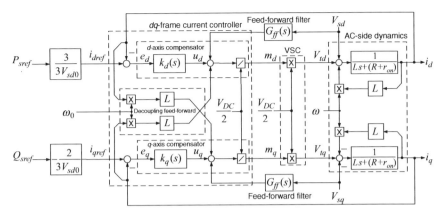

FIGURE 11.5 Control block diagram of the STATCOM dq-frame current controller (see Section 8.4.1 for details).

approach as the one adopted for the controlled DC-voltage power port (Section 8.6). In contrast, however, in the STATCOM Q_s, which is usually a free control variable in the controlled DC-voltage power port, is controlled through a closed-loop mechanism that regulates the PCC voltage. The PCC voltage regulation is based on the model of Figure 11.2 or its simplified version (Fig. 11.4). Since $V_{sq} = 0$, the control of P_s and Q_s is equivalent to the control of i_d and i_q, respectively (see equations (4.83) and (4.84)).

The structure of the VSC dq-frame current controller shown in Figure 8.10, is repeated here as Figure 11.5. There is, however, a minor difference between the control systems of Figures 8.10 and 11.5: if the AC system is stiff, the PLL dynamics are decoupled from those of the other system variables. Thus, once the PLL start-up transients are passed, the angular velocity of the dq-frame settles to the constant value ω_0. This condition holds for the current controller of Figure 8.10, where $\omega = \omega_0$ appears as a constant parameter in both the plant coupling terms and the controller decoupling terms. However, in the case of the STATCOM, ω is a dynamic variable in the plant model, as illustrated in Figure 11.5; this is due to the AC system weakness and the PLL dynamics. Thus, to decouple the plant d- and q-axis dynamics, theoretically, ω (that is, an output of the PLL) should be used in the controller decoupling terms, rather than ω_0. However, accepting a suboptimal (but quite effective) decoupling, we use ω_0 in the STATCOM current-control scheme, as shown in Figure 11.5.

With reference to Figure 11.5, compensators $k_d(s)$ and $k_q(s)$ are

$$k_d(s) = k_q(s) = \frac{k_p s + k_i}{s}. \tag{11.45}$$

Selecting k_p and k_i as

$$k_p = \frac{L}{\tau_i},$$ (11.46)

$$k_i = \frac{R + r_{on}}{\tau_i},$$ (11.47)

we obtain the following closed-loop transfer functions for d and q-axis current controllers:

$$G_i(s) = \frac{I_d(s)}{I_{dref}(s)} = \frac{I_q(s)}{I_{qref}(s)} = \frac{1}{\tau_i s + 1},$$ (11.48)

where the design choice τ_i is the desired time constant of the closed-loop step response.

11.7   COMPENSATOR DESIGN FOR PCC VOLTAGE CONTROLLER

The compensator of the PCC voltage controller, that is, $k_{Vac}(s)$, is designed based on the block diagram of Figure 11.2. The plant transfer function $G_q(s)$ can be derived from linearization of the system nonlinear equations, as discussed in Section 11.4. However, if the PLL dynamics are ignored, a simple, low-order model is deduced for $G_q(s)$ and, instead the block diagram of Figure 11.4 can be employed. Based on Figure 11.4, the plant transfer function is a pure gain, and $k_{Vac}(s)$ in its simplest form is a PI compensator. The bandwidth of the closed-loop PCC voltage control system is usually selected to be adequately smaller than that of the closed-loop current controllers, that is, $1/\tau_i$. Parameters of $k_{Vac}(s)$ are selected based on the phase margin and bandwidth requirements, as elaborated in Example 11.1.

11.8   MODEL EVALUATION

As explained in Section 11.5, the compensator for the AC voltage control loop can be designed based on a simplified model that ignores the dynamics of the PLL, shunt harmonic filter(s), and load(s). Based on these simplifying assumptions, we approximated the high-order plant model by a pure gain (Fig. 11.4). In this section, we evaluate fidelity of the simplified model and accuracy of the results.

To conduct accurate examinations, in general, a detailed switched model of the STATCOM can be developed in a digital time-domain simulation environment, based on Figure 11.1. In our case, however, the response pattern may be difficult to examine in the presence of switching harmonics, especially if the variables change in a vicinity of their respective steady-state values. Thus, we develop a simplified simulation model of the STATCOM, shown in Figure 11.6, in which the VSC AC side is represented by its averaged model, that is, by three, linear, dependent voltage sources. This model avoids the switching harmonics that exist in the actual response. However, it does not

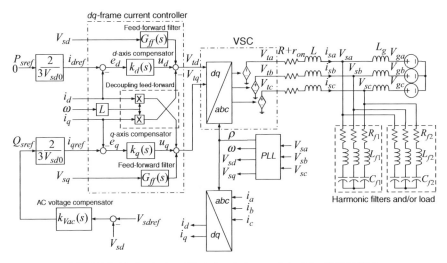

FIGURE 11.6 A simplified simulation model of the STATCOM.

alter the system dynamic characteristics since the switching harmonics are of high frequencies and well beyond the bandwidth of controllers.

The model of the VSC in Figure 11.6 is based on the following equations, as outlined in Section 8.4.1 and Figure 11.5:

$$V_{td}(t) = \frac{V_{DC}}{2} m_d(t), \tag{11.49}$$

$$V_{tq}(t) = \frac{V_{DC}}{2} m_q(t), \tag{11.50}$$

and

$$m_d = \frac{2}{V_{DC}}\left(u_d - L\omega i_q + V_{sd}\right), \tag{11.51}$$

$$m_q = \frac{2}{V_{DC}}\left(u_q + L\omega i_d + V_{sq}\right). \tag{11.52}$$

Comparing (11.49) with (11.51), and (11.50) with (11.52), we deduce

$$V_{td} = \left(u_d - L\omega i_q + V_{sd}\right), \tag{11.53}$$

$$V_{tq} = \left(u_q + L\omega i_d + V_{sq}\right). \tag{11.54}$$

The control signals of the dependent sources are then generated from V_{td} and V_{tq}, based on a dq- to abc-frame transformation. In the steady state, V_{td} and V_{tq} become DC quantities and thus V_{tabc} is a pure sinusoidal waveform. In reality, however, V_{tabc} is a modulated waveform of which only the per-switching-cycle average is a sinusoidal function of time. Note that, other than the VSC, the STAT-COM of Figure 11.6 includes all other elements (i.e., PLL, filters, compensators, feed-forward terms, etc.) of the actual STATCOM of Figure 11.1. The voltage at the PCC is controlled via Q_{sref}. However, the STATCOM exchanges a small amount of real power, corresponding to the VSC losses, with the PCC. Thus, P_{sref} is set to zero in the STATCOM of Figure 11.6. In the STATCOM of Figure 11.1, P_{sref} is a small value to regulate the DC-bus voltage.

Based on the model of Figure 11.6, dynamic performance of the STATCOM can be evaluated under various test signals and load conditions, with the impact of switching harmonics masked. The model of Figure 11.6 can also reveal potential instabilities due to both the harmonic filters and the load dynamics. Moreover, the STATCOM response can be compared with that of the simplified model of Figure 11.4 that is used for the compensator design. The model of Figure 11.6 preserves the main dynamic characteristics of the STATCOM and the AC system.

In the following examples, we evaluate the developed model and design of the STATCOM by comparing the models of Figures 11.4 and 11.6, and also by comparing both foregoing models with the detailed switched model of the STATCOM that is based on the system of Figure 11.1.

EXAMPLE 11.1   Compensator Design for AC Voltage Controller

Consider the STATCOM model of Figure 11.6 in which the objective is to regulate the PCC voltage at $V_{sd0} = \widehat{V}_g = 0.391$ kV. The STATCOM includes two filters but no loads. The system parameters are

- $L = 200$ μH, $R = 2.38$ mΩ, and $r_{on} = 0.88$ mΩ, for the VSC,
- $\widehat{V}_g = 0.391$ kV, $\omega_0 = 377$ rad/s, and $L_g = 50$ μH, for the AC system, and
- $R_{f1} = 1.0$ mΩ, $L_{f1} = 19$ μH, $C_{f1} = 508$ μF, $R_{f2} = 1.0$ mΩ, $L_{f2} = 19$ μH, and $C_{f2} = 450$ μF for the filters.

Transfer functions of current-control compensators and feed-forward filters are

$$k_d(s) = k_q(s) = \frac{0.2s + 3.26}{s} \quad [\Omega],$$

$$G_{ff}(s) = \frac{1}{0.002s + 1}.$$

Transfer function of the PLL compensator (see Figure 8.5) is

$$H(s) = \frac{683,790 \left(s^2 + 568,516\right) \left(s^2 + 166s + 6889\right)}{s \left(s^2 + 1508s + 568,516\right) \left(s^2 + 964s + 232,324\right)} \quad [(\text{rad/s})/(\text{kV})].$$

It should be noted that in the simulation model of Figure 11.6, i_{abc} and V_{sabc} are sampled based on a gain of $1/1000$; that is, the feedback, feed-forward, and control signals are expressed in kA and kV. Therefore, each dependent voltage source has a gain of 1000 to compensate for the attenuation ratio.

Based on the given parameters, the closed-loop transfer functions of d-and q-axis current controllers are

$$G_i(s) = \frac{1000}{s + 1000}.$$

Then, based on the block diagram of Figure 11.4, the compensator

$$k_{Vac}(s) = \frac{2000}{s} \quad [\text{kA}]$$

results in a gain crossover frequency of $\omega_c = 64$ rad/s and a phase margin of about 86°. We note that ω_c is about 15 times smaller than the bandwidth of $G_i(s)$, and therefore, the loop gain is effectively an integrator. Consequently, the closed-loop transfer function is predominantly of the first order and its time constant is approximately equal to $1/\omega_c = 15$ ms. Although this design may seem too conservative, it should be noted that the grid inductance is not precisely known, and is often subject to a large tolerance. Moreover, we have ignored the dynamics of the filters, loads, and PLL. Thus, by selecting such a large phase margin, we can ensure that the closed-loop system remains stable in the presence of unmodeled dynamics and parameter uncertainties. The resultant closed-loop system can regulate the PCC voltage in less than one cycle of the power frequency, as will be seen later.

Initially, the STATCOM is under a steady-state condition and $V_{sd} = V_{sdref} = 391$ V. At $t = 0.4$ s, V_{sdref} is subjected to a stepwise change from 391 to 450 V. Figures 11.7(a) and (b) illustrate the response of V_{sd} to the disturbance.

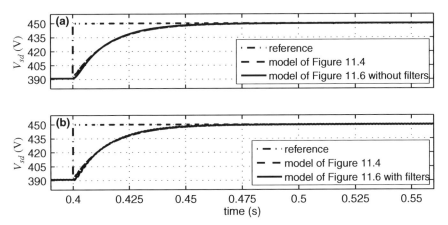

FIGURE 11.7   Step response of V_{sd} based on the model or Figure 11.6 (a) without harmonic filters and (b) with harmonic filters; Example 11.1.

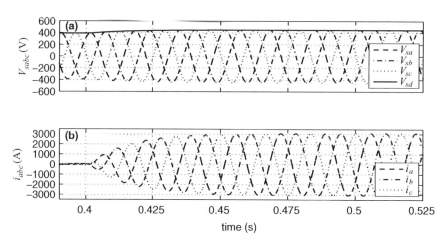

FIGURE 11.8   Responses of the PCC voltage and the STATCOM AC-side current when the harmonic filters are not in service; Example 11.1.

Figure 11.7(a) (solid line) illustrates the response of V_{sd} obtained from the model of Figure 11.6, when the harmonic filters are disconnected. Figure 11.7(b) (solid line) illustrates the response when the harmonic filters are in service. For the sake of comparison, Figures 11.7(a) and (b) (dashed line) also depict the response obtained from the simplified control model of Figures 11.4. As Figures 11.7(a) and (b) illustrate, V_{sd} tracks V_{sdref} based on a first-order exponential function and reaches the steady state in about 75 ms. This conforms to our expectation that the time constant of the step response is about 15 ms. Moreover, it is observed that there is an insignificant difference between the STATCOM response corresponding to the case where the filters are disconnected and the case where the filters are in service. Figures 11.7(a) and (b) also show that the response based on the control model of Figure 11.4 (dashed line) closely agrees with that obtained from the simulation model of Figure 11.6 (solid line).

Figure 11.8 illustrates the waveforms of the PCC voltage and the STATCOM AC-side current, corresponding to the case where the harmonic filters are not in service. As Figure 11.8(a) shows, V_{sabc} smoothly increases and reaches the steady state while its peak value resides on the waveform of V_{sd}. The reason is that, based on (4.77) and since $V_{sq} \approx 0$, the amplitude of the PCC line-to-neutral voltage is equal to V_{sd}. Figure 11.8(b) illustrates that subsequent to the voltage reference change, the STATCOM line current increases from zero to about 3000 A (peak value), corresponding to the reactive power that must be injected into the grid to increase the PCC voltage from 391 to 450 V (peak value).

Figure 11.9 illustrates the responses of the PCC voltage and STATCOM AC-side current when the harmonic filters are in service. Figure 11.9(a) shows that the behavior of V_{sabc} is the same as that of the case without harmonic filters (Fig. 11.8(a)). However, a comparison between Figures 11.9(b) and 11.8(b) reveals that when the harmonic filters are in the circuit, the STATCOM line current is nonzero before the step change in the PCC voltage command. The reason is that

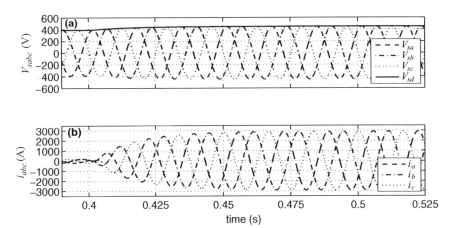

FIGURE 11.9   Responses of the PCC voltage and the STATCOM AC-side current when the harmonic filters are in service; Example 11.1.

the harmonic filters are capacitive at the AC system frequency and thus supply the grid with a small amount of reactive power. Consequently, to maintain V_{sd} at 391 V, the STATCOM absorbs the reactive power of the filters, at the expense of a small AC-side current. Note that Figures 11.7–11.9 provide clear illustrations of the system variables and waveforms, as the model of Figure 11.6 does not include the switching distortions.

EXAMPLE 11.2   STATCOM Performance Based on Switched Model

Consider the STATCOM of Figure 11.1 that employs the three-level NPC of Figure 6.18. The converter parameters are $2C = 19250$ μF, $V_d = 1.0$ V, and $f_s = 1680$ Hz. The three-level NPC also adopts the third-harmonic injected pulse-width modulation (PWM) strategy. Two harmonic filters are paralleled with the STATCOM at the PCC. The filters are the same as those in Example 11.1 and exhibit low impedances to the 27th- and 29th-order harmonics of the AC system frequency. The DC-bus voltage reference V_{DCref} is set to 1750 V. The PCC voltage reference, V_{sdref}, is set to 391 V, corresponding to the line-to-line voltage of 480 Vrms. The controllers of the three-level NPC partial DC-side voltage balancer are

$$K(s) = 0.7 \quad [(\text{kV})^{-1}],$$

$$F(s) = \frac{s^2 + (3\omega_0)^2}{(s + 3\omega_0)^2} = \frac{s^2 + 1131^2}{s^2 + 2262s + 1131^2}.$$

(See Section 6.7.2 and Figure 6.17 or Section 8.5 and Figure 8.20 for details). All the other parameters and controllers are the same as those in Example 11.1. Figure 11.10 illustrates the system response to stepwise changes in V_{sdref} and

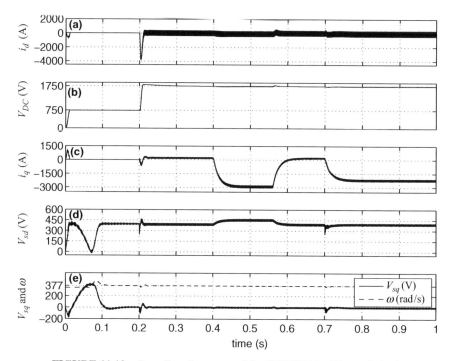

FIGURE 11.10   Overall performance of the STATCOM of Example 11.2.

sudden energization of a series RL load, based on a detailed switched model of the STATCOM of Figure 11.1.

Initially, the STATCOM gating pulses are blocked, all the controllers are inactive, and the load is disconnected; however, the PLL is in service. Since the STATCOM is connected to the PCC, the DC-side capacitors are charged via the VSC antiparallel diodes, and V_{DC} increases to about 750 V as Figure 11.10(b) shows. During this uncontrolled charging phase, the STATCOM AC-side current assumes a fairly large amplitude, and therefore i_d and i_q exhibit relatively large spikes (Figs. 11.10(a) and (c)). Moreover, the PLL attempts to lock to V_{sabc}, and therefore V_{sd}, V_{sq}, and ω undergo large excursions, as Figures 11.10(d) and (e) illustrate. Figures 11.10(d) and (e) also show that the PLL reaches the steady state in 0.15 s, and V_{sq} and ω settle down at zero and 377 rad/s, respectively.

At $t = 0.2$ s, the STATCOM gating pulses are unblocked and the controllers are enabled. Thus, the DC-bus voltage controller decreases i_d to a negative value (Fig. 11.10(a)), to transfer real power from the AC system to the DC side. This action results in an increase of V_{DC}, as Figure 11.10(b) illustrates. The abrupt flow of real power to the VSC DC side, at $t = 0.2$ s, results in a transient voltage drop at the PCC (Fig. 11.10(d)). Consequently, the PCC voltage controller commands a negative i_q to regulate V_{sd}, as Figure 11.10(c) illustrates. Figures 11.10(e) show that V_{sq} and ω undergo excursions at $t = 0.2$ s. Figures 11.10(b) and (d) indicate that V_{DC} and V_{sd} are rapidly regulated at their respective reference values. Figure 11.10(c) illustrates that i_q assumes a positive, small, steady-state value. The reason is that the harmonic filters supply reactive power to the PCC. Thus, to regulate V_{sd}, the STATCOM absorbs the reactive power of the filters.

At $t = 0.4$ s, V_{sdref} is changed stepwise from 391 to 450 V. Consequently, the PCC voltage controller commands a negative i_q (Fig. 11.10(c)), and V_{sd} increases as a first-order exponential function (Fig. 11.10(d)). The disturbance has insignificant impacts on V_{DC}, V_{sq}, and ω, as Figures 11.10(b) and (e) illustrate. At $t = 0.56$ s, V_{sdref} is changed back to 391 V. Thus, i_q and V_{sd} exponentially approach the same respective steady-state values as they possessed from $t = 0.2$ to 0.4 s (Fig. 11.10(c) and (d)). Figures 11.10(b) and (e) illustrate that the disturbance has insignificant impacts on V_{DC}, V_{sq}, and ω.

At $t = 0.7$ s, a three-phase series RL load is connected to the PCC. The load configuration is the same as that of Figure 9.4, but its inductance and resistance are $L_1 = 137\,\mu$H and $R_1 = 83$ mΩ, respectively. Based on the given parameters, and assuming a line-to-line voltage of 480 Vrms, the load real and reactive power are calculated as 2.0 MW and 1.247 MVAr, respectively. Subsequent to the load energization, V_{sd} drops (Fig. 11.10(d)) and V_{sq} and ω are disturbed (Fig. 11.10(e)). Thus, the PCC voltage controller responds accordingly and i_q approaches a negative steady-state value corresponding to the load (and filters) reactive power (Fig. 11.10(c)). The disturbance has an insignificant impact on V_{DC}, as Figure 11.10(b) illustrates.

Figure 11.11 provides a closer look at the STATCOM response at about $t = 0.4$ s. Figure 11.11(a) illustrates that V_{sd} is distorted by the PWM switching harmonics. However, its per-switching-cycle average changes based on a first-order exponential function that closely agrees with the response of V_{sd} shown in Figures 11.7(a) and (b). Figures 11.11(b) shows that subsequent to the voltage increase, i_q assumes a large negative steady-state value to supply reactive power to the AC system. Figure 11.11(c) illustrates that the amplitude of the STAT-COM line current is equal to the absolute value of $i_q(t)$. This can be explained based on (4.77) and the fact that $i_d \approx 0$. Figure 11.11(c) also shows that the PCC voltage and the STATCOM line current are $90°$ phase shifted. The reason is that the STATCOM exchanges a negligible real power with the AC system. Figure 11.11(d) illustrates the partial DC-side voltages of the three-level NPC (see Fig. 6.18 for schematic of a three-level NPC converter). As Figure 11.11(d) shows, the partial DC-side voltages, V_1 and V_2, include triple power-frequency ripples whose amplitudes grow as the STATCOM line current increases. However, the DC components of V_1 and V_2 are kept equal by the VSC DC-side voltage equalizing scheme (see Section 6.7.2 and Figure 6.17, or Section 8.5 and Figure 8.20 for details).

Figure 11.12 shows the system performance when the load is energized, that is, at $t = 0.7$ s. Figure 11.12(a) illustrates that following the energization, the load real and reactive power increase from zero to, respectively, 2.0 MW

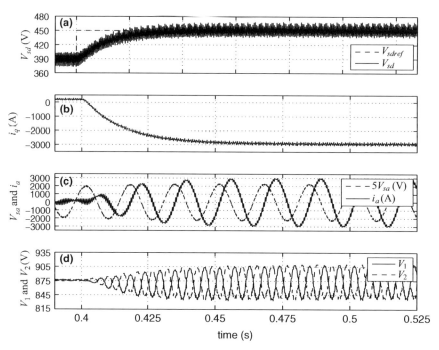

FIGURE 11.11   Response of the STATCOM to step change in V_{sdref}; Example 11.2.

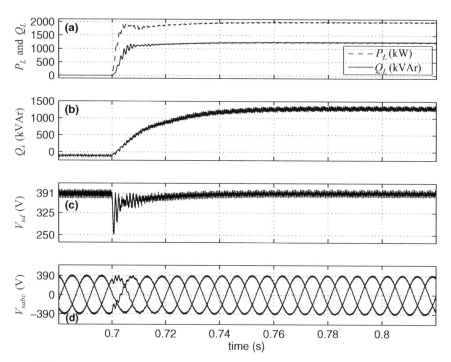

FIGURE 11.12   Response of the STATCOM to load energization; Example 11.2.

and 1.247 MVAr. Therefore, the AC voltage controller responds and the STAT-COM delivers reactive power to the AC system, as Figure 11.12(b) illustrates. Figure 11.12(c) shows that the disturbance results in a 30% voltage drop in V_{sd}; however, V_{sd} reverts to its nominal value in less than one cycle of the AC system frequency. Figure 11.12(d) shows that V_{sabc} remains tightly regulated in spite of the disturbance. The high-frequency components on the waveforms of V_{sa}, V_{sb}, and V_{sc} are due to the dynamic interactions between the filter capacitors, the combined effect of the AC system, filter, and STATCOM inductances.

12 Back-to-Back HVDC Conversion System

12.1 INTRODUCTION

This chapter deals with the control of the VSC-based back-to-back high-voltage DC (HVDC) conversion system. Developments of this chapter are based on the *dq*-frame models and controls for the *real-/reactive-power controller* and the *controlled DC-voltage power port*, presented in Chapter 8. Although we concentrate on the *dq*-frame control, the $\alpha\beta$-frame control approach of Chapter 7 is also applicable to the HVDC system.

12.2 HVDC SYSTEM STRUCTURE

Figure 12.1 illustrates a schematic diagram of a VSC-based back-to-back HVDC system [98]. The HVDC conversion system is composed of two back-to-back connected VSC systems. Both VSC systems employ the three-level NPC as their power converters, labeled as NPC1 and NPC2. Each three-level NPC has a DC-side capacitive voltage divider with two nominally identical capacitors (Fig. 6.18). However, the capacitors of one three-level NPC are not necessarily identical to those of the other one. As discussed in Sections 6.7.2 and 8.5, the partial DC-side voltages of each three-level NPC must be equalized by means of a corresponding DC-side voltage equalizing scheme. Figure 12.1 also shows that NPC1 (NPC2) is interfaced with an AC system, that is, Grid1 (Grid2), through an interface transformer, TR1 (TR2), at the point of common coupling PCC1 (PCC2). As Figure 12.1 illustrates, each grid is represented by a three-phase voltage source, V'_{gabc1} (V'_{gabc2}), and an internal inductance, L'_{i1} (L'_{i2}). To mitigate switching voltage/current harmonics, a shunt, tuned filter is connected to PCC1 (PCC2). Figure 12.1 also shows an optional AC voltage regulator for PCC1 (PCC2). The AC voltage regulator utilizes the feedback signal V_{sabc1} (V_{sabc2}) and issues the command Q_{sref1} (Q_{sref2}). The analysis and design of the AC voltage regulator is based on the methodology that was presented in Chapter 11 in the context of the static compensator (STATCOM).

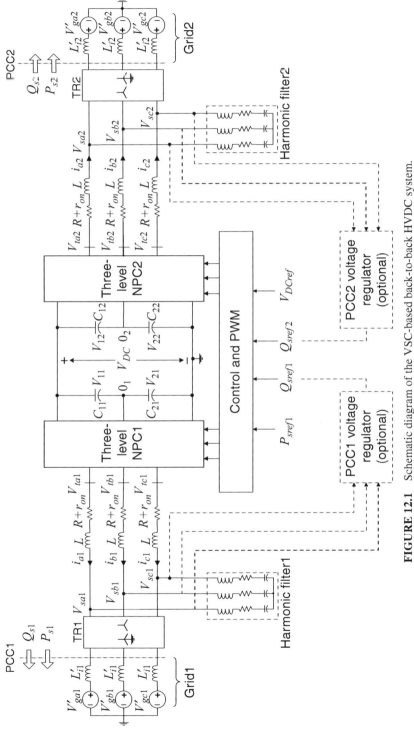

FIGURE 12.1 Schematic diagram of the VSC-based back-to-back HVDC system.

335

It should be noted that in the HVDC system of Figure 12.1, the DC-side midpoints of NPC1 and NPC2, that is, nodes 0_1 and 0_2, are not linked. As discussed in Chapter 6, the midpoint current of the three-level NPC has a third-harmonic component that distorts the partial DC-side voltages of the three-level NPC. Thus, if two midpoints are connected, the midpoint current of one NPC also affects the partial DC-side voltages of the other three-level NPC since the net DC-bus voltage is the same for both three-level NPCs (Fig. 12.1). This results in the generation of interharmonics through the pulse width modulation (PWM) process, especially if Grid1 and Grid2 are at different frequencies, for example, 50 and 60 Hz.

The operational strategy of the back-to-back HVDC system of Figure 12.1 can be itemized as follows:

- The HVDC system enables bidirectional real-power exchange between Grid1 and Grid2, based on the command P_{sref1}.
- Q_{sref1} (Q_{sref2}) is set to a prespecified value and not changed during the operation. Alternatively, it can be closed-loop controlled to regulate the corresponding AC voltage at PCC1 (PCC2).
- The net DC-bus voltage must be regulated at a prespecified value, that is, V_{DCref}, while the partial DC-side voltages of each three-level NPC must be kept equal, that is, $V_{11} = V_{21}$ and $V_{12} = V_{22}$.

In our developments, we may review and discuss the HVDC system components based on generic block diagrams. However, when we develop specific formulations for one of the two VSC systems, we index the variables and parameters by either "1" or "2", corresponding to Grid1 or Grid2, respectively.

12.3   HVDC SYSTEM MODEL

12.3.1   Grid and Interface Transformer Models

Figure 12.2(a) provides a close-up of the interface transformer and the grid for one side of the HVDC converter system of Figure 12.1. The transformer high-to-low (line-to-line) voltage ratio is N. In the HVDC system of Figure 12.1, the synchronization and feed-forward signals, that is, V_{sabc}, are obtained from the transformer low-voltage side. Therefore, the transformer leakage inductance L'_l, referred to the high-voltage side, effectively is lumped with the grid inductance L'_i, and the effective grid inductance is $L'_g = L'_l + L'_i$ (Fig. 12.2(a)). The neutral point of the grid is connected to the ground and serves as the potential reference node for the transformer high-voltage side (Fig. 12.2(a)). However, the potential reference node for the low-voltage side of the transformer is the midpoint of the corresponding three-level NPC, that is, node 0_1 or 0_2 in Figure 12.1. Thus, the star point of the transformer low-voltage side assumes a voltage V_{null} with reference to the corresponding DC-side midpoint.

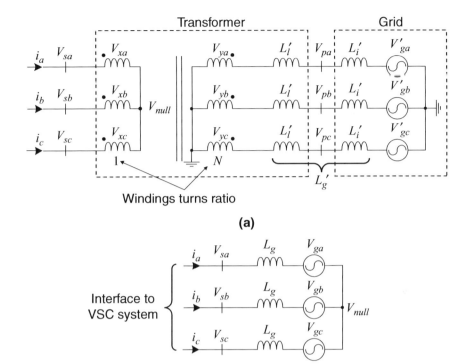

FIGURE 12.2 (a) Schematic diagram of the interface transformer and grid. (b) Equivalent circuit where parameters and variables are referred to the low-voltage side of the interface transformer.

With reference to Figure 12.2(a), the following equations hold if the transformer leakage inductance is ignored:

$$V_{xa} = \frac{1}{N} V_{ya} \approx \frac{1}{N} V_{pa}, \tag{12.1}$$

$$V_{xb} = \frac{1}{N} V_{yb} \approx \frac{1}{N} V_{pb}, \tag{12.2}$$

$$V_{xc} = \frac{1}{N} V_{yc} \approx \frac{1}{N} V_{pc}. \tag{12.3}$$

V_{sabc} is expressed as

$$V_{sa} = V_{xa} + V_{null}, \tag{12.4}$$

$$V_{sb} = V_{xb} + V_{null}, \tag{12.5}$$

$$V_{sc} = V_{xc} + V_{null}. \tag{12.6}$$

Figure 12.2(b) illustrates a circuit equivalent to that of Figure 12.2(a), where the grid voltage and its equivalent inductance are referred to the transformer low-voltage side. The equivalent circuit of Figure 12.2(b) is a more convenient representation for the analysis of the HVDC system. Considering the transformer turns ratio and winding configuration, one concludes that the amplitude of V_{gabc} is $1/N$ times that of V'_{gabc}, and $L_g = L'_g/N^2 = (L'_i + L'_l)/N^2$.

12.3.2   Back-to-Back Converter System Model

Figure 12.3(a) illustrates that the back-to-back HVDC converter system can be considered as the composition of two constant-frequency VSC systems: the left-hand-side VSC system is a real-/reactive-power controller (Section 8.3, Fig. 8.3), and the right-hand-side VSC system is a controlled DC-voltage power port (Section 8.6, Fig. 8.21). As Figure 12.3(a) shows, the real-/reactive-power controller and the controlled DC-voltage power port are interfaced with Grid1 and Grid2, respectively. The real-/reactive-power controller and the controlled DC-voltage power port are connected in parallel from their DC-side terminals.

In the HVDC system of Figure 12.3(a), the real-/reactive-power controller can independently control the real and reactive power exchanged with Grid1, that is, P_{s1} and Q_{s1}. The DC-side voltage of a real-/reactive-power controller must be supported by a DC voltage source. In the HVDC system of Figure 12.3(a), the right-hand side VSC system, that is, the controlled DC-voltage power port, provides the DC voltage support to the real-/reactive-power controller. In a steady state, P_{s1} is approximately equal to the negative of the DC power $P_{ext} = V_{DC}i_{ext}$, which is imposed on the controlled DC-voltage power port. However, Q_{s1} is merely the result of energy exchange among the three phases of VSC1 and does not correspond to any power exchange with the converter DC side. The principles of operation and control of the real-/reactive-power controller are detailed in Sections 8.3 and 8.4. In this chapter, we assume that Q_{s1} is regulated at a prespecified value. However, as outlined in Chapter 11, Q_{s1} can be independently controlled in a closed-loop system to regulate V_{sabc1}, in case the grid inductance L_{g1} is significant.

In the HVDC system of Figure 12.3(a), the controlled DC-voltage power port, that is, the right-hand-side VSC system, provides a regulated DC voltage to the real-/reactive-power controller, that is, the left-hand-side VSC system. The DC-bus voltage, V_{DC}, is regulated by the control of P_{s2}, that is, the real power that is exchanged with Grid2. Thus, P_{s2} must be kept equal to P_{ext} to maintain the power exchanged with the DC-bus capacitor at zero, such that the DC-bus voltage remains constant. The end result is that the negative of P_{s1}, which is a free control variable in the real-/reactive-power controller, is transferred to Grid2 through the controlled DC-voltage power port. The principles of operation and control of the controlled DC-voltage power port are presented in Section 8.6. Similar to the real-/reactive-power controller, in the controlled DC-voltage power port Q_{s2} does not contribute to the exchange of energy with the VSC DC side; therefore, it is regulated at an arbitrary value, for example, zero for unity power-factor operation. Alternatively, as outlined in

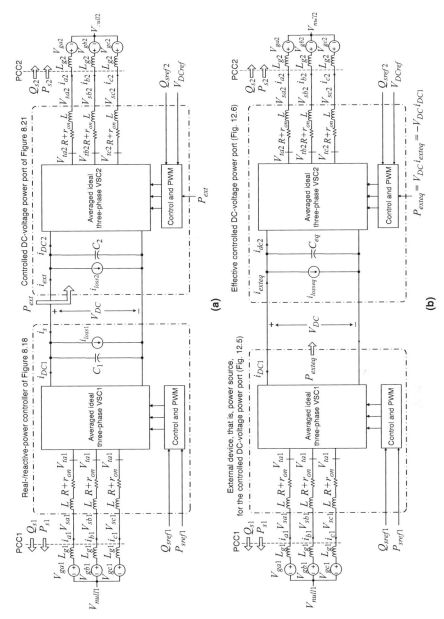

FIGURE 12.3 (a) Back-to-back HVDC system based on a real-/reactive-power controller and a controlled DC-voltage power port; (b) Equivalent system where the two DC-bus capacitors and power-loss current sources are merged.

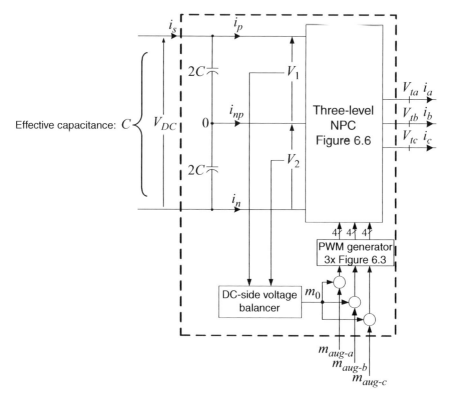

FIGURE 12.4   Block diagram of the three-level NPC.

Chapter 11, Q_{s2} can be independently controlled in a closed-loop system to regulate V_{sabc2}, in case the grid inductance L_{g2} is significant.

As illustrated in Figure 12.1, converters of the back-to-back HVDC system are of the three-level NPC type (Fig. 12.4), to allow an adequately large DC-bus voltage. In Figure 12.3(a), however, these converters are represented by equivalent two-level VSCs. This representation is justified based on the unified model of the two-level VSC and three-level NPC, presented in Section 6.7.4. It was shown that a VSC can be modeled as an ideal (lossless) two-level VSC whose DC side is in parallel with a current source corresponding to the converter power loss, and an effective DC-bus capacitance; the effective capacitance is half the capacitance of each of the DC-side capacitors of the three-level NPC (Fig. 12.4). As discussed in Section 8.5, an internal closed-loop control mechanism is required for the three-level NPC to balance the voltages of the corresponding DC-side capacitors.

Figure 12.3(b) illustrates a modified schematic diagram of the HVDC converter system, equivalent to that of Figure 12.3(a), that is more convenient for the analysis and control design. In the equivalent system of Figure 12.3(b), the DC-bus capacitor and the power-loss current source in the real-/reactive-power controller are lumped with their corresponding counterparts in the controlled DC-voltage power port. Thus, the

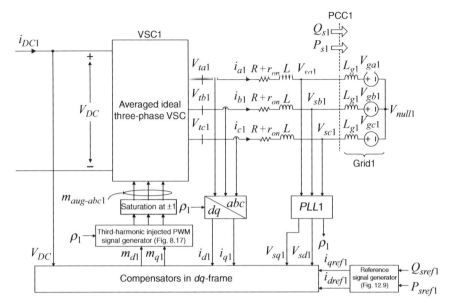

FIGURE 12.5 Schematic diagram of the real-/reactive-power controller in the HVDC system of Figure 12.3(b).

HVDC system of Figure 12.3(b) is composed of (i) a real-/reactive-power controller with no DC-side capacitor and (ii) a controlled DC-voltage power port that has the effective DC-side capacitance $C_{eq} = C_1 + C_2$ in parallel with the effective current source $i_{losseq} = i_{loss1} + i_{loss2}$. The new real-/reactive-power controller and controlled DC-voltage power port are shown in Figures 12.5 and 12.6, respectively, and constitute the base for our subsequent developments.

Based on the model of Figure 12.3(b), the real-/reactive-power controller (Fig. 12.5) behaves as an external device and imposes the DC power $P_{exteq} = V_{DC}i_{exteq} = -V_{DC}i_{DC1}$ on the DC side of the controlled DC-voltage power port (Fig. 12.6). Thus, the main function of the controlled DC-voltage power port is to regulate the DC voltage for the real-/reactive-power controller. As detailed in Section 8.6, the DC-bus voltage regulation is achieved by adjusting P_{s2} in a closed-loop scheme that feedbacks V_{DC}^2 and commands P_{sref2}, as Figure 12.6 illustrates. For the controlled DC-voltage power port, P_{exteq} and the power-loss component $P_{losseq} = V_{DC}i_{losseq}$ are both disturbance inputs. P_{losseq} is comparatively small and fairly constant, and thus it can be compensated by an integrator in the compensator $K_V(s)$. However, P_{exteq} is highly variable within the power rating of the HVDC system and therefore has a significant dynamic impact on V_{DC}^2. Hence, a measure of P_{exteq} is included in the control loop, as a feed-forward signal (Fig. 12.6).

Based on the model of Figure 12.3(b) and the foregoing discussion, the back-to-back HVDC system is operated as follows:

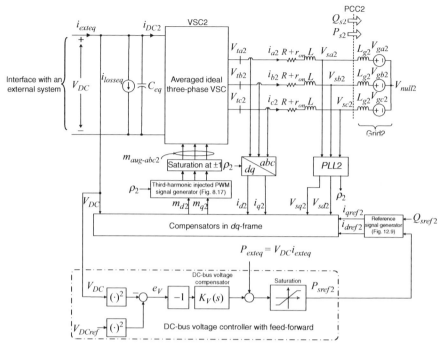

FIGURE 12.6 Schematic diagram of the controlled DC-voltage power port in the HVDC system of Figure 12.3(b).

- The power exchange between Grid1 and Grid2 is controlled by P_{s1} for which P_{sref1} is the reference value. However, P_{sref2} is an internal control variable that is commanded by the DC-bus voltage controller of the HVDC system. Consequently, P_{s2} is a by-product of the control and thus not explicitly controllable. A positive (negative) P_{s1} corresponds to the transfer of real power from Grid2 to Grid1 (Grid1 to Grid2).

- At both AC sides of the HVDC system, Q_{s1} and Q_{s2} can be regulated to arbitrary values within the ratings of VSC1 and VSC2, through their respective reference commands, Q_{sref1} and Q_{sref2}. Alternatively, as discussed in Chapter 11, it is possible to determine Q_{sref1} and Q_{sref2} in a closed-loop fashion, to regulate V_{sabc1} and V_{sabc2}, respectively. However, the latter option is not considered in this chapter.

12.4   HVDC SYSTEM CONTROL

12.4.1   Phase-Locked Loop (PLL)

Each VSC system in the HVDC converter system of Figure 12.3(b), that is, the real-/reactive-power controller and the controlled DC-voltage power port, is controlled in

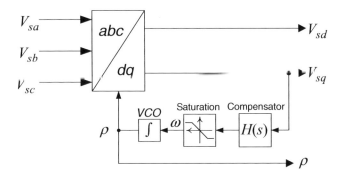

FIGURE 12.7 Schematic diagram of the PLL.

a *dq*-frame. As Figures 12.5 and 12.6 show,

$$i_{dq} = \vec{i}\, e^{-j\rho(t)}, \tag{12.7}$$

$$V_{sdq} = \vec{V_s}\, e^{-j\rho(t)}, \tag{12.8}$$

$$\vec{m} = m_{dq}\, e^{j\rho(t)}, \tag{12.9}$$

where $\rho(t)$ is the phase angle of the corresponding PCC voltage. For each grid, $\rho(t)$ is obtained from a phase-locked loop (PLL) (see Section 8.3.4). Figure 12.7 illustrates the PLL block diagram. The PLL transforms V_{sabc} to V_{sdq} and dynamically adjusts the rotational speed of the *dq*-frame, ω, to fulfill $V_{sq} = 0$. This is achieved based on the control law

$$\frac{d\rho}{dt} = \omega(t) = H(p)V_{sq}(t), \quad \omega_{min} \le \omega \le \omega_{max}, \tag{12.10}$$

where $p = d(\,\cdot\,)/dt$ is the differentiation operator and $H(s)$ is the transfer function of the PLL compensator. Thus, $H(p)V_{sq}(t)$ represents the zero-state response of $H(s)$ to the input $V_{sq}(t)$[1]. ω_{min} and ω_{max} are the lower and upper limits of ω, respectively.

To analyze the PLL, let us consider a balanced grid voltage. Thus, without loss of generality, we assume that V'_{gabc} is

$$V'_{ga} = \left(N\widehat{V_s} \right) \cos\left(\omega_0 t + \theta_0 \right), \tag{12.11}$$

---

[1]Throughout this chapter, we use $F(p)x(t)$, where $F(s)$ is a linear transfer function and $x(t)$ is an arbitrary time-domain signal, to represent the zero-state response of $F(s)$ to the input $x(t)$. In other words, $\mathcal{L}\{F(p)x(t)\} = F(s)X(s)$.

$$V'_{gb} = \left(N\widehat{V}_s\right) \cos\left(\omega_0 t + \theta_0 - \frac{2\pi}{3}\right), \tag{12.12}$$

$$V'_{gc} = \left(N\widehat{V}_s\right) \cos\left(\omega_0 t + \theta_0 - \frac{4\pi}{3}\right). \tag{12.13}$$

Let us also assume that the grid is relatively stiff, that is, L'_g is relatively small, and therefore $V_{pabc} \approx V'_{gabc}$, Fig. 12.2. Then, based on (12.1)–(12.3), we deduce

$$V_{xa} = \widehat{V}_s \cos(\omega_0 t + \theta_0), \tag{12.14}$$

$$V_{xb} = \widehat{V}_s \cos\left(\omega_0 t + \theta_0 - \frac{2\pi}{3}\right), \tag{12.15}$$

$$V_{xc} = \widehat{V}_s \cos\left(\omega_0 t + \theta_0 - \frac{4\pi}{3}\right). \tag{12.16}$$

Substituting for V_{xabc} in (12.4)–(12.6), from (12.14)–(12.16), we deduce

$$V_{sa} = \widehat{V}_s \cos(\omega_0 t + \theta_0) + V_{null}, \tag{12.17}$$

$$V_{sb} = \widehat{V}_s \cos\left(\omega_0 t + \theta_0 - \frac{2\pi}{3}\right) + V_{null}, \tag{12.18}$$

$$V_{sc} = \widehat{V}_s \cos\left(\omega_0 t + \theta_0 - \frac{4\pi}{3}\right) + V_{null}. \tag{12.19}$$

The space phasor corresponding to V_{sabc} can be derived based on (4.2) as

$$\overrightarrow{V_s} = \frac{2}{3}\left[e^{j0}V_{sa} + e^{j\frac{2\pi}{3}}V_{sb} + e^{j\frac{4\pi}{3}}V_{sc}\right] = \widehat{V}_s e^{j(\omega_0 t + \theta_0)}. \tag{12.20}$$

Note that, since $(e^{j0} + e^{j\frac{2\pi}{3}} + e^{j\frac{4\pi}{3}}) \equiv 0$, V_{null} does not appear in $\overrightarrow{V_s}$. Substituting for $\overrightarrow{V_s}$ from (12.20) in (12.8), and decomposing the resultant into real and imaginary components, we obtain

$$V_{sd} = \widehat{V}_s \cos[\omega_0 t + \theta_0 - \rho(t)], \tag{12.21}$$

$$V_{sq} = \widehat{V}_s \sin[\omega_0 t + \theta_0 - \rho(t)]. \tag{12.22}$$

Then, $V_{sq} = 0$ corresponds to $\rho(t) = \omega_0 t + \theta_0$. Considering that, in general, the two grids are not synchronized or can operate at different frequencies, we have

$$\rho_1(t) = \omega_{01} t + \theta_{01}, \tag{12.23}$$

$$\rho_2(t) = \omega_{02} t + \theta_{02}, \tag{12.24}$$

where θ_{01} and θ_{02} are the initial phase angles of V_{sabc1} and V_{sabc2}, respectively.

As detailed in Section 8.3.5, $H(s)$ must include an integral term. If the grid is stiff, V_{pabc} and V_{sabc} are independent of i_{abc}. Therefore, once the PLL reaches the steady state, V_{sq} settles at zero, and $\omega(t)$ becomes constant and equal to the grid angular frequency ω_0. However, if the grid is not stiff, V_{pabc} and V_{sabc} are also functions of i_{abc}, and thus V_{sq} and $\omega(t)$ are subject to excursions when i_d or i_q changes (see equations (11.16) and (11.18)). This, in turn, implies that the PLL dynamics are coupled with the VSC system dynamics. As an approximation, however, we assume a stiff grid condition and consider the PLL dynamics to be independent of those of the VSC system. This subject and the approximation are discussed in Chapter 11.

It was also discussed in Section 8.3.5 that in addition to its integral characteristic, $H(s)$ must exhibit a considerable gain drop at frequency $2\omega_0$, to accommodate operation under grid voltage imbalances. A voltage imbalance is often due to asymmetrical faults, for example, a line-to-ground fault occurring at the PCC (close-in fault) or within the power grid (remote fault). In Section 12.5, we analyze the performance of the HVDC system subjected to a close-in asymmetrical fault.

12.4.2 *dq*-Frame Current-Control Scheme

In the HVDC system of Figure 12.3(b), both VSC systems, that is, the real-/reactive-power controller and the controlled DC-voltage power port, are controlled based on the current-mode control strategy. As detailed in Sections 8.4 and 8.4.1, the current-mode control strategy is employed for controlling the real and reactive power that each VSC system exchanges with the corresponding AC system (grid). Thus, we adopt the current-mode control strategy to control (P_{s1}, Q_{s1}) in the real-/reactive-power controller of Figure 12.5 and to control (P_{s2}, Q_{s2}) in the controlled DC-voltage power port of Figure 12.6.

Figure 12.8 illustrates a block diagram of the *dq*-frame current-control scheme, which is a replicate of Figure 8.10. As Figure 12.8 shows, the *d*- and *q*-axis compensators, $k_d(s)$ and $k_q(s)$, process the error signals $e_d = i_{dref} - i_d$ and $e_q = i_{qref} - i_q$ and deliver the outputs u_d and u_q, respectively. Then, u_d and u_q are augmented by feed-forward signals $V_{sd} - L\omega i_q$ and $V_{sq} + L\omega i_d$, and the signals V_{tdref} and V_{tqref} are generated. V_{tdref} and V_{tqref} are equivalent to *d*- and *q*-axis components of the VSC averaged AC-side terminal voltage; by the subscript *ref* we distinguish them from their counterparts in the actual AC-side terminal voltage, that is, V_{td} and V_{tq}. The feed-forward compensation is (i) to decouple dynamics of i_d and i_q, (ii) to enhance the disturbance rejection capability of the closed-loop system, (iii) to ensure a bumpless system start-up, and (iv) to decouple dynamics of i_d and i_q from those of the grid. For the feed-forward compensation, ω, that is, the rotational speed of the *dq*-frame, is obtained from the PLL. If the PLL dynamics are ignored, ω can be replaced by the (constant) value ω_0.

Figure 12.8 also shows that to generate m_d and m_q, V_{tdref} and V_{tqref} are divided by $V_{DC}(t)/2$. This operation can be considered as a feed-forward compensation, and has to compensate for the VSC conversion gain $V_{DC}(t)/2$ and to ensure that V_{td} and V_{tq} (in the actual converter AC-side terminal voltage) are an accurate reproduction of, respectively, V_{tdref} and V_{tqref} (produced by the current-control scheme of Fig. 12.8)

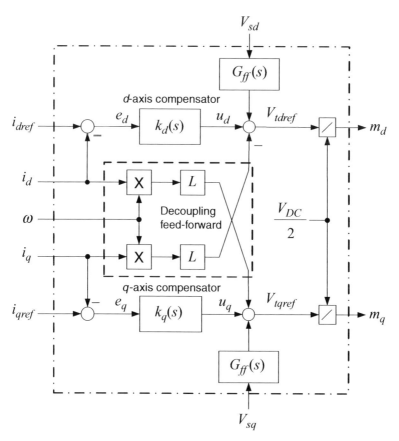

FIGURE 12.8 Block diagram of the *dq*-frame current-control scheme for the real-/reactive-power controller and the controlled DC-voltage power port.

in spite of DC-bus voltage fluctuations. In Figure 12.8, $V_{DC}(t)$ can be replaced by its steady-state value, that is, V_{DCref}, if the DC-bus voltage is relatively constant. Otherwise, its dynamically measured value should be used. Both these options and their results are discussed in Section 12.5.5.

As detailed in Section 8.4.1, $k_d(s)$ is a proportional-integral (PI) compensator,

$$k_d(s) = \frac{k_p s + k_i}{s},\tag{12.25}$$

where k_p and k_i are the proportional and integral gains, respectively. Selecting

$$k_p = \frac{L}{\tau_i},\tag{12.26}$$

$$k_i = \frac{R + r_{on}}{\tau_i},\tag{12.27}$$

we have

$$\frac{I_d(s)}{I_{dref}(s)} = G_i(s) = \frac{1}{\tau_i s + 1}, \tag{12.28}$$

where the time constant τ_i is a design choice. Equation (12.28) indicates that if k_p and k_i are selected based on (12.26) and (12.27), the step response of $i_d(t)$ to $i_{dref}(t)$ is a first-order exponential function with time constant τ_i. This is desirable since the first-order exponential response is smooth and exhibits no steady-state error or overshoot. τ_i specifies the speed of response and is usually selected in the range of 0.5–5 ms, depending on the VSC switching frequency, the desired speed of response, the DC-bus voltage level, and the types of transients. The same compensator as $k_d(s)$ can also be adopted for the q-axis compensator, that is, $k_q(s)$. Thus,

$$\frac{I_q(s)}{I_{qref}(s)} = G_i(s) = \frac{1}{\tau_i s + 1}. \tag{12.29}$$

Based on (4.83) and (4.84), the real and reactive power that the VSC system exchanges with the corresponding PCC are $P_s = (3/2)\left[V_{sd}i_d + V_{sq}i_q\right]$ and $Q_s = (3/2)\left[-V_{sd}i_q + V_{sq}i_d\right]$, respectively. If the PCC voltage is balanced, then in steady state $V_{sq} = 0$, and V_{sd} becomes equal to the amplitude of the PCC line-to-neutral voltage. Therefore, P_s and Q_s become proportional to i_d and i_q, respectively. Thus,

$$P_s = \frac{3}{2}\widehat{V}_s i_d, \tag{12.30}$$

$$Q_s = -\frac{3}{2}\widehat{V}_s i_q, \tag{12.31}$$

where $\widehat{V}_s$ is the peak value of the line-to-neutral voltage of the PCC and assumed to be a constant parameter.[2] Figure 12.8 illustrates that i_{dref} and i_{qref} are the reference inputs to the VSC current-control scheme. Thus, in view of (12.30) and (12.31), i_{dref} and i_{qref} are calculated based on the desired real and reactive power, that is, P_{sref} and Q_{sref}, as illustrated in the block diagram of Figure 12.9. Then, i_d and i_q track i_{dref} and i_{qref}, based on the closed-loop transfer functions of (12.28) and (12.29). Substituting for $i_d(t) = G_i(p)i_{dref}(t)$ and $i_q(t) = G_i(p)i_{qref}(t)$ in (12.30) and (12.31), we obtain

$$P_s(t) = G_i(p)P_{sref}(t), \tag{12.32}$$

$$Q_s(t) = G_i(p)Q_{sref}(t), \tag{12.33}$$

---

[2] In this chapter, for the sake of compactness in formulations and without loss of generality, we assume that Grid1 and Grid2 have the same nominal voltage.

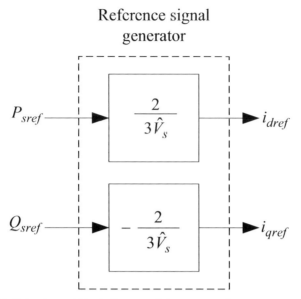

FIGURE 12.9   Control block diagram of the current reference signal generator.

where $P_{sref} = (3/2)\widehat{V}_s i_{dref}$ and $Q_{sref} = -(3/2)\widehat{V}_s i_{qref}$, as illustrated in Figure 12.9. Equations (12.32) and (12.33) can be expressed in the Laplace domain as

$$\frac{P_s(s)}{P_{sref}(s)} = \frac{Q_s(s)}{Q_{sref}(s)} = G_i(s), \tag{12.34}$$

that is, P_s and Q_s track their respective reference commands based on the transfer function $G_i(s)$, which is a first-order transfer function with a time constant of τ_i, (12.28) and (12.29).

12.4.3   PWM Gating Signal Generator

Figure 12.10 illustrates a schematic diagram of the PWM gating signal generator for each VSC of the HVDC system of Figure 12.3(b). The PWM signal generator receives m_d and m_q from the current-control scheme of Figure 12.8 and generates m_{abc} from a dq- to abc-frame transformation for which the angle ρ is provided by the corresponding PLL. Then, m_{abc} is augmented by a fraction of its third-order harmonic to yield $m_{aug-abc}$. Finally, $m_{aug-abc}$ is augmented by a DC offset, m_0, and delivered to the gating signal generator (Fig. 6.3). m_0 is the output of the DC-side voltage equalizing scheme of the three-level NPC and is employed to balance the partial DC-side voltages of the three-level NPC. The DC-side voltage equalizing scheme was extensively discussed in Chapters 6–8 and is briefly reviewed in the next subsection for ease of reference.

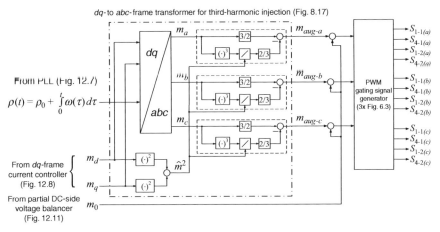

FIGURE 12.10 Block diagram of PWM gating signal generator for the three-level NPC.

12.4.4   Partial DC-Side Voltage Equalization

As discussed in Chapter 6, a three-level NPC with a capacitive voltage divider at its DC side requires an internal closed-loop control scheme to balance its partial DC-side voltages. Figure 12.11 shows a schematic diagram of the DC-side voltage equalizing scheme that is required for each three-level NPC in the HVDC system of Figure 12.1. With reference to the three-level NPC of Figure 12.4, the DC-side voltage equalizing scheme of Figure 12.11 processes the error signal $e = V_1 - V_2$ through the compensator $K(s)$ and provides the control signal u. Since $V_1 - V_2$ is distorted by a relatively large third-order harmonic component, filter $F(s)$ is required in the feedback loop. As detailed in Section 8.5, Figure 8.20, the loop gain is proportional to P_s. Therefore, to ensure the stability of the closed-loop system under both the rectifying and the inverting modes of operation, u is multiplied by the sign of P_s, as Figure 12.11 shows, and m_0 is generated for the PWM signal generator of Figure 12.10. The procedures for designing $F(s)$ and $K(s)$ were outlined in Example 7.3.

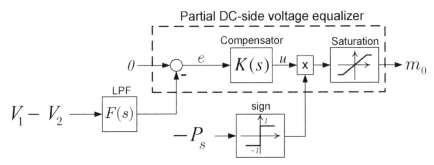

FIGURE 12.11 Control block diagram of the partial DC-side voltage equalizing scheme for the three-level NPC.

12.4.5   Power Flow Control

As explained in Section 12.3.2, P_{s1}, that is, the real power that the real-/reactive-power controller of Figure 12.5 delivers to Grid1, is the main control variable of the back-to-back HVDC system and corresponds to a power transfer from Grid2 to Grid1. As discussed in Section 12.4.2, P_{s1} is controlled by P_{sref1} based on a current-mode control strategy. The transfer function from P_{sref1} to P_{s1} is a first-order filter as described by (12.34). It was also discussed in Section 12.3.2 that P_{s1} results in a DC power component, P_{exteq}, that is imposed on the DC bus of the controlled DC-voltage power port (Fig. 12.3(b)). P_{exteq} acts as a disturbance input to the controlled DC-voltage power port, but its impact can be considerably mitigated through a feed-forward compensation mechanism. Hence, we formulate P_{exteq} in terms of P_{s1} and the system parameters. An estimate of P_{exteq} is used in the feed-forward scheme.

Based on (7.91) and assuming a negligible resistive power loss, the power that is exchanged with the AC-side of VSC1, that is, P_{t1}, is

$$P_{t1} = P_{s1} + \left(\frac{2L}{3\widehat{V}_s^2}\right) P_{s1} \frac{dP_{s1}}{dt} + \left(\frac{2L}{3\widehat{V}_s^2}\right) Q_{s1} \frac{dQ_{s1}}{dt}. \tag{12.35}$$

Since VSC1 is lossless, its DC-side power, that is, $P_{DC1} = i_{DC1} V_{DC}$, is equal to P_{t1}. Thus,

$$P_{DC1} = P_{s1} + \left(\frac{2L}{3\widehat{V}_s^2}\right) P_{s1} \frac{dP_{s1}}{dt} + \left(\frac{2L}{3\widehat{V}_s^2}\right) Q_{s1} \frac{dQ_{s1}}{dt}. \tag{12.36}$$

Figure 12.3(b) illustrates that $i_{exteq} = -i_{DC1}$. Therefore, $P_{exteq} = V_{DC} i_{exteq}$ is equal to $-P_{DC1}$, and we deduce

$$P_{exteq} = -P_{s1} - \left(\frac{2L}{3\widehat{V}_s^2}\right) P_{s1} \frac{dP_{s1}}{dt} - \left(\frac{2L}{3\widehat{V}_s^2}\right) Q_{s1} \frac{dQ_{s1}}{dt}. \tag{12.37}$$

P_{exteq} is imposed on the DC bus of the controlled DC-voltage power port of Figure 12.6. The controlled DC-voltage power port then transfers P_{exteq} to AC side of VSC2 and, thus, to Grid2, in order to maintain the power balance and to regulate V_{DC}. Based on (12.37), during transients P_{exteq} can be larger or lower than $-P_{s1}$, depending on the polarity of P_{s1} (and Q_{s1}) and whether it is increasing or decreasing. However, in an HVDC system, P_{sref1} or Q_{sref1} are usually changed as ramp functions rather than stepwise. Therefore, the derivative terms dP_s/dt and dQ_s/dt are small and may be ignored in (12.37). Thus,

$$P_{exteq} \approx -P_{s1} = -G_i(s)P_{sref1}. \tag{12.38}$$

Equation (12.38) suggests that a measure of P_{exteq} can be estimated by passing the reference signal P_{sref1} through a filter whose transfer function is $G_i(s)$. This estimation method is preferable over the calculation of P_{exteq} based on $P_{exteq} = -i_{DC1}V_{DC}$. The reasons are (i) to measure i_{DC1}, one has to utilize a (Hall effect) current transducer that can be costly, and (ii) the measured i_{DC1} includes switching ripple that must be filtered since only the average value of P_{exteq} is of interest. The disadvantage of estimating P_{exteq} is that (12.38) is developed based on the assumption of a balanced grid voltage. If PCC1 is subjected to an asymmetrical fault, for example, a line-to-ground fault, then P_{exteq} is a periodic function of time that consists of a DC component and a sinusoidal component. While the DC component is exchanged with the converter DC side, the sinusoidal component results in periodic fluctuations superimposed on the DC-bus voltage. We will further demonstrate in Section 12.5.4 that subsequent to a fault inception, the (absolute value of the) DC component of P_{exteq} drops stepwise from its prefault value to a smaller value, even if P_{sref1} is kept constant. However, (12.38) fails to reveal such a change in P_{exteq}. Consequently, the feed-forward compensation based on (12.38) cannot mitigate DC-bus voltage disturbances due to faults or unbalanced voltage grid conditions.

12.4.6 DC-Bus Voltage Regulation

In the HVDC system of Figure 12.3(b), the DC-bus voltage regulation is a function of the controlled DC-voltage power port of Figure 12.6 and is achieved by controlling P_{s2}. As discussed in Section 12.4.2, P_{s2} is controlled through a current-control mechanism for which P_{sref2} and $G_i(s)$ are the reference command and the transfer function, respectively. P_{sref2} is, in turn, the output of the DC-bus voltage controller and is also augmented by a measure of P_{exteq} as a feed-forward signal (Fig. 12.6). The model and control methodology for the controlled DC-voltage power port are extensively discussed in Section 8.6, but are briefly reviewed here for ease of reference.

Similar to (12.36), with respect to Figure 12.6, the DC-side power of VSC2, that is, $P_{DC2} = V_{DC}i_{DC2}$, is expressed as

$$P_{DC2} = P_{s2} + \left(\frac{2L}{3\widehat{V}_s^2}\right) P_{s2}\frac{dP_{s2}}{dt} + \left(\frac{2L}{3\widehat{V}_s^2}\right) Q_{s2}\frac{dQ_{s2}}{dt}. \qquad (12.39)$$

The power-balance equation for the DC side of VSC2 requires that

$$P_{exteq} - P_{losseq} - \frac{dW_{Ceq}}{dt} = P_{DC2}, \qquad (12.40)$$

where $P_{exteq} = V_{DC}i_{exteq}$, $P_{losseq} = V_{DC}i_{losseq}$, and W_{Ceq} is the energy stored in the DC-bus capacitor. Substituting for $W_{Ceq} = (1/2)C_{eq}V_{DC}^2$ and also for P_{s2} from (12.39) in (12.40), we obtain

$$\left(\frac{C_{eq}}{2}\right)\frac{dV_{DC}^2}{dt} = P_{exteq} - P_{losseq} - \left(\frac{2L}{3\widehat{V}_s^2}\right) Q_{s2}\frac{dQ_{s2}}{dt} - P_{s2} - \left(\frac{2L}{3\widehat{V}_s^2}\right) P_{s2}\frac{dP_{s2}}{dt}.$$

$$(12.41)$$

Equation (12.41) describes a first-order plant for which V_{DC}^2 is the output, P_{s2} is the control input, and P_{losseq}, P_{exteq}, and Q_{s2} are the disturbance inputs. Figure 12.6 illustrates that $-K_V(s)$ processes the error signal $e_V = V_{DCref}^2 - V_{DC}^2$ and issues P_{sref2}[3]. P_{sref2} is then tracked by P_{s2} based on the transfer function $G_i(s) = 1/(\tau_i s + 1)$. P_{losseq} is a relatively small value and a function of the operating point, as discussed in Section 5.2.3, but it typically changes over a narrow range of tolerance and, therefore, can be considered as a constant disturbance that can be rejected via an integral term included in $K_V(s)$. However, P_{exteq} can assume any value between the negative rated power and the positive rated power of the HVDC system, and can cause V_{DC}^2 to significantly deviate from V_{DCref}^2. Therefore, as shown in Figure 12.6, a measure of P_{exteq} is added as a feed-forward signal to the output of $K_V(s)$. It should be noted that with typical system parameters, the impact of Q_{s2} on V_{DC}^2 is insignificant and requires no compensation.

With respect to the control input P_{s2}, the plant described by (12.41) is nonlinear due to the term $P_{s2} dP_{s2}/dt$. To design $K_V(s)$, we linearize (12.41) as[4]

$$\frac{d\tilde{V}_{DC}^2}{dt} = -\frac{2}{C_{eq}} \left[\tilde{P}_{s2} + \left(\frac{2L P_{s2ss}}{3\hat{V}_s^2} \right) \frac{d\tilde{P}_{s2}}{dt} \right], \tag{12.42}$$

where superscript $\sim$ and index ss represent small-signal perturbations and steady-state values, respectively. Taking Laplace transform from both sides of (12.42), we obtain the transfer function $G_V(s) = \tilde{V}_{DC}^2/\tilde{P}_{s2}$ as

$$G_V(s) = \frac{\tilde{V}_{DC}^2(s)}{\tilde{P}_{s2}(s)} = -\left(\frac{2}{C_{eq}} \right) \frac{\tau s + 1}{s}, \tag{12.43}$$

where the time constant τ is

$$\tau = \frac{2L P_{s2ss}}{3\hat{V}_s^2}. \tag{12.44}$$

Replacing the derivative terms by zero in (12.41), we deduce

$$P_{s2ss} = P_{exteqss} - P_{losseq}. \tag{12.45}$$

If P_{sref1} is constant, then $P_{s1} = P_{sref1}$ is in steady state. Hence, based on (12.38), we have $P_{exteqss} = -P_{sref1}$, and (12.45) can be rewritten as

$$P_{s2ss} = -P_{sref1} - P_{losseq}. \tag{12.46}$$

---

[3]The negative sign of the compensator is to compensate for the plant negative gain.
[4]We usually do not include disturbances in a small-signal analysis since we are concerned about the closed-loop stability in which disturbance signals play no role.

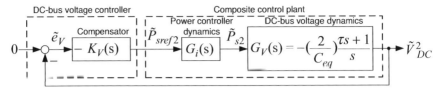

FIGURE 12.12   Control block diagram of the DC-bus voltage regulator.

If $|P_{sref1}| \gg P_{losseq}$, (12.46) can be approximated as

$$P_{s2ss} \approx -P_{sref1}. \tag{12.47}$$

Substituting for P_{s2ss} in (12.44), from (12.47), we conclude

$$\tau = -\frac{2L P_{sref1}}{3\widehat{V}_s^2}. \tag{12.48}$$

Based on (12.48), τ is proportional to the (constant) real-power flow command P_{sref1}. As (12.48) suggests, if P_{sref1} is small, τ is negligible and the plant $G_V(s)$ is predominantly an integrator. As the absolute value of P_{sref1} is increased, $|\tau|$ increases and can cause a considerable phase shift to the loop gain. If P_{sref1} is negative, that is, the power flows from Grid1 to Grid2 (Fig. 12.3(b)), τ is positive and adds to the loop gain phase. However, if P_{sref1} is positive, corresponding to a power flow from Grid2 to Grid1, then τ is negative and imposes a phase delay on the loop gain. A negative τ, which corresponds to a non-minimum-phase zero for $G_V(s)$, has a destabilizing effect on the closed-loop system. Thus, to account for the worst-case scenario, we design $K_V(s)$ for the positive rated value of P_{sref1}. Note that if P_{sref1} is positive, P_{exteq} is negative based on (12.38), and therefore, the controlled DC-voltage power port of Figure 12.6 operates in the rectifying mode of operation. As discussed in Section 8.6, in the rectifying mode of operation, the controlled DC-voltage power port is a non-minimum-phase plant.

Figure 12.12 illustrates a control block diagram of the DC-bus voltage controller, based on the linearized model of (12.43). As Figure 12.12 shows, $K_V(s)$ commands $\widetilde{P}_{sref2}(s)$. However, the plant control input is $\widetilde{P}_{s2}$, which is related to $\widetilde{P}_{sref2}(s)$ through the transfer function $G_i(s)$. Therefore, the transfer function of the effective plant is $G_i(s)G_V(s)$. $K_V(s)$ must include an integral term to compensate for the constant disturbance P_{losseq} and to ensure zero steady-state error.

12.5   HVDC SYSTEM PERFORMANCE UNDER AN ASYMMETRICAL FAULT

Thus far, we have discussed the control of the back-to-back HVDC system of Figure 12.1 (or its equivalent system, i.e., Fig. 12.3(b)) under balanced voltage grid conditions. However, if one of the PCCs (or both) is subjected to an asymmetrical fault, for example, a line-to-ground fault, the HVDC system has to accommodate the grid voltage imbalance and continue its operation, at least temporarily. In this section, we analyze the performance of the HVDC system under unbalanced grid conditions.

12.5.1 PCC Voltage Under an Asymmetrical Fault

When the PCC is subjected to an asymmetrical fault, its voltage becomes unbalanced. Based on the theory of symmetrical components [99], an unbalanced three-phase voltage can be formulated as the superposition of a positive-sequence component, a negative-sequence component, and a zero-sequence component. The formal mathematical formulation is

$$
\begin{aligned}
V_{sa}(t) &= \underbrace{a\widehat{V}_s\cos\left(\omega_0 t + \theta_0\right)}_{} & \underbrace{+\, b\widehat{V}_s\cos\left(\omega_0 t + \theta_0 + \psi\right)}_{} & + & \underbrace{V_s^0(t)}_{}\,, \\
V_{sb}(t) &= a\widehat{V}_s\cos\left(\omega_0 t + \theta_0 - \frac{2\pi}{3}\right) & +\, b\widehat{V}_s\cos\left(\omega_0 t + \theta_0 + \psi - \frac{4\pi}{3}\right) & + & V_s^0(t)\,, \\
V_{sc}(t) &= \underbrace{a\widehat{V}_s\cos\left(\omega_0 t + \theta_0 - \frac{4\pi}{3}\right)}_{\text{positive-sequence}} & \underbrace{+\, b\widehat{V}_s\cos\left(\omega_0 t + \theta_0 + \psi - \frac{2\pi}{3}\right)}_{\text{negative-sequence}} & + & \underbrace{V_s^0(t)}_{\text{zero-sequence}}\,,
\end{aligned}
$$

$$(12.49)$$

where $\widehat{V}_s$ and θ_0 are the amplitude and the phase angle of the line-to-neutral voltage of the otherwise sound PCC, respectively. ψ is the phase angle of the negative-sequence component, relative to the positive-sequence component, when the PCC voltage is unbalanced. Parameters a and b are, respectively, the amplitudes of the positive- and negative-sequence components of (unbalanced) V_{sabc}, relative to $\widehat{V}_s$, and thus specify the degree of imbalance. For example, a balanced three-phase voltage is a special case of (12.49), where $a = 1$, $b = 0$, and $V_s^0(t) \equiv 0$.

Based on (4.2), the space phasor corresponding to V_{sabc} is

$$
\overrightarrow{V}_s = \underbrace{a\widehat{V}_s e^{j(\omega_0 t + \theta_0)}}_{\overrightarrow{V}_s{}^+} + \underbrace{b\widehat{V}_s e^{-j(\omega_0 t + \theta_0 + \psi)}}_{\overrightarrow{V}_s{}^-}. \tag{12.50}
$$

Note that $V_s^0(t)$ does not appear in the expression for $\overrightarrow{V}_s$ since $\left(e^{j0} + e^{j\frac{2\pi}{3}} + e^{j\frac{4\pi}{3}}\right) \equiv 0$.

Equation (12.50) indicates that $\overrightarrow{V}_s$ is composed of two components: (i) the counterclockwise rotating space phasor $\overrightarrow{V}_s{}^+ = a\widehat{V}_s e^{j(\omega_0 t + \theta_0)}$, which we refer to as *positive-sequence space phasor*, and (ii) the clockwise rotating space phasor

$\overrightarrow{V_s}^- = b\widehat{V_s}e^{-j(\omega_0 t + \theta_0 + \psi)}$, which we refer to as *negative-sequence space phasor*. The positive- and negative-sequence components of V_{sabc} can be retrieved by applying (4.7) to, respectively, $\overrightarrow{V_s}^+$ and $\overrightarrow{V_s}^-$. This yields

$$V_{sa}^+(t) = Re\left\{\overrightarrow{V_s}^+ e^{-j0}\right\}, \tag{12.51}$$

$$V_{sb}^+(t) = Re\left\{\overrightarrow{V_s}^+ e^{-j\frac{2\pi}{3}}\right\}, \tag{12.52}$$

$$V_{sc}^+(t) = Re\left\{\overrightarrow{V_s}^+ e^{-j\frac{4\pi}{3}}\right\}, \tag{12.53}$$

and

$$V_{sa}^-(t) = Re\left\{\overrightarrow{V_s}^- e^{-j0}\right\}, \tag{12.54}$$

$$V_{sb}^-(t) = Re\left\{\overrightarrow{V_s}^- e^{-j\frac{2\pi}{3}}\right\}, \tag{12.55}$$

$$V_{sc}^-(t) = Re\left\{\overrightarrow{V_s}^- e^{-j\frac{4\pi}{3}}\right\}, \tag{12.56}$$

where $Re\{\cdot\}$ is the real-part operator. It should, however, be noted that V_{sabc} cannot be retrieved from $\overrightarrow{V_s}$ unless $V_s^0(t)$ is known.[5]

EXAMPLE 12.1   PCC Voltage Under A Line-to-Ground Fault

Figure 12.13 illustrates a scenario in which the PCC is subjected to a line-to-ground fault. Consider that the fault is due to a short circuit between phase c of the PCC and the ground, that is, $V_{pc} \equiv 0$. Thus, based on (12.1)–(12.3), we have

$$V_{xa} \approx \frac{1}{N} V_{pa},$$

$$V_{xb} \approx \frac{1}{N} V_{pb},$$

$$V_{xc} \approx 0.$$

$$\tag{12.57}$$

---

[5]The zero-sequence component of the three-phase signal $f_{abc}(t)$ is given as $f^0(t) = \left[f_a(t) + f_b(t) + f_c(t)\right]/3$.

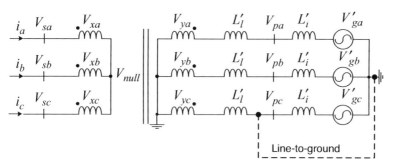

FIGURE 12.13   Schematic diagram of the interface transformer and AC Systems under a line-to-ground fault.

Assuming that $V_{pa} \approx V'_{ga}$ and $V_{pb} \approx V'_{gb}$, and substituting for V'_{ga} and V'_{gb} in (12.57), respectively, from (12.11) and (12.12), we deduce

$$V_{xa} \approx \widehat{V}_s \cos\left(\omega_0 t + \theta_0\right),$$

$$V_{xb} \approx \widehat{V}_s \cos\left(\omega_0 t + \theta_0 - \frac{2\pi}{3}\right),$$

$$V_{xc} \approx 0.$$

$$(12.58)$$

Based on (12.4)–(12.6) and (12.58), we have

$$V_{sa} \approx \widehat{V}_s \cos\left(\omega_0 t + \theta_0\right) + V_{null},$$

$$V_{sb} \approx \widehat{V}_s \cos\left(\omega_0 t + \theta_0 - \frac{2\pi}{3}\right) + V_{null},$$

$$V_{sc} \approx V_{null},$$

$$(12.59)$$

which constitute an unbalanced three-phase voltage. Based on (4.2), we deduce

$$\overrightarrow{V_s} = \underbrace{\frac{2}{3}\widehat{V}_s e^{j(\omega_0 t + \theta_0)}}_{\overrightarrow{V_s}^+} + \underbrace{\frac{1}{3}\widehat{V}_s e^{-j(\omega_0 t + \theta_0 - \frac{\pi}{3})}}_{\overrightarrow{V_s}^-}. \qquad (12.60)$$

Comparing (12.50) and (12.60), one concludes that $a = 2/3$, $b = 1/3$, and $\psi = -\pi/3$ under the line-to-ground fault condition. Based on (12.49), V_{sabc} can be expressed as

$$V_{sa}(t) = \frac{2}{3}\widehat{V}_s \cos{(\omega_0 t + \theta_0)} \qquad + \frac{1}{3}\widehat{V}_s \cos{\left(\omega_0 t + \theta_0 - \frac{\pi}{3}\right)} + \underbrace{V_{null}(t)}\,,$$

$$V_{sb}(t) = \frac{2}{3}\widehat{V}_s \cos{\left(\omega_0 t + \theta_0 - \frac{2\pi}{3}\right)} + \frac{1}{3}\widehat{V}_s \cos{\left(\omega_0 t + \theta_0 - \frac{5\pi}{3}\right)} + V_{null}(t)\,,$$

$$V_{sc}(t) = \underbrace{\frac{2}{3}\widehat{V}_s \cos{\left(\omega_0 t + \theta_0 - \frac{4\pi}{3}\right)}}_{\textit{positive-sequence}} + \underbrace{\frac{1}{3}\widehat{V}_s \cos{(\omega_0 t + \theta_0 - \pi)}}_{\textit{negative-sequence}} + \underbrace{V_{null}(t)}_{\textit{zero-sequence}}\,,$$

$$(12.61)$$

where V_{null} is the zero-sequence component of V_{sabc}.

12.5.2 Performance of PLL Under an Asymmetrical Fault

Substituting for $\overrightarrow{V}_s$ from (12.50) in (12.8), and decomposing the result into its real and imaginary components, we obtain

$$V_{sd} = a\widehat{V}_s \cos{[\omega_0 t + \theta_0 - \rho(t)]} + b\widehat{V}_s \cos{[\omega_0 t + \theta_0 + \psi + \rho(t)]}, \quad (12.62)$$

$$V_{sq} = a\widehat{V}_s \sin{[\omega_0 t + \theta_0 - \rho(t)]} - b\widehat{V}_s \sin{[\omega_0 t + \theta_0 + \psi + \rho(t)]}. \quad (12.63)$$

The PLL function is to synchronize the dq-frame to V_{sabc}, that is, to ensure that $\rho(t) = \omega_0 t + \theta_0$ in the steady state. Thus, assuming a small-signal perturbation, that is, $\rho(t) \approx \omega_0 t + \theta_0$, we can rewrite (12.63) as

$$V_{sq} = a\widehat{V}_s \sin{[\omega_0 t + \theta_0 - \rho(t)]} - b\widehat{V}_s \sin{[2(\omega_0 t + \theta_0) + \psi]}$$

$$\approx a\widehat{V}_s [\omega_0 t + \theta_0 - \rho(t)] - b\widehat{V}_s \sin{[2(\omega_0 t + \theta_0) + \psi]}. \quad (12.64)$$

Substituting for V_{sq} in (12.10), we find

$$\frac{d\rho}{dt} = a\widehat{V}_s H(p)[\omega_0 t + \theta_0 - \rho(t)] - b\widehat{V}_s H(p)\sin{[2(\omega_0 t + \theta_0) + \psi]}, \quad (12.65)$$

where $p = d(\cdot)/dt$ is the differentiation operator.

Equation (12.65) can be associated with a classical feedback mechanism for which $\omega_0 t + \theta_0$ is the reference input, $\rho(t)$ is the output, and $a\widehat{V}_s H(s)/s$ is the open-loop transfer function. Input $d(t) = -b\widehat{V}_s \sin{[2(\omega_0 t + \theta_0) + \psi]}H(p)$ represents a disturbance to the control loop. Figure 12.14 shows a general control block diagram of the PLL that covers both balanced and unbalanced conditions. Under normal operating condition, $(a, b) = (1, 0)$, and the block diagram of Figure 12.14 becomes an equivalent to its counterpart under balanced grid conditions, that is, Figure 8.4. However, when the PCC is subjected to an asymmetrical fault, the coefficient a becomes less than unity and the coefficient b assumes a nonzero value; consequently, (i) the PLL

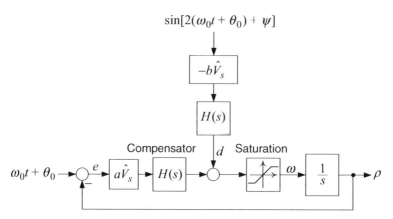

FIGURE 12.14 Control block diagram of the PLL: (a) $a = 1$ and $b = 0$ for a balanced grid condition; (b) $a < 1$ and $b \neq 0$ for an unbalanced grid condition, for example, due to an asymmetrical fault.

loop gain magnitude drops by $100(1 - a)\%$, for example, 33% for a line-to-ground fault, and (ii) $\omega(t)$ gets distorted by a sinusoidal disturbance that would not exist under a balanced grid condition. The magnitude drop may slightly reduce the closed-loop speed of response but poses no instability issues. However, the disturbance input results in fluctuations in ω and ρ at the frequency $2\omega_0$. These fluctuations, in turn, generate additional harmonics in the VSC terminal voltage and current. Thus, to attenuate the disturbance, $H(s)$ must exhibit an appreciable gain drop at $2\omega_0$. This can be achieved, for example, by inclusion of the complex-conjugate zeros $\pm j(2\omega_0)$ in $H(s)$. In Example 8.1, a procedure to design $H(s)$ was presented.

12.5.3 Performance of dq-Frame Current-Control Scheme Under an Asymmetrical Fault

In this section, we analyze the performance of the d- and q-axis current-control scheme under an unbalanced PCC voltage, subsequent to an asymmetrical fault. For both the real-/reactive-power controller of Figure 12.5 and the controlled DC-voltage power port of Figure 12.6, the model of the AC side is given by (8.5), repeated here as (12.66):

$$L\frac{d\vec{i}}{dt} = -(R + r_{on})\,\vec{i} + \vec{V_t} - \vec{V_s}. \tag{12.66}$$

Expressing each space phasor of (12.66) in terms of dq-frame components, that is, $\vec{f} = (f_d + jf_q)e^{j\rho}$, we obtain

$$L\frac{di_d}{dt} = (L\omega)i_q - (R + r_{on})i_d + V_{td} - V_{sd}, \tag{12.67}$$

$$L\frac{di_q}{dt} = -(L\omega)i_d - (R + r_{on})i_q + V_{tq} - V_{sq}, \qquad (12.68)$$

where $\omega = d\rho/dt$. If the grid voltage is balanced, V_{sd} and V_{sq} are given by (12.21) and (12.22), which, since $\rho(t) = \omega_0 t + \theta_0$, can be rewritten as

$$V_{sd} = \widehat{V}_s, \qquad (12.69)$$

$$V_{sq} = 0. \qquad (12.70)$$

Under an asymmetrical fault condition, V_{sd} and V_{sq} are given by (12.62) and (12.63), which can be written as

$$V_{sd} = a\widehat{V}_s + b\widehat{V}_s \cos[2(\omega_0 t + \theta_0) + \psi], \qquad (12.71)$$

$$V_{sq} = -b\widehat{V}_s \sin[2(\omega_0 t + \theta_0) + \psi]. \qquad (12.72)$$

A comparison between (12.72) and (12.70) reveals that when an asymmetrical fault occurs, V_{sq} assumes a sinusoidal ripple component while retaining its zero DC component. However, a comparison between (12.71) and (12.69) indicates that V_{sd} assumes a sinusoidal ripple component in addition to a level drop in its DC component, from $\widehat{V}_s$ to $a\widehat{V}_s$. As indicated by (12.67) and (12.68), if not properly compensated, the ripple components of V_{sd} and V_{sq} result in fluctuations in i_d and i_q; the fluctuations, in turn, entail imbalance and generation of a positive-sequence third-order harmonic in i_{abc}.[6] The impact of V_{sd} and V_{sq} on i_d and i_q, as explained above, can be mitigated by a feed-forward mechanism as follows.

Figure 12.8 illustrates that (filtered) measures of V_{sd} and V_{sq} are added to the control signals u_d and u_q; the transfer function of the filters is $G_{ff}(s)$. Thus, any change in V_{sd} and/or V_{sq} is rapidly communicated to V_{tdref} and/or V_{tqref}, such that the voltage drop across the interface reactors remains merely a function of i_d and i_q.

Since the frequency of the ripple components of V_{sd} and V_{sq} is $2\omega_0$, the bandwidth of $G_{ff}(s)$ must be adequately larger than $2\omega_0$, in order for the feed-forward system to be effective. Otherwise, the gain drop and the phase delay associated with $G_{ff}(j2\omega_0)$ result in deviations of the measured values of V_{sd} and V_{sq} from their actual values, and render the feed-forward compensation ineffective. In addition, the switching frequency must be adequately large to enable the VSC to faithfully reproduce the ripple components of V_{sd} and V_{sq}. If these conditions are satisfied, i_d and i_q contain no ripple components due to fluctuations in V_{sd} and V_{sq}, and i_{abc} remains more or less balanced and distortion-free under the fault. A detailed explanation of the feed-forward compensation and its technical merits was presented in Section 3.4.

---

[6]This will be formally proven in Section 12.5.7.

12.5.4   Dynamics of DC-Bus Voltage Under an Asymmetrical Fault

Consider the HVDC system of Figure 12.3(b) and assume that both PCCs are subjected to asymmetrical faults. For the compactness of formulations and without loss of generality, we assume that the frequencies of both grids are equal, that is, ω_0. However, their initial phase angles can be different, that is, $\theta_{01} \neq \theta_{02}$. We further assume that both sides of the HVDC system exchange power with their corresponding grids at unity power factor, that is, $Q_{sref} = Q_s = 0$. Based on the developments of Sections 12.4.1 and 12.5.2, we can express the dq-frame components of each PCC voltage as

$$V_{sd} = a\widehat{V}_s + b\widehat{V}_s \cos\left[2(\omega_0 t + \theta_0) + \psi\right], \tag{12.73}$$

$$V_{sq} = -b\widehat{V}_s \sin\left[2(\omega_0 t + \theta_0) + \psi\right], \tag{12.74}$$

where coefficients a and b depend on the type of fault and the degree of voltage imbalance. For example, $a = 1$ and $b = 0$ represent a sound grid, (12.69) and (12.70), whereas $a = 2/3$ and $b = 1/3$ represent a grid that is subjected to a line-to-ground fault, (12.71) and (12.72), and so on. Based on (4.83) and since $i_q = 0$, the (actual) real power that is exchanged with the PCC is

$$P_{s\text{-}act} = \frac{3}{2}\left\{a\widehat{V}_s + b\widehat{V}_s \cos\left[2(\omega_0 t + \theta_0) + \psi\right]\right\} i_d. \tag{12.75}$$

Substituting for $i_d = 2P_s/(3\widehat{V}_s)$ in (12.75), we have

$$P_{s\text{-}act} = aP_s + bP_s \cos\left[2(\omega_0 t + \theta_0) + \psi\right]. \tag{12.76}$$

It should be noted that $P_{s\text{-}act}$ represents the real power that is actually exchanged with the grid and is not the same as P_s. Based on (12.30), $P_s = (3/2)\widehat{V}_s i_d$ defines a quantity that is equivalent to $P_{s\text{-}act}$ only if the PCC voltage is balanced; that is, $P_{s\text{-}act} = P_s$ when $a = 1$ and $b = 0$. However, under an unbalanced grid condition P_s is not equal to $P_{s\text{-}act}$ but, as (12.76) indicates, appears as a factor in the expression for the DC component (or the average value) of $P_{s\text{-}act}$. The DC component of $P_{s\text{-}act}$, that is, aP_s, is proportional to and smaller in absolute value than P_s, since $a \leq 1$. Moreover, under an unbalanced grid condition b is nonzero and, thus, $P_{s\text{-}act}$ includes an AC ripple component in addition to its DC component. The ripple component is a sinusoidal function of time whose amplitude and frequency are bP_s and $2\omega_0$, respectively. Similar conclusions can be made also for $Q_{s\text{-}act}$ and Q_s.

The VSC DC-side power is given by

$$P_{DC} = P_t$$

$$= P_{s\text{-}act} + P_L, \tag{12.77}$$

where P_L, the instantaneous power of the interface reactors, is given by (4.41) as

$$P_L = \frac{3L}{2} Re \left\{ \frac{d\vec{i}}{dt} \vec{i}^* \right\}$$

$$= \frac{3L}{2} Re \left\{ \frac{d(i_{dq}e^{j\rho})}{dt} (i_{dq}e^{j\rho})^* \right\}$$

$$= \frac{3L}{4} \left(\frac{di_d^2}{dt} + \frac{di_q^2}{dt} \right)$$

$$= \frac{3L}{4} \frac{di_d^2}{dt}, \text{ since } i_q = 0. \tag{12.78}$$

Since $i_d = 2P_s/(3\widehat{V}_s)$, (12.78) can be rewritten as

$$P_L = \frac{L}{3\widehat{V}_s^2} \frac{dP_s^2}{dt}. \tag{12.79}$$

Substituting for $P_{s\text{-}act}$ and P_L in (12.77), respectively, from (12.76) and (12.79), we deduce the VSC terminal power as

$$P_{DC} = \left(aP_s + \frac{L}{3\widehat{V}_s^2} \frac{dP_s^2}{dt} \right) + bP_s \cos\left[2(\omega_0 t + \theta_0) + \psi\right]. \tag{12.80}$$

With reference to the HVDC system of Figure 12.3(b), the DC-bus voltage dynamics can be formulated based on the principle of power balance as

$$\left(\frac{C_{eq}}{2} \right) \frac{dV_{DC}^2}{dt} = P_{exteq} - P_{losseq} - P_{DC2}$$

$$\approx P_{exteq} - P_{DC2}$$

$$\approx -P_{DC1} - P_{DC2}. \tag{12.81}$$

Substituting for P_{DC1} and P_{DC2} from the generic equation (12.80) in (12.81), we have

$$\left(\frac{C_{eq}}{2} \right) \frac{dV_{DC}^2}{dt} = -\left(a_1 P_{s1} + \frac{L}{3\widehat{V}_s^2} \frac{dP_{s1}^2}{dt} \right) - \left(a_2 P_{s2} + \frac{L}{3\widehat{V}_s^2} \frac{dP_{s2}^2}{dt} \right)$$

$$- \{b_1 P_{s1} \cos\left[2(\omega_0 t + \theta_{01}) + \psi_1\right] + b_2 P_{s2} \cos\left[2(\omega_0 t + \theta_{02}) + \psi_2\right]\}, \tag{12.82}$$

or equivalently

$$
\frac{dV_{DC}^2}{dt} = -\left(\frac{2}{C_{eq}}\right)\left(a_1 P_{s1} + \frac{L}{3\widehat{V}_s^2}\frac{dP_{s1}^2}{dt}\right) - \left(\frac{2}{C_{eq}}\right)\left(a_2 P_{s2} + \frac{L}{3\widehat{V}_s^2}\frac{dP_{s2}^2}{dt}\right)
$$
$$
- \left(\frac{1}{C_{eq}}\right)\left[b_1 P_{s1} e^{j(2\theta_{01}+\psi_1)} + b_2 P_{s2} e^{j(2\theta_{02}+\psi_2)}\right]e^{j2\omega_0 t}
$$
$$
- \left(\frac{1}{C_{eq}}\right)\left[b_1 P_{s1} e^{-j(2\theta_{01}+\psi_1)} + b_2 P_{s2} e^{-j(2\theta_{02}+\psi_2)}\right]e^{-j2\omega_0 t}. \qquad (12.83)
$$

Based on (12.82) or (12.83), V_{DC}^2 includes a steady-state AC component with frequency $2\omega_0$, if b_1 or b_2 is nonzero. Let us approximate V_{DC}^2 as

$$
V_{DC}^2 \approx y_0(t) + y_\alpha(t)\cos(2\omega_0 t) - y_\beta(t)\sin(2\omega_0 t)
$$
$$
= y_0(t) + \frac{1}{2}\left[y_\alpha(t) + jy_\beta(t)\right]e^{j2\omega_0 t} + \frac{1}{2}\left[y_\alpha(t) - jy_\beta(t)\right]e^{-j2\omega_0 t}, \quad (12.84)
$$

where the time-varying coefficients $y_0(t)$, $y_\alpha(t)$, and $y_\beta(t)$ settle at respective constant values in the steady state. Formulation of V_{DC}^2 as (12.84) is based on the assumption that $y_0(t)$, $y_\alpha(t)$, and $y_\beta(t)$ change considerably slower than $\cos(2\omega_0 t)$ and $\sin(2\omega_0 t)$. Taking derivative of (12.84), we obtain

$$
\frac{dV_{DC}^2}{dt} = \frac{dy_0}{dt}
$$
$$
+ \left[\frac{1}{2}\left(\frac{dy_\alpha}{dt} + j\frac{dy_\beta}{dt}\right) + j\omega_0\left(y_\alpha + jy_\beta\right)\right]e^{j2\omega_0 t}
$$
$$
+ \left[\frac{1}{2}\left(\frac{dy_\alpha}{dt} - j\frac{dy_\beta}{dt}\right) - j\omega_0(y_\alpha - jy_\beta)\right]e^{-j2\omega_0 t}. \qquad (12.85)
$$

Substituting for dV_{DC}^2/dt from (12.85) in (12.83) and equating the terms that are multiples of $e^{j0\omega_0 t}$, $e^{j2\omega_0 t}$, and $e^{-j2\omega_0 t}$ to their counterparts from the other side of the resultant equation, we deduce

$$
\frac{dy_0}{dt} = -\left(\frac{2}{C_{eq}}\right)\left(a_1 P_{s1} + \frac{L}{3\widehat{V}_s^2}\frac{dP_{s1}^2}{dt}\right) - \left(\frac{2}{C_{eq}}\right)\left(a_2 P_{s2} + \frac{L}{3\widehat{V}_s^2}\frac{dP_{s2}^2}{dt}\right),
$$

$$(12.86)$$

and

$$\left[\frac{1}{2}\left(\frac{dy_\alpha}{dt} + j\frac{dy_\beta}{dt}\right) + j\omega_0\left(y_\alpha + jy_\beta\right)\right] =$$
$$-\left(\frac{1}{C_{eq}}\right)\left[b_1 P_{s1}e^{j(2\theta_{01}+\psi_1)} + b_2 P_{s2}e^{j(2\theta_{02}+\psi_2)}\right], \qquad (12.87)$$

$$\left[\frac{1}{2}\left(\frac{dy_\alpha}{dt} - j\frac{dy_\beta}{dt}\right) - j\omega_0\left(y_\alpha - jy_\beta\right)\right] =$$
$$-\left(\frac{1}{C_{eq}}\right)\left[b_1 P_{s1}e^{-j(2\theta_{01}+\psi_1)} + b_2 P_{s2}e^{-j(2\theta_{02}+\psi_2)}\right]. \qquad (12.88)$$

Equation (12.86) describes the dynamics of the DC component of V_{DC}^2, that is, y_0. Equation (12.87) embeds two equations describing the dynamics of the direct and quadrature components of the AC component of V_{DC}^2, that is, y_α and y_β. It can be shown that (12.88) is equivalent to (12.87) and thus superfluous. Therefore, we proceed with our developments by decomposing (12.87) into real and imaginary components, as

$$\frac{dy_\alpha}{dt} = 2\omega_0 y_\beta - \frac{2}{C_{eq}}[b_1 P_{s1}\cos(2\theta_{01} + \psi_1) + b_2 P_{s2}\cos(2\theta_{02} + \psi_2)], \quad (12.89)$$

$$\frac{dy_\beta}{dt} = -2\omega_0 y_\alpha - \frac{2}{C_{eq}}[b_1 P_{s1}\sin(2\theta_{01} + \psi_1) + b_2 P_{s2}\sin(2\theta_{02} + \psi_2)]. \quad (12.90)$$

Equations (12.89) and (12.90) indicate that the dynamics of y_α and y_β are coupled but independent of those of y_0. Likewise, y_0 is independent of y_α or y_β. Let us rewrite (12.84) as

$$V_{DC}^2 = y_0(t) + \hat{y}(t)\cos[2\omega_0 t + \eta(t)], \qquad (12.91)$$

where

$$\hat{y} = \sqrt{y_\alpha^2 + y_\beta^2},$$
$$\eta = tan^{-1}\left(\frac{y_\beta}{y_\alpha}\right).$$
$$(12.92)$$

Equation (12.92) indicates that the amplitude and the phase angle of the AC component of V_{DC}^2 are functions of y_α and y_β. The steady-state values of y_α and y_β can be determined by solving (12.89) and (12.90), where dy_α/dt and dy_β/dt are set to zero.

The result is

$$y_{\beta ss} = \frac{1}{C_{eq}\omega_0}[b_1 P_{s1ss} \cos(2\theta_{01} + \psi_1) + b_2 P_{s2ss} \cos(2\theta_{02} + \psi_2)], \quad (12.93)$$

$$y_{\alpha ss} = -\frac{1}{C_{eq}\omega_0}[b_1 P_{s1ss} \sin(2\theta_{01} + \psi_1) + b_2 P_{s2ss} \sin(2\theta_{02} + \psi_2)], \quad (12.94)$$

where subscript ss denotes the steady-state value. It then follows from the substitution of $y_{\alpha ss}$ and $y_{\beta ss}$ in (12.92), from (12.93) and (12.94), that

$$\widehat{y}_{ss} = \frac{\sqrt{(b_1 P_{s1ss})^2 + (b_2 P_{s2ss})^2 + 2b_1 b_2 P_{s1ss} P_{s2ss} \cos[2(\theta_{01} - \theta_{02}) + (\psi_1 - \psi_2)]}}{C_{eq}\omega_0}.$$

$$(12.95)$$

It is preferable to express $\widehat{y}_{ss}$ in terms of P_{sref1}, that is, the power-flow command of the HVDC system. To that end, we first set the left-hand side of (12.86) to zero and obtain

$$P_{s2ss} = -\left(\frac{a_1}{a_2}\right)\underbrace{P_{s1ss}}_{P_{sref1}} = -\left(\frac{a_1}{a_2}\right)P_{sref1}. \quad (12.96)$$

Then, we substitute in (12.95) for $P_{s1ss} = P_{sref1}$ and $P_{s2ss} = -(a_1/a_2)P_{sref1}$, and deduce

$$\widehat{y}_{ss} = \frac{\sqrt{(a_1 b_2)^2 + (a_2 b_1)^2 - 2(a_1 b_2)(a_2 b_1)\cos[2(\theta_{01} - \theta_{02}) + (\psi_1 - \psi_2)]}}{a_2 C_{eq}\omega_0}|P_{sref1}|.$$

$$(12.97)$$

Equation (12.97) indicates that the steady-state amplitude of the AC component of V_{DC}^2, that is, $\widehat{y}_{ss}$, depends on the parameters a and b and thus on the type of the fault. $\widehat{y}_{ss}$ is also a function of AC Systems' initial phase angles. Furthermore, $\widehat{y}_{ss}$ is proportional to the absolute value of the power flow command P_{sref1} and inversely proportional to the effective DC-bus capacitance. The AC component of V_{DC}^2 results in a periodic overvoltage for which the DC-bus capacitor and the VSC switch cells must be rated. To calculate the steady-state overvoltage, let us rewrite (12.91) as

$$V_{DC} = \sqrt{y_{0ss} + \widehat{y}_{ss} \cos(2\omega_0 t + \eta)}$$

$$= \sqrt{y_{0ss} \left[1 + \left(\frac{\widehat{y}_{ss}}{y_{0ss}} \right) \cos(2\omega_0 t + \eta) \right]}$$

$$= \sqrt{y_{0ss}} \sqrt{1 + \left(\frac{\widehat{y}_{ss}}{y_{0ss}} \right) \cos(2\omega_0 t + \eta)}. \qquad (12.98)$$

As discussed in the next section, the DC-bus voltage must regulate y_0, rather than V_{DC}^2, and the AC component of V_{DC}^2 is remarkably attenuated in the feedback loop. In other words, the objective of the DC-bus voltage controller is to ensure that $y_{0ss} = V_{DCref}^2$. Hence, (12.98) can be rewritten as

$$V_{DC} = V_{DCref} \sqrt{1 + \left(\frac{\widehat{y}_{ss}}{V_{DCref}^2} \right) \cos(2\omega_0 t + \eta)}. \qquad (12.99)$$

It is plausible to assume that $\widehat{y}_{ss} \cos(2\omega_0 t + \eta) \ll V_{DCref}^2$. Thus, (12.99) can be approximated as

$$V_{DC} \approx V_{DCref} \left[1 + \left(\frac{\widehat{y}_{ss}}{2V_{DCref}^2} \right) \cos(2\omega_0 t + \eta) \right]$$

$$= V_{DCref} + V_{ov} \cos(2\omega_0 t + \eta), \qquad (12.100)$$

where the overvoltage, V_{ov}, is formulated as

$$V_{ov} = \frac{\widehat{y}_{ss}}{2V_{DCref}}$$

$$= \frac{\sqrt{(a_1 b_2)^2 + (a_2 b_1)^2 - 2(a_1 b_2)(a_2 b_1) \cos[2(\theta_{01} - \theta_{02}) + (\psi_1 - \psi_2)]}}{2a_2 V_{DCref} C_{eq} \omega_0} |P_{sref1}|.$$

$$(12.101)$$

12.5.5 Generation of Low-Order Harmonics Under an Asymmetrical Fault

As discussed in Section 12.4.2, the effectiveness of the current-control scheme of Figure 12.8 depends on faithful reproduction of the control signals V_{tdref} and V_{tqref} by the VSC. The idea was to translate V_{tdref} and V_{tqref} into modulating signals m_d and m_q such that actual AC-side voltage components V_{td} and V_{tq} are as close to V_{tdref} and V_{tqref} as possible, despite dynamic fluctuations of the DC-bus voltage. This goal was pursued by dividing V_{tdref} and V_{tqref} by the dynamic value $V_{DC}(t)/2$, to generate

m_d and m_q (see Fig. 12.8). It was also mentioned that if the DC-bus voltage is fairly constant, V_{tdref} and V_{tqref} can be divided by the constant gain $V_{DCref}/2$ to yield m_d and m_q. Hereinafter, we refer to the former strategy as the *PWM with DC-bus voltage feed-forward* and to the latter as the *PWM without DC-bus voltage feed-forward*.

One scenario in which the DC-bus voltage continuously fluctuates is when the HVDC system is subjected to an symmetrical fault, as detailed in Section 12.5.4. In this section, we demonstrate that an asymmetrical fault at one of the AC sides of the HVDC system results in the injection of low-order current harmonics into the other (sound) AC system if the PWM without DC-bus voltage feed-forward is employed. By contrast, the harmonics can be effectively mitigated by means of the PWM with DC-bus voltage feed-forward.

12.5.5.1 PWM without DC-Bus Voltage Feed-Forward

Consider the HVDC system of Figure 12.3(a) (or Fig. 12.3(b)) whose, for example, PCC1 is subjected to an asymmetrical fault while PCC2 is undisturbed. Based on (5.10)–(5.12), the PWM modulating signals and the AC-side terminal voltages of the VSC are related as

$$V_{ta2}(t) = \frac{V_{DC}}{2} m_{a2}(t),$$

$$V_{tb2}(t) = \frac{V_{DC}}{2} m_{b2}(t),$$

$$V_{tc2}(t) = \frac{V_{DC}}{2} m_{c2}(t).$$

(12.102)

Since VSC2 operates under a balanced grid condition, $V_{tabc2}(t)$ must also constitute a balanced three-phase waveform. Let $V_{tabc2ref}(t)$ be the three-phase equivalent of the control signals V_{td2ref} and V_{tq2ref} (Fig. 12.8), such that

$$V_{ta2ref}(t) = \widehat{V}_{t2} \cos{(\omega_0 t + \delta_2)},$$

$$V_{tb2ref}(t) = \widehat{V}_{t2} \cos{\left(\omega_0 t + \delta_2 - \frac{2\pi}{3} \right)},$$

$$V_{tc2ref}(t) = \widehat{V}_{t2} \cos{\left(\omega_0 t + \delta_2 - \frac{4\pi}{3} \right)},$$

(12.103)

where $\widehat{V}_{t2}$ and δ_2 are the amplitude and the initial phase angle of $V_{tabc2ref}(t)$, respectively. Then, based on the PWM without DC-bus voltage feed-forward, that is, if V_{DC} is approximated by V_{DCref} in the current-control scheme of Figure 12.8, m_{abc2}

assumes the form

$$m_{a2}(t) = \left(2\widehat{V}_{t2}/V_{DCref}\right) \cos{(\omega_0 t + \delta_2)},$$

$$m_{b2}(t) = \left(2\widehat{V}_{t2}/V_{DCref}\right) \cos{\left(\omega_0 t + \delta_2 - \frac{2\pi}{3}\right)},$$

$$m_{c2}(t) = \left(2\widehat{V}_{t2}/V_{DCref}\right) \cos{\left(\omega_0 t + \delta_2 - \frac{4\pi}{3}\right)}. \tag{12.104}$$

The VSC actual AC-side terminal voltage is then calculated based on (12.102), for V_{DC} and m_{abc2} described, respectively, by (12.100) and (12.104). Thus,

$$V_{ta2}(t) = \widehat{V}_{t2} \cos{(\omega_0 t + \delta_2)} + \widehat{V}_{t2} \left(V_{ov}/V_{DCref}\right) \cos{(2\omega_0 t + \eta)}\cos{(\omega_0 t + \delta_2)},$$

$$V_{tb2}(t) = \widehat{V}_{t2} \cos{\left(\omega_0 t + \delta_2 - \frac{2\pi}{3}\right)}$$
$$+ \widehat{V}_{t2} \left(V_{ov}/V_{DCref}\right) \cos{(2\omega_0 t + \eta)}\cos{\left(\omega_0 t + \delta_2 - \frac{2\pi}{3}\right)},$$

$$V_{tc2}(t) = \widehat{V}_{t2} \cos{\left(\omega_0 t + \delta_2 - \frac{4\pi}{3}\right)}$$
$$+ \widehat{V}_{t2} \left(V_{ov}/V_{DCref}\right) \cos{(2\omega_0 t + \eta)}\cos{\left(\omega_0 t + \delta_2 - \frac{4\pi}{3}\right)}, \tag{12.105}$$

or equivalently

$$V_{ta2}(t) = V_{ta2ref}(t) + \widehat{V}_{t2} \left(V_{ov}/V_{DCref}\right) \cos{(2\omega_0 t + \eta)}\cos{(\omega_0 t + \delta_2)},$$

$$V_{tb2}(t) = V_{tb2ref}(t) + \widehat{V}_{t2} \left(V_{ov}/V_{DCref}\right) \cos{(2\omega_0 t + \eta)}\cos{\left(\omega_0 t + \delta_2 - \frac{2\pi}{3}\right)},$$

$$V_{tc2}(t) = \underbrace{V_{tc2ref}(t)}_{\text{desired}} + \underbrace{\widehat{V}_{t2} \left(V_{ov}/V_{DCref}\right) \cos{(2\omega_0 t + \eta)}\cos{\left(\omega_0 t + \delta_2 - \frac{4\pi}{3}\right)}}_{\text{unwanted}},$$

$$\tag{12.106}$$

where V_{ov}, expressed by (12.101), is the amplitude of the DC-bus double-frequency voltage oscillations due to the fault at PCC1.

Equation (12.106) indicates that V_{tabc2} is not an accurate reproduction of the desired voltage $V_{tabc2ref}$, as it also includes an additional (unwanted) component. Using the identity $\cos x \cos y = (1/2)\left[\cos(x+y) + \cos(x-y)\right]$, we deduce that the unwanted component consists of a negative-sequence fundamental component and a positive-sequence third-order harmonic. The negative-sequence component results in an AC-side voltage imbalance, while the positive-sequence third-order harmonic results in a third-order current harmonic. Based on (12.106), the amplitudes of both the fundamental and third-order voltage harmonics are proportional to the ratio V_{ov}/V_{DCref} that, as indicated by (12.101), can be limited to a small value by selecting a large capacitance for C_{eq}. Low-order voltage/current harmonic generation under unbalanced grid conditions is discussed in Ref. [100].

12.5.5.2 PWM with DC-Bus Voltage Feed-Forward   The impact of the double-frequency component of V_{DC} on the VSC AC-side terminal voltage/current can be effectively mitigated if the PWM with DC-bus voltage feed-forward is utilized, that is, a measure of $V_{DC}(t)$ is used for the calculation of m_d and m_q (see Fig. 12.8). Thus,

$$m_{a2}(t) = \left[2\widehat{V}_{t2}/V_{DC}(t) \right] \cos{(\omega_0 t + \delta_2)},$$

$$m_{b2}(t) = \left[2\widehat{V}_{t2}/V_{DC}(t) \right] \cos{\left(\omega_0 t + \delta_2 - \frac{2\pi}{3} \right)},$$

$$m_{c2}(t) = \left[2\widehat{V}_{t2}/V_{DC}(t) \right] \cos{\left(\omega_0 t + \delta_2 - \frac{4\pi}{3} \right)}. \tag{12.107}$$

Substituting for m_{abc2} in (12.102), from (12.107), one obtains

$$V_{ta2}(t) = \widehat{V}_{t2} \cos{(\omega_0 t + \delta_2)} = V_{ta2ref},$$

$$V_{tb2}(t) = \widehat{V}_{t2} \cos{\left(\omega_0 t + \delta_2 - \frac{2\pi}{3} \right)} = V_{tb2ref},$$

$$V_{tc2}(t) = \widehat{V}_{t2} \cos{\left(\omega_0 t + \delta_2 - \frac{4\pi}{3} \right)} = V_{tc2ref}. \tag{12.108}$$

Equation (12.108) indicates that V_{tabc2} is an exact reproduction of the desired voltage $V_{tabc2ref}$. This outcome is based on two prerequisites: (i) the measurement bandwidth of $V_{DC}(t)$ is adequately larger than $2\omega_0$ and (ii) the PWM switching frequency (in rad/s) is adequately larger than $2\omega_0$. The first requirement ensures that fluctuations of $V_{DC}(t)$ are communicated to $m_{abc2}(t)$ with insignificant attenuation or phase delay. The second requirement is to ensure that $m_{abc2}(t)$ is faithfully tracked and reproduced by the VSC, with low distortions. In practice, the two foregoing conditions, especially the second one, are not fully satisfied, and therefore, the impact of an symmetrical fault at one PCC on the other PCC cannot be entirely eliminated.

Comparing (12.103) with (12.107), we derive

$$m_{a2}(t) = \frac{2}{V_{DC}(t)} V_{ta2ref}, \tag{12.109}$$

$$m_{b2}(t) = \frac{2}{V_{DC}(t)} V_{tb2ref}, \tag{12.110}$$

$$m_{c2}(t) = \frac{2}{V_{DC}(t)} V_{tc2ref}. \tag{12.111}$$

Multiplying both sides of (12.109)–(12.111) by $(2/3)e^{j0}$, $(2/3)e^{j2\pi/3}$, and $(2/3)e^{j4\pi/3}$, respectively, and adding both sides of the resultant equations, based on (4.2) we deduce

$$\overrightarrow{m}_2(t) = \frac{2}{V_{DC}(t)} \overrightarrow{V}_{t2ref}(t). \tag{12.112}$$

Multiplying both sides of (12.112) by $e^{-j\rho}$ and expressing each space phasor in terms of its d- and q-axis components based on $f_{dq} = \overrightarrow{f} e^{-j\rho}$, we obtain

$$m_{d2}(t) = \frac{2}{V_{DC}(t)} V_{td2ref}(t), \tag{12.113}$$

$$m_{q2}(t) = \frac{2}{V_{DC}(t)} V_{tq2ref}(t), \tag{12.114}$$

which represent the dq-frame realization of the DC-bus voltage feed-forward. This realization is illustrated in the block diagram of Figure 12.8.

12.5.6 Steady-State Power-Flow Under an Asymmetrical Fault

For a sound HVDC system, the power components that are exchanged with the PCCs are DC quantities. However, if one of the two PCCs is subjected to an asymmetrical fault, the real power (and also the reactive power) exchanged at that PCC includes a pulsating ripple in addition to the DC component (equation (12.76)). While the DC power component of the faulted PCC is equalized by the real power that is exchanged with the other (sound) PCC, the pulsating power component causes DC-bus voltage fluctuations, (12.100). In this section, we formulate the steady-state real power at the two PCCs, in terms of the real-power command, when either or both PCCs are subjected to an asymmetrical fault.

Based on (12.76), the actual real power at the PCCs is

$$P_{s1-act} = a_1 P_{s1ss} + b_1 P_{s1ss} \cos\left[2\omega_{01}t + (2\theta_{01} + \psi_1)\right], \tag{12.115}$$

$$P_{s2-act} = a_2 P_{s2ss} + b_2 P_{s2ss} \cos\left[2\omega_{02}t + (2\theta_{02} + \psi_2)\right]. \tag{12.116}$$

In a steady state, $P_{s1ss} = P_{sref1}$ and $P_{s2ss} = -(a_1/a_2)P_{sref1}$ based on (12.96). Therefore, (12.115) and (12.116) can be rewritten as

$$P_{s1-act} = a_1 P_{sref1} + b_1 P_{sref1} \cos\left[2\omega_{01}t + (2\theta_{01} + \psi_1)\right], \tag{12.117}$$

$$P_{s2-act} = -a_1 P_{sref1} - \left(\frac{a_1}{a_2}\right) b_2 P_{sref1} \cos\left[2\omega_{02}t + (2\theta_{02} + \psi_2)\right]. \tag{12.118}$$

Equations (12.117) and (12.118) indicate that the DC power components of the two PCCs are equal, no matter if the PCCs are sound or faulted. This implies that the DC-bus voltage has a constant average value in the steady state. However, under an asymmetrical fault condition when $b \neq 0$, the real power exchanged between the HVDC system and the faulted PCC also includes a sinusoidal ripple component. Equations (12.117) and (12.118) show that the amplitude of the ripple component is proportional to the real-power set point, P_{sref1}.

Equations (12.117) and (12.118) also show that if an asymmetrical fault occurs at PCC1, that is, at the PCC corresponding to the real-/reactive-power controller half of the HVDC system of Figure 12.3(b), then $a_1 < 1$ and the absolute value of the DC power component of each PCC is reduced from $|P_{sref1}|$ to $a_1|P_{sref1}|$. Then, while the amplitude of the VSC1 line current does not change, the amplitude of the VSC2 line current reduces to a_1 times its pre-fault value. However, if the fault takes place at PCC2, that is, at the PCC corresponding to the controlled DC-voltage power port half of the HVDC system, the amplitude of the VSC1 line current remains unchanged, whereas the amplitude of the VSC2 line current increases to $1/a_2$ times its prefault value. The reason is that, in this case, the DC component of the power remains the same as $|P_{sref1}|$. Let us assume that the amplitude of the VSC2 (and VSC1) AC-side current is limited to its rated value.[7] Then, the maximum DC power component that VSC2 can transfer from PCC2 to the DC bus and vice versa is $a_2 P_{rated}$, where P_{rated} is the power rating of VSC1 and VSC2. Consequently, if $|P_{sref1}| > a_2 P_{rated}$, the balance of power is lost and the DC-bus voltage drifts. The solution to this problem is to introduce an external protective control loop that limits $|P_{sref1}|$ under such a condition.

EXAMPLE 12.2   DC-Bus Overvoltage Under a Line-to-Ground Fault

Consider a back-to-back HVDC system with the equivalent system of Figure 12.3(b), with $C_{eq} = 500 \ \mu F$, $\omega_{01} = \omega_{02} = 377$ rad/s, and $V_{DCref} = 35$ kV, and also assume that the power-flow set point is $P_{sref1} = 24$ MW while $Q_{sref1} = Q_{sref2} = 0$. We wish to calculate the steady-state DC-bus overvoltage, as well as the actual real power that is exchanged with each PCC, when one of the PCCs is subjected to a line-to-ground fault.

If a line-to-ground fault occurs at PCC1, we have $(a_1, b_1) = (2/3, 1/3)$, $(a_2, b_2) = (1, 0)$, and $\psi_1 = 0$. Then, based on (12.100) and (12.101), we obtain

$$V_{ov} \approx 0.606 \ [kV],$$

$$V_{DC} \approx 35 + 0.606 \cos(754t + \eta) \quad [kV], \tag{12.119}$$

which corresponds to about 1.7% overvoltage. It then follows from (12.117) and (12.118) that

$$P_{s1-act} = 16 + 8 \cos(754t + 2\theta_{01}) \ [MW],$$

$$P_{s2-act} = -16 \quad [MW]. \tag{12.120}$$

---

[7] In practice, the saturation limit on the current amplitude is picked 10–20% higher than the rated value, to allow both dynamic excursions and temporary overloads.

Equation (12.120) indicates that compared to a sound operating condition, a lower amount of (average) real power, that is, 16 MW, flows from Grid2 to Grid1. The reason is that the positive-sequence voltage of PCC1 has dropped to 2/3 of its prefault value (see Example 12.1), while $i_{d1} = 2P_{sref1}/(3\widehat{V}_s)$ has not been changed.

However, if the line-to-ground fault takes place at PCC2, then $(a_1, b_1) = (1, 0)$, $(a_2, b_2) = (2/3, 1/3)$, and $\psi_2 = 0$; thus,

$$V_{ov} \approx 0.909 \ [\text{kV}],$$

$$V_{DC} \approx 35 + 0.909 \cos(754t + \eta) \quad [\text{kV}], \tag{12.121}$$

which corresponds to about 2.6% overvoltage. Furthermore,

$$P_{s1-act} = 24 \ [\text{MW}],$$

$$P_{s2-act} = -24 - 12\cos(754t + 2\theta_{02}) \quad [\text{MW}]. \tag{12.122}$$

In this case, as (12.122) shows, the (average) power flow from Grid2 to Grid1 remains the same as $P_{sref1} = 24$ MW since PCC1 is sound. This, however, is at the expense of 1.5 times increase in i_{d2} compared to a sound condition. The reason is that, due to the fault at PCC2, the positive-sequence voltage of PCC2 becomes 2/3 of the prefault value, while based on the principle of power balance the (average) real power exchanged with PCC2 is dictated to be $P_{sref1} = 24$ MW.

12.5.7 DC-Bus Voltage Control Under an Asymmetrical Fault

Thus far, we have discussed that V_{DC}^2 is a DC quantity under balanced grid conditions. However, it includes a sinusoidal component with the frequency $2\omega_0$, in addition to its DC (average) component y_0, when one or both PCCs are subjected to an asymmetrical fault. In this section, we demonstrate that the DC-bus voltage controller of Figure 12.6 should regulate y_0 rather than V_{DC}^2, to prevent generation of voltage and current harmonics at the sound grid. This can be achieved if the DC-bus voltage compensator $K_V(s)$ (Fig. 12.6) exhibits a large gain drop at frequency $2\omega_0$. If this condition is not satisfied, under asymmetrical faults the HVDC AC current waveforms experience low-order harmonic distortions and imbalance.

12.5.7.1 Steady-State Analysis   Consider the HVDC system of Figure 12.3(a) for which frequencies of both grids are equal to ω_0. The DC-bus voltage is controlled through the control scheme shown at the bottom of Figure 12.6. Assume that PCC1 is subjected to an asymmetrical fault, and the system is under a steady-state condition. Therefore, based on (12.91), V_{DC}^2 is expressed as

$$V_{DC}^2 = y_{0ss} + \widehat{y}_{ss}\cos(2\omega_0 t + \eta). \tag{12.123}$$

Consequently, the error signal $e_V = V_{DCref}^2 - V_{DC}^2$, the output of $K_V(s)$, and the real-power command P_{sref2} also include sinusoidal and DC components. P_{sref2} can be expressed as

$$P_{sref2} = |K_V(j0)|\left(y_{0ss} - V_{DCref}^2\right) + |K_V(j2\omega_0)|\widehat{y}_{ss}\cos(2\omega_0 t + \vartheta), \quad (12.124)$$

where

$$\vartheta = \eta + \angle K_V(j2\omega_0). \tag{12.125}$$

As Figure 12.9 illustrates, the d-axis current reference i_{dref2} is calculated from P_{sref2}, as

$$i_{dref2} = \left(\frac{2}{3\widehat{V}_s}\right)|K_V(j0)|\left(y_{0ss} - V_{DCref}^2\right) + \left(\frac{2}{3\widehat{V}_s}\right)|K_V(j2\omega_0)|\widehat{y}_{ss}\cos(2\omega_0 t + \vartheta).$$

$$\tag{12.126}$$

Based on (12.28) and (12.125), the response of i_{d2} to i_{dref2} is

$$i_{d2ss} = A_0 + A_1\cos(2\omega_0 t + \varrho), \tag{12.127}$$

where

$$A_0 = \left(\frac{2}{3\widehat{V}_s}\right)|K_V(j0)|\underbrace{|G_i(0j)|}_{=1}\left(y_{0ss} - V_{DCref}^2\right)$$

$$= \left(\frac{2}{3\widehat{V}_s}\right)|K_V(j0)|\left(y_{0ss} - V_{DCref}^2\right), \tag{12.128}$$

$$A_1 = \left(\frac{2}{3\widehat{V}_s\sqrt{1 + (2\tau_i\omega_0)^2}}\right)|K_V(j2\omega_0)|\widehat{y}_{ss}, \tag{12.129}$$

$$\varrho = \vartheta - \tan^{-1}(2\tau_i\omega_0). \tag{12.130}$$

The space phasor corresponding to i_{abc2} is

$$\overrightarrow{i_{2ss}} = \left(i_{d2} + j\underbrace{i_{q2}}_{=0}\right)e^{j\rho_2}$$

$$= \left[A_0 + A_1\cos(2\omega_0 t + \varrho)\right]e^{j(\omega_0 t + \theta_{02})}. \tag{12.131}$$

Based on the identity $\cos\theta = (1/2)\left(e^{j\theta} + e^{-j\theta}\right)$, we can rewrite (12.131) as

$$\overrightarrow{i_{2ss}} = A_0 e^{j(\omega_0 t + \theta_{02})} + \underbrace{\frac{A_1}{2} e^{-j(\omega_0 t - \theta_{02} + \varrho)}}_{imbalance} + \underbrace{\frac{A_1}{2} e^{j(3\omega_0 t + \theta_{02} + \varrho)}}_{third\ harmonic}, \qquad (12.132)$$

Equation (12.132) indicates that $\overrightarrow{i_{2ss}}$ and thus i_{abc2} includes a negative-sequence fundamental component and a third-order harmonic component, in addition to the positive-sequence component. The negative-sequence component causes i_{abc2} to be unbalanced, and the third-order harmonic results in distortion.

Equation (12.132) shows that the amplitudes of both the negative-sequence and the third-order harmonic components are proportional to A_1. Thus, one must make A_1 as small as possible to mitigate imbalance and harmonic distortion of i_{abc2}. Based on (12.97) and (12.129), one can select larger DC-bus capacitors, that is, larger C_{eq}, to limit $\hat{y}_{ss}$ and thus A_1. However, due to the cost, weight, and footprint, C_{eq} is practically limited.

An elegant and potentially less expensive method for limiting A_1 is to design $K_V(s)$ such that it exhibits a considerable gain attenuation at frequency $2\omega_0$, that is, $|K_V(j2\omega_0)| \ll 1$. This is possible, for example, by including one pair of complex-conjugate zeros at $s = \pm 2j\omega_0$ in $K_V(s)$. In case where the frequencies of the two grids are not equal, two pairs of complex-conjugate poles may be located, one at $s = \pm 2j\omega_{01}$ and the other at $s = \pm 2j\omega_{02}$; alternatively, one pair of complex-conjugate zeros may be located at $s = \pm 2j\sqrt{\omega_{01}\omega_{02}}$ if ω_{01} and ω_{02} are fairly close.

As discussed earlier in this chapter, under a balanced condition $V_{DC}^2 (= y_0)$ is a pure DC quantity in steady state. However, if the HVDC system is subjected to an asymmetrical fault, y_0 constitutes only the DC component (average value) of V_{DC}^2. Note that if the sinusoidal component of V_{DC}^2 is significantly attenuated in the feedback signal—to avoid imbalance and distortion of the AC-side current—then y_0 will be the only component of V_{DC}^2 that can be controlled. Substituting in (12.127) for A_0 from (12.128) and assuming that $A_1 = 0$, we have

$$i_{d2ss} = \left(\frac{2}{3\widehat{V}_s}\right) |K_V(j0)| \left(y_{0ss} - V_{DCref}^2\right). \qquad (12.133)$$

Since i_{d2ss} is a limited nonzero variable, $(y_{0ss} - V_{DCref}^2) \longrightarrow 0$ if $|K_V(j0)| \longrightarrow \infty$. This implies that y_0 can be regulated at V_{DCref} with zero steady-state error, if $K_V(s)$ exhibits an infinite DC gain, for example, through a pole at $s = 0$.

Based on the foregoing steady-state analysis, we conclude that the DC-bus voltage compensator, $K_V(s)$, must include one pair of complex-conjugate zeros at $s = \pm j2\omega_0$, and one pole at $s = 0$. Other poles and/or zeros of $K_V(s)$ must be identified through a design procedure, based on the dynamic model of the plant as presented in Section 12.5.7.2.

12.5.7.2 Dynamic Analysis   In the HVDC system of Figure 12.3(a), the DC-bus voltage, V_{DC}^2, is regulated by the controlled DC-voltage power port of Figure 12.6, through P_{s2}. As discussed in Section 12.5.7.1, the DC-bus voltage controller must regulate the DC component of V_{DC}^2, that is, y_0, rather than V_{DC}^2 itself. Otherwise, under unbalanced grid conditions, for example, due to asymmetrical faults, low-order harmonics and imbalance are introduced in the corresponding AC-side current.

Dynamics of y_0 are described by (12.86). In (12.86), y_0 is the output, P_{s1} is the disturbance input, and P_{s2} is the control input, where P_{s1} and P_{s2} are responses of d-axis current controllers of VSC1 and VSC2 to reference signals P_{sref1} and P_{sref2}, respectively. The objective of the control system is to regulate y_0 at V_{DCref}^2. A model of the plant suitable for the control design is deduced by linearizing (12.86) about a steady-state operating point. Thus,

$$\frac{d\tilde{y}_0}{dt} = -\left(\frac{2}{C_{eq}}\right)\left[a_2\tilde{P}_{s2} + \left(\frac{2LP_{s2ss}}{3\widehat{V}_s^2}\right)\frac{d\tilde{P}_{s2}}{dt}\right], \tag{12.134}$$

where superscript $\sim$ and index ss, respectively, represent small-signal perturbations and steady-state values of the variables. Substituting for P_{s2ss} from (12.96) in (12.134), we deduce

$$\frac{d\tilde{y}_0}{dt} = -\left(\frac{2}{C_{eq}}\right)\left[a_2\tilde{P}_{s2} - \left(\frac{2a_1LP_{sref1}}{3a_2\widehat{V}_s^2}\right)\frac{d\tilde{P}_{s2}}{dt}\right], \tag{12.135}$$

where the a_1 and a_2 are unity when PCC1 and PCC2 are sound, and are less than unity when PCCs are subjected to asymmetrical faults (see Example 12.1). The Laplace transform of (12.135) provides the transfer function $G_V(s) = \tilde{y}_0/\tilde{P}_{s2}$ as

$$G_V(s) = \frac{\tilde{y}_0(s)}{\tilde{P}_{s2}(s)} = -a_2\left(\frac{2}{C_{eq}}\right)\frac{\tau s + 1}{s}, \tag{12.136}$$

where the time constant τ is

$$\tau = -\left(\frac{a_1}{a_2^2}\right)\left(\frac{2LP_{sref1}}{3\widehat{V}_s^2}\right). \tag{12.137}$$

Equations (12.136) and (12.137) constitute a linearized model for controlling (the DC component of) the DC-bus voltage of a back-to-back HVDC system subjected to an unbalanced grid condition. It should be noted that (12.43) and (12.48), which describe the plant dynamics for a balanced grid condition, are the special cases of (12.136) and (12.137) where a_1 and a_2 are both are equal to 1.

Figure 12.15 illustrates a control block diagram of the DC-bus voltage control loop based on the model of the plant for an unbalanced grid condition, that is, (12.136) and (12.137). The control block diagram of Figure 12.15 is structurally the same as its counterpart under a balanced grid condition of Figure 12.12. In both control systems of Figures 12.15 and 12.12, P_{s2} is the plant input. However, under a balanced grid condition, P_{s2} controls the DC-bus voltage, that is, V_{DC}^2, whereas under an unbalanced

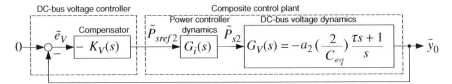

FIGURE 12.15 Control block diagram of the DC-bus voltage regulator for a back-to-back HVDC system subjected to an asymmetrical fault.

grid condition, P_{s2} controls merely the DC component of the DC-bus voltage, that is, y_0. Note that y_0 is not directly measurable but is embedded in V_{DC}^2. As discussed in Section 12.5.7.1, the double-frequency ripple component of V_{DC}^2, if penetrated into P_{sref2} and P_{s2}, results in imbalance and harmonic distortion of the corresponding AC-side current. Therefore, $K_V(s)$ must exhibit a low gain at $2\omega_0$, to attenuate the ripple component of V_{DC}^2 and recover y_0. This can be achieved by including one pair of complex-conjugate zeros at $s = \pm j\omega_0$, in $K_V(s)$. Moreover, to regulate y_0 at V_{DCref}^2 with zero steady-state error, $K_V(s)$ must also include at least one pole at $s = 0$. Other poles and/or zeros of $K_V(s)$ must be located through a design procedure, accounting for the desired closed-loop bandwidth, stability margins, and so on, and based on the dynamic model of the plant described by (12.136) and (12.137).

Based on (12.137), if P_{sref1} is positive, then τ, the time constant corresponding to the zero of $G_V(s)$, is negative and $G_V(s)$ represents a non-minimum-phase plant. If τ is negative, $\angle G_V(j\omega)$ decreases as the frequency increases. This, in turn, indicates that the control-loop phase margin is reduced when P_{sref1} is set to a positive value and (real) power flows from Grid2 to Grid1. Based on (12.137), a larger P_{sref1} results in a larger absolute value of τ and thus a lower phase margin. Equation (12.137) also indicates that when P_{sref1} is positive, the largest reduction of $\angle G_V(j\omega)$ corresponds to the case where PCC2 is subjected to an asymmetrical fault while PCC1 is sound, that is, when $a_2 < 1$ and $a_1 = 1$. The destabilizing effect of this phase reduction is, however, outweighed by the drop in the loop gain magnitude due to the factor a_2 (see (12.136) and Fig. 12.15). For example, if PCC2 is subjected to a line-to-ground fault while PCC1 is sound, that is, $a_2 = 2/3$ and $a_1 = 1$, then τ increases to 2.25 times its prefault value, whereas the loop gain reduces to 66% of its prefault magnitude. To design $K_V(s)$, in addition to the normal rated operating points, the operating points corresponding to possible types of asymmetrical fault must also be considered. The guidelines for compensator design are provided in Examples 7.4, 8.1, and 8.4, and are not repeated in this chapter. The next example demonstrates the performance of the back-to-back HVDC system, under normal and faulted grid conditions.

EXAMPLE 12.3   Performance of Back-to-Back HVDC System

Consider the HVDC system of Figure 12.1 for which both AC sides are identical. The rating of the HVDC system is 36 MVA, that is, either a maximum of 36 MW real power can be transferred from one grid to the other one if both sides operate at unity power factor, or each side of the HVDC system can independently exchange a maximum of ± 36 MVAr with the corresponding

TABLE 12.1   HVDC System Parameters; Example 12.3

Parameter	Value	Comment
HVDC system power rating	36 MVA	$\sqrt{P_s^2 + Q_s^2} < 36$
Grid voltage	138 kV (l-l rms)	V'_{gabc}
Grid inductance	0.2 mH	L'_i
Grid frequency	60 Hz	$\omega_0 = 377$ rad/s
Transformer rating	36 MVA	
Transformer voltage ratio	138/17.9 kV	Delta/Y
Transformer leakage inductance		
referred to high-voltage side	112 mH	L'_l
Interface reactor inductance	8.5 mH[8]	L
Interface reactor resistance	75 mΩ	R
Capacitance of DC-side capacitors	500 μF	C_1 and C_2
Switch cell on-state resistance	6.5 mΩ	
Switch cell on-state voltage drop	11.5 V	
Equivalent DC-bus capacitance	500 μF	C_{eq}
DC-bus voltage	35 kV	V_{DCref}

grid when no real power is exchanged. The exchange of a reactive power of $|Q_s| < 36$ MVAr will reduce the real-power rating of the HVDC system, based on $P_{max} = \sqrt{36^2 - Q_s^2}$.

Figure 12.1 shows that each VSC of the HVDC system is a three-level NPC with a capacitive voltage divider (Fig. 12.4). The third-harmonic injected PWM strategy is adopted for the three-level NPCs. Switching frequencies of both three-level NPCs are 1680 Hz. However, PWM carrier waveforms of the two converters are synchronized neither to the corresponding grids nor to each other. For each three-level NPCs, the voltages of the DC-side capacitors are equalized by means of a corresponding partial DC-side voltage equalizing scheme (Fig. 12.11). The total DC-bus voltage of the HVDC system is regulated at $V_{DCref} = 35$ kV. The on-state resistance and voltage drop of each switch cell are about 6.5 mΩ and 11.5 V, respectively.[9] Thus, the effective internal resistance of each VSC is approximately $r_{on} = 2 \times 6.5 = 13$ mΩ. Table 12.1 provides the parameters of the HVDC system.

As Figure 12.1 illustrates, two harmonic filters—one in parallel with PCC1 and the other in parallel with PCC2—are provided for the HVDC system. The function of each harmonic filter is to mitigate the harmonic distortion of the corresponding PCC voltage. Each harmonic filter is composed of two series-resonant RLC circuits as shown in Figure 12.16: one resonant circuit exhibits a low impedance at the 27th-order harmonic and the other resonant circuit exhibits a low impedance at the 29th-order harmonic. For both resonant circuits, $L_{f1} = L_{f2} = 1.93$ mH, $r_{f1} = r_{f2} = 25$ mΩ, and $R_{f1} = R_{f2} = 25$ Ω.

---

[8]Depending on the design specifics and/or operational requirements, this reactance can assume a smaller value or even be replaced by a more elaborate filter configuration.

[9]These data are estimated based on the assumption that each switch cell is composed of nine series-connected 2500 V/2000 A IGBTs, 5SNR-20H2500, from ABB.

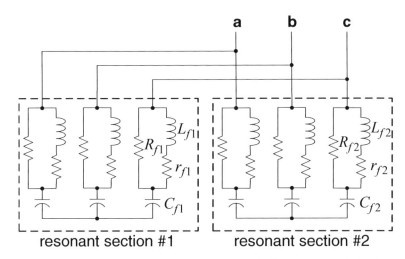

FIGURE 12.16   Schematic diagram of harmonic filters of Example 12.3.

However, $C_{f1} = 5.0$ μF for the circuit tuned to the 27th-order harmonic, whereas $C_{f2} = 4.3$ μF for the other resonant circuit.

Assuming that a closed-loop time constant of $\tau_i = 1.0$ ms is desired for current control, based on (12.25)–(12.27), the compensators of dq-frame current controllers (Fig. 12.8) are

$$k_d(s) = k_q(s) = \frac{8.5s + 88}{s} \quad [\Omega]. \tag{12.138}$$

The transfer function of the feed-forward filters is $G_{ff}(s) = 1/(8 \times 10^{-6}s + 1)$. Transfer functions of the compensator and filter of the DC-side voltage equalizing scheme (Fig. 12.11) are

$$K(s) = 0.05 \quad [(\text{kV})^{-1}],$$

$$F(s) = \frac{s^2 + (3\omega_0)^2}{(s + 3\omega_0)^2} = \frac{s^2 + 1131^2}{s^2 + 2262s + 1131^2}.$$

(See Section 7.4 and Fig. 7.16 for analysis, and Example 7.3 for design guidelines.) Transfer function of the PLL compensator (Fig. 12.7) is

$$H(s) = \frac{18{,}340 \left(s^2 + 754^2\right) \left(s^2 + 166s + 6889\right)}{s \left(s^2 + 1508s + 754^2\right) \left(s^2 + 964s + 232{,}324\right)} \quad [(\text{rad/s})/(\text{kV})].$$

(See Section 8.3.5 and Example 8.1 for design guidelines.). The compensator of the DC-bus voltage controller has the transfer function

$$K_V(s) = 0.022 \left(\frac{s^2 + 754^2}{s^2 + 75.4s + 754^2}\right) \left(\frac{s + 50.9}{s}\right) \quad [\Omega^{-1}].$$

Note that both $H(s)$ and $K_V(s)$ include integral terms and zeros at $\pm 754\,j$ rad/s. Figure 12.17 illustrates response of the HVDC system to the start-up process, a ramp change in the power-flow command, and the reversal of the power-flow direction.

Initially, the HVDC system is disconnected from both grids and the DC capacitors are discharged, the gating pulses of the two VSCs are blocked, all controllers are inactive, and P_{sref1}, Q_{sref1}, and Q_{sref2} are all set to zero. At $t - 0$ s, the transformers are connected to the corresponding grids through 1200 Ω preinsertion resistors (not shown in Fig. 12.1). Consequently, the DC-side capacitors of the VSCs are smoothly charged and the DC-bus voltage, V_{DC}, reaches about 25 kV (Fig. 12.17(d)). At $t = 0.15$ s, the preinsertion resistors are bypassed and TR1 and TR2 are directly connected to the grids. At $t = 0.2$ s, while the power-flow command P_{sref1} is kept at zero (Fig. 12.17(a)), the gating signals are unblocked, the controllers are activated, and the DC-bus voltage reference is ramped up to 35 kV. Thus, V_{DC} tracks V_{DCref} and settles at 35 kV, at about $t = 0.35$ s. Figure 12.17(b) shows that during the start-up, the real power exchanged with PCC1 remains at zero since $P_{sref1} = 0$. However, to increase V_{DC} from 25 to 35 kV, power must be delivered to the DC-bus capacitors. Consequently, as Figure 12.17(c) illustrates, P_{s2-act} becomes slightly negative to facilitate the flow of power from Grid2 to the capacitors.

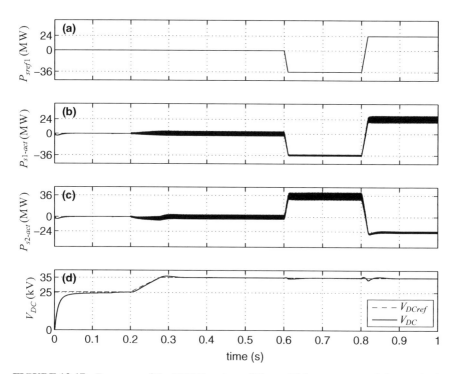

FIGURE 12.17   Response of the HVDC system of Figure 12.1 to start-up and changes in the power-flow command.

At $t = 0.6$ s, P_{sref1} is ramped down from zero to -36 MW, corresponding to a 36 MW real-power transfer from Grid1 to Grid2 (Fig. 12.17(a)). Likewise, $P_{s1\text{-}act}$ follows P_{sref1} from zero to -36 MW (Fig. 12.17(b)) and $P_{s2\text{-}act}$ increases from zero to slightly less than 36 MW (Fig. 12.17(c)); the discrepancy is due to the power losses of the VSCs. Figure 12.17(d) illustrates that due to the intervention of the power feed-forward signal, that is, P_{exteq} in the lower part of Figure 12.6, the deviation of V_{DC} from 35 kV is insignificant. If P_{sref1} is changed stepwise rather than slowly, a large DC-bus voltage is required to prevent the VSCs from entering into the overmodulation mode of operation (see Sections 7.3.4 and 7.3.5 for analysis).

Figure 12.17 also illustrates the response of the HVDC system to a reversal in the power-flow direction. As Figure 12.17(a) and (b) illustrates, at $t = 0.8$ s P_{sref1} is ramped up from -36 MW to 24 MW and is tracked by $P_{s1\text{-}act}$. Thus, $P_{s2\text{-}act}$ changes from about 36 MW to -24 MW, as Figure 12.17(c) shows. Figure 12.17(d) shows that the DC-bus voltage responds to the power-flow reversal with a relatively small deviation from V_{DCref}.

Figure 12.18 provides a close-up of the response of the HVDC system to the power-flow reversal. Figure 12.18(a) shows that P_{sref1} is changed from -36 to 24 MW with a rate of 3.6 MW/ms. Figure 12.18(a) also shows that $P_{s1\text{-}act}$ tracks P_{sref1} with zero steady-state error. However, since $P_{s1\text{-}act}(s)/P_{sref1}(s) = G_i(s)$ is a first-order transfer function, $P_{s1\text{-}act}$ is slightly smaller than P_{sref1} during the

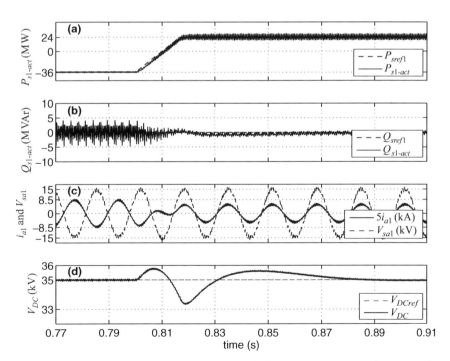

FIGURE 12.18 Response of the HVDC system of Figure 12.1 to a power-flow reversal at $t = 0.8$ s.

period in which P_{sref1} is ramped up. Figure 12.18(b) illustrates that despite the change in P_{s1-act}, Q_{s1-act} remains fairly regulated at its set point of $Q_{sref1} = 0$. This is due to the decoupling mechanism employed in the dq-frame current-control scheme of Figure 12.8. Figure 12.18(c) illustrates the waveforms of i_{a1} and V_{sa1}. Figure 12.18(c) indicates that (i) until $t = 0.8$ s, i_{sa1} and V_{sa1} are 180° phase shifted, since P_{s1-act} is negative, (ii) between $t = 0.8$ and 0.81 s, P_{s1-act} reduces in absolute value, so does the amplitude of i_{a1}, and (iii) after $t = 0.81$ s, P_{s1-act} becomes positive and increases; hence, i_{a1} and V_{sa1} become cophasal, and the amplitude of i_{a1} increases. Figure 12.18(d) shows that V_{DC} exhibits an overshoot of about 2.7%, followed by an undershoot of about −5% due to the power-flow reversal.

It should be noted that P_s is a mathematical quantity defined by (12.30) and is equivalent to P_{s-act}, that is, the actual real power that is exchanged with the grid, only under a balanced grid condition. However, under unbalanced grid conditions, as indicated by (12.76), P_s constitutes a fraction of the DC (average) component of P_{s-act}.

Figure 12.19 illustrates the response of the HVDC system to line-to-ground faults. At $t = 1.05$ s, phase c of PCC1 is subjected to a line-to-ground fault while PCC2 is sound. Consequently, P_{s1-act} converts from a constant 24-MW waveform to a composite waveform that consists of a 16-MW DC component and a double-power-frequency sinusoidal ripple with an amplitude of about 8

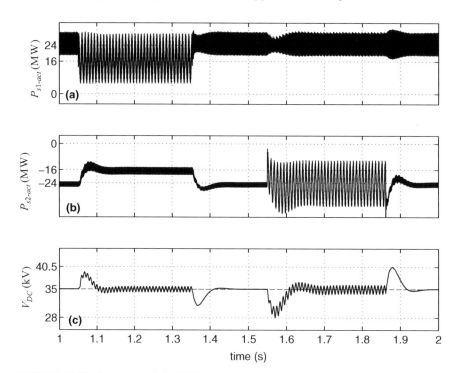

FIGURE 12.19   Response of the HVDC system of Figure 12.1 to line-to-ground faults at PCC1 and PCC2.

MW (Fig. 12.19(a)). Thus, to keep the average of the DC-bus voltage regulated, the DC-bus voltage controller changes $P_{s2\text{-}act}$ from -24 MW to -16 MW, as Figure 12.19(b) shows, to balance the DC component of $P_{s1\text{-}act}$. Figure 12.19(c) illustrates that during the fault the DC-bus voltage assumes a sinusoidal component; moreover, V_{DC} undergoes an overshoot of about 14% following the fault inception. The overshoot, however, decays to zero in about 100 ms and the average value of V_{DC} is settled at 35 kV. The fault is cleared at $t = 1.35$ s. Thereafter, $P_{s1\text{-}act}$, $P_{s2\text{-}act}$, and V_{DC} resume their respective pre-fault forms, as Figure 12.19 illustrates.

At $t = 1.55$ s, phase c of PCC2 is subjected to a line-to-ground fault, but PCC1 remains sound; the fault is cleared at $t = 1.85$ s. During the fault $P_{s1\text{-}act}$ remains constant at 24 MW (Fig. 12.19(a)), whereas $P_{s2\text{-}act}$ assumes a sinusoidal ripple component with an amplitude of 12 MW, in addition to a DC component of about -24 MW (equal to $P_{s1\text{-}act}$) (Fig. 12.19(b)). Figure 12.19(c) shows that V_{DC} includes a sinusoidal ripple component superimposed on an average value of about 35 kV and is disturbed at the instants of fault inception and clearance. Figure 12.19(c) also shows that the overshoot of V_{DC} is about 20% and damped in less than 100 ms.

Figure 12.20 illustrates the VSC line current waveforms and the PCC voltages for the time period over which PCC1 is subjected to the line-to-ground

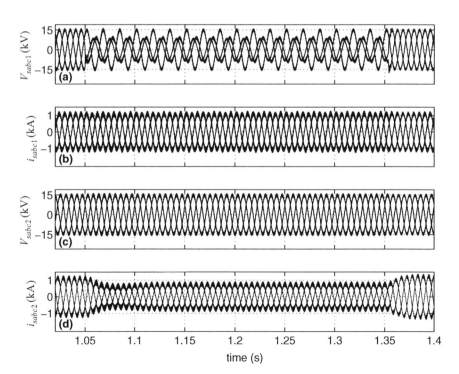

FIGURE 12.20   Line current and PCC voltage waveforms of the HVDC system of Figure 12.1 when the fault occurs at PCC1.

fault. As Figure 12.20(a) shows, during the fault V_{sabc1} is unbalanced. However, due to feed-forward of V_{sd1} and V_{sq1} (see Fig. 12.8), i_{abc1} remains balanced (Fig. 12.20(b)). Moreover, since P_{sref1} is not changed during the faults period, i_{d1} remains constant. Therefore, since $\widehat{i}_1 = \sqrt{i_{d1}^2 + i_{q1}^2}$ and $i_{q1} = 0$, the amplitude of i_{abc1} does not change with respect to the prefault condition, as Figure 12.20(b) illustrates. Figure 12.20(c) and (d) shows that V_{sabc2} and i_{abc2} are balanced since PCC2 is sound. However, to maintain the balance of (average) real power, the DC-bus voltage controller reduces the absolute value of P_{s2-act} from 24 to 16 MW (see Fig. 12.19(b)), through the reduction of the amplitude of i_{abc2} (Fig. 12.20(d)).

Figure 12.21 illustrates the same variables shown in Figure 12.21, but for the case where PCC2 is subjected to the line-to-ground fault. In this case, V_{sabc1} and i_{abc} remain unchanged with respect to the prefault condition, whereas V_{sabc2} is unbalanced. However, in contrast to the case of the fault at PCC1, the amplitude of i_{abc2} increases following the fault inception (Fig. 12.21(d)). The reason is that during the fault at PCC2, the DC component of P_{s2-act} remains equal to the prefault value of P_{s2-act} as shown in Figure 12.19(b). However, the amplitude of the positive-sequence component of V_{sabc2} drops due to the imbalance. Consequently, to transfer the same average power as the prefault condition, i_{abc2} is increased proportionally by the DC-bus voltage controller.

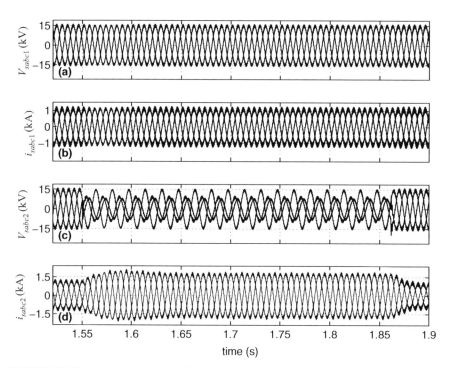

FIGURE 12.21   Line current and PCC voltage waveforms of the HVDC system of Figure 12.1 when the fault occurs at PCC2.

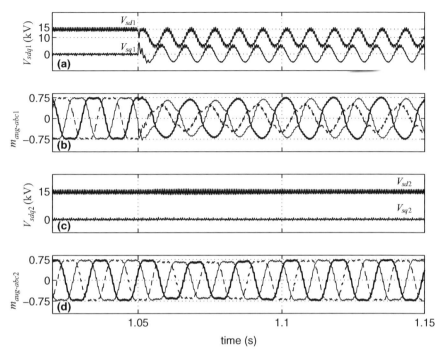

FIGURE 12.22 d- and q-axis components of PCC voltages, and PWM modulating signals of the HVDC system of Figure 12.1 when PCC1 is faulted.

Figure 12.22 illustrates the waveforms of d- and q-axis components of PCC voltages, and the PWM modulating signals of the two VSCs, when PCC1 is subjected to the line-to-ground fault. Figure 12.22(a) shows that subsequent to the fault, V_{sd1} and V_{sq1} assume double-frequency sinusoidal components. Moreover, while the the DC component of V_{sq1} is regulated by the corresponding PLL at zero, the DC component of V_{sd1} drops by 33%. Prior to the fault, the modulating signals of VSC1 constitute a balanced set of waveforms with an injected third-order harmonic (Fig. 12.22(b)). However, following the fault at $t = 1.05$ s, since V_{sd1} and V_{sq1} are incorporated as feed-forward signals (see Fig. 12.8), $m_{aug-abc2}$ includes both the second- and third-order harmonics, as shown in Figure 12.22(b). The feed-forward action, however, ensures that the corresponding line current, that is, i_{abc1}, remains balanced and free of distortions (see Fig. 12.20(b)). Figure 12.22(c) and (d) illustrates that since PCC2 is sound, irrespective of the fault V_{sd2} and V_{sq2} are constant functions of time and $m_{aug-abc2}$ is a balanced three-phase signal.

Figure 12.23 illustrates the DC-bus voltage and the partial DC-side voltages of VSC1 and VSC2 when PCC1 is subjected to the line-to-ground fault. Figure 12.23(a) illustrates that subsequent to the fault, V_{DC} assumes a double-frequency ripple component with an amplitude of about 600 V. Figure 12.23(b) and (c) shows that the partial DC-side voltages of VSC1 and VSC2 are periodic but asymmetrical. The reason is that the voltages of the DC-side capacitors in-

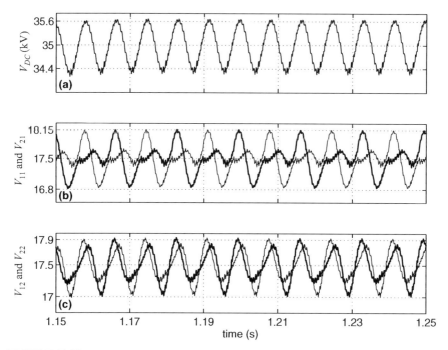

FIGURE 12.23   Total DC-bus voltage and the partial DC-side voltages of the HVDC system of Figure 12.1 when PCC1 is subjected to a line-to-ground fault.

clude both the second- and third-order harmonics. The second-order harmonic is due to the sinusoidal ripple of the net DC-bus voltage. The third-order harmonic is a result of the third-order harmonic of the three-level NPC midpoint current (see Section 6.7 for details). However, as Figures 12.23(b) and (c) illustrate, the DC (average) voltages of the capacitors of each three-level NPC are equalized by the corresponding partial DC-side voltage equalizing scheme.

13 Variable-Speed Wind-Power System

13.1 INTRODUCTION

This chapter deals with the principles of operation and control of a class of grid-connected variable-speed[1] wind-power systems, that is, the doubly-fed asynchronous generator-based wind-power system.[2]

Chapters 7 and 8 introduced the *controlled DC-voltage power port*, and Chapter 10 introduced the *variable-frequency VSC system*. This chapter demonstrates that a grid-connected variable-speed wind-power system, more specifically a doubly-fed asynchronous machine-based wind-power system, in principle, consists of a variable-frequency VSC system and a controlled DC-voltage power port. The variable-frequency VSC system controls the turbine-generator set of the wind-power system, whereas the controlled DC-voltage power port provides the grid interface.

13.2 CONSTANT-SPEED AND VARIABLE-SPEED WIND-POWER SYSTEMS

Wind-power systems are broadly classified as either the *constant-speed* or the *variable-speed* systems. The following subsections provide a brief review of the two classes.

13.2.1 Constant-Speed Wind-Power Systems

Figure 13.1 illustrates a simplified schematic diagram of a constant-speed wind-power system. The wind-power system is composed of a wind turbine that is mechanically coupled, via a gearbox, to an asynchronous generator. Wind turbines naturally rotate at relatively low speeds. Thus, the gearbox is employed to elevate the machine rotor speed, close to the machine synchronous speed. Figure 13.1 also shows that in a constant-speed wind-power system the asynchronous machine is directly interfaced

---

[1] The adjectives *variable speed* and *constant speed* are defined in Section 13.2.
[2] In the technical literature, this class is commonly referred to as the doubly-fed induction generator (DFIG) based wind-power system.

Voltage-Sourced Converters in Power Systems, by Amirnaser Yazdani and Reza Iravani
Copyright © 2010 John Wiley & Sons, Inc.

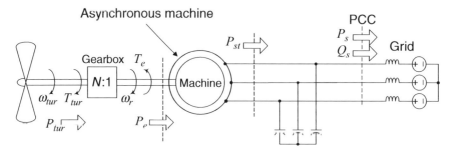

FIGURE 13.1   Schematic diagram of the constant-speed wind-power system.

with the utility grid, and thus the machine synchronous frequency is imposed by the grid. The rotor speed of an asynchronous machine is typically within ±3–8% of the synchronous speed and is thus fairly constant. It should be noted that since the machine operates in the generating mode, its rotor speed is slightly higher than the synchronous speed. An asynchronous machine absorbs reactive power. Therefore, as Figure 13.1 illustrates, the constant-speed wind-power system is equipped with shunt compensation capacitors to maintain the voltage profile and to ensure stable operation, particularly under nonstiff grid conditions.

The constant-speed wind-power system of Figure 13.1 is structurally simple and rugged. However, since the rotor speed is constant, fluctuations in the wind speed and the turbine power are directly transferred to the asynchronous machine and translate into power/voltage fluctuations. This also subjects the drive train and the machine to excessive mechanical and electrical stresses. Moreover, if the grid is not adequately stiff, as in the case of remote wind system installations, the current fluctuations typically cause voltage excursions and flicker [101]. A more noticeable demerit of a constant-speed wind-power system is its relatively poor energy capturing capability and low capacity factor.[3] We will take a closer look at these characteristics in Section 13.3.

13.2.2   Variable-Speed Wind-Power Systems

Figures 13.2(a)–(c) illustrate simplified schematic diagrams of three dominant types of variable-speed wind-power systems. Figure 13.2(a) shows the schematic diagram of a variable-speed wind-power system based on the asynchronous machine. The machine frequency and rotor speed are adjusted by a power-electronic converter system that also enables the flow of real power from the variable-frequency machine to the constant-frequency utility grid.

Figure 13.2(b) illustrates a schematic diagram of a variable-speed wind-power system based on the doubly-fed asynchronous machine. In the wind-power system of

---

[3]The capacity factor of a generator, for example, a wind-power system, is defined as the ratio of its actual delivered energy over a period of time to the energy that could have been delivered over the same period, if the generator had operated at its rated power. Obviously, the capacity factor is also a function of the demand. For a wind-power system, we assume that the rated power is demanded. Typically, a wind-power system has a capacity factor of about 20–40%.

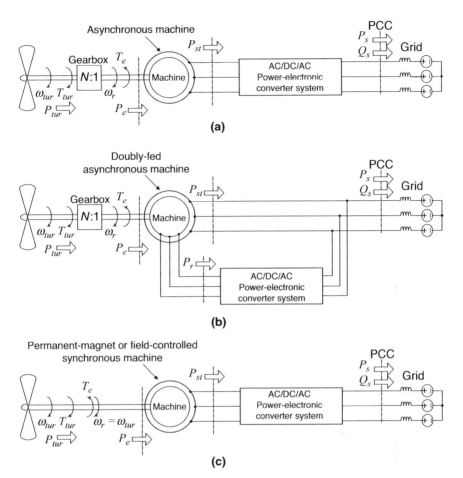

FIGURE 13.2 Schematic diagrams of three main classes of variable-speed wind-power systems: (a) based on the asynchronous machine and power-electronic converter, (b) based on the doubly-fed asynchronous machine and reduced rating power-electronic converter, and (c) based on the gearless-driven synchronous machine and power-electronic converter.

Figure 13.2(b), a power-electronic converter system adjusts the excitation frequency of the machine rotor circuit. The power-electronic converter system also permits a bidirectional power exchange between the machine rotor circuit and the utility grid. Figure 13.2(b) also shows that the machine stator is directly connected to the utility grid and thus its synchronous frequency is directly dictated by the grid frequency. However, the rotor speed can be varied through the adjustment of the rotor frequency. This electrical configuration is explained in more detail in Section 10.3.2.

Figure 13.2(c) illustrates a schematic diagram of a variable-speed wind-power system based on the synchronous machine. The principle of operation of the wind-power system of Figure 13.2(c) is conceptually the same as that of the system of Figure 13.2(a). In the system of Figure 13.2(c), the power-electronic converter system adjusts the frequency of the stator circuit excitation to permit a variable rotor speed. The structural difference between the systems of Figures 13.2(a) and (c) is that the gearbox can be eliminated in the configuration of Figure 13.2(c) if a low-speed (high-pole) synchronous machine is used. The synchronous machine can be of either the field-controlled type [102] or the permanent-magnet type [103].

13.3 WIND TURBINE CHARACTERISTICS

Operation of a wind turbine can be characterized by its mechanical power, as given by [104, 105]

$$P_{tur} = 0.5\rho A V_W^3 C_p(\lambda, \beta) \quad [W], \tag{13.1}$$

where ρ is the air mass density in kg/m^3, $A = \pi r^2$ is the turbine swept area in m^2, r is the turbine radius in m, and V_W is the wind speed in m/s. Function $C_p(\lambda, \beta)$ is called the *performance coefficient* or *power efficiency* and is smaller than 0.59 (Betz limit) [105]. β and λ are the blade *pitch angle* [106] (in degrees) and the *tip-speed ratio* (dimensionless), respectively. The tip-speed ratio is defined as

$$\lambda = \frac{r\omega_{tur}}{V_W}, \tag{13.2}$$

where ω_{tur} is the turbine angular speed in rad/s. Equation (13.2) indicates that λ is the ratio of the tangential speed of the blades' tips to the wind speed.

Equation (13.1) indicates that for a given turbine, the power is the product of two terms. The first term, $0.5\rho A V_W^3$, is proportional to the cube of the wind speed while the second term, $C_p(\lambda, \beta)$, is a variable quantity. The former cannot be influenced as the wind speed cannot be controlled. However, the latter, that is, $C_p(\lambda, \beta)$, can be manipulated by λ and/or β. However, based on (13.2), λ itself is a function of V_W and ω_{tur}. Therefore, the control of C_p indeed boils down to the control of ω_{tur} and β.

$C_p(\lambda, \beta)$ is a static, highly nonlinear, function of λ and β, for which analytical expressions are available [104, 107]. Figure 13.3 depicts $C_p(\lambda)$ versus λ, for two

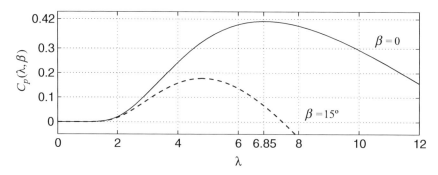

FIGURE 13.3 Performance-coefficient versus tip-speed-ratio characteristic curve of a wind turbine.

values of pitch angle, $\beta = 0$ and $\beta = 15°$. Figure 13.3 shows that $C_p(\lambda = 0) \approx 0$, and that when λ increases C_p also increases until it reaches a peak value at $\lambda = \lambda_{opt}$ (for $\beta = 0$, $\lambda_{opt} \approx 6.85$ in the characteristic of Fig. 13.3); thereafter, a further increase in λ results in a drop in $C_p(\lambda)$. It should be noted that the peak value of $C_p(\lambda)$, that is, $C_p(\lambda_{opt})$, is the largest when β is zero; as β is increased, the peak value of $C_p(\lambda)$ drops (Fig. 13.3), so does the turbine power, P_{tur}. In most wind-power systems, β (i) is set to zero if the electrical power is below the rated value, (ii) is actively controlled to limit the turbine power in case the power exceeds the rated value, and (iii) is set to its maximum value, for example, 90°, to stop power generation under extreme wind conditions.

The differences between a constant-speed wind-power system and a variable-speed counterpart can be explained on the basis of properties of $C_p(\lambda)$. In a constant-speed wind-power system, there is no control over λ since ω_{tur} cannot be varied. Consequently, λ becomes a function of V_W, and $C_p(\lambda)$ cannot necessarily assume its maximum value. This results in a nonoptimum turbine power over a wide range of wind speeds. However, as (13.2) indicates, if ω_{tur} is adjusted in proportion to V_W, then λ can be kept constant and equal to λ_{opt} to maximize $C_p(\lambda)$. Thus, for any wind speed, the turbine power assumes its maximum value based on (13.1) in which $C_p(\lambda)$ is maximized.

The power–speed characteristic of a wind turbine can be described based on (13.1). To simplify the analysis, we first introduce a per-unit system as follows:

- The machine nominal power is selected as the base for electrical and mechanical power, P_b.
- The machine nominal electrical frequency is selected as the base for the machine rotor speed, ω_b. Hence, ω_b/N is the base for the turbine speed, where N is the gearbox ratio.
- Based on the aforementioned power and voltage base values, the base for the machine electrical torque is calculated as $T_b = P_b/\omega_b$. Thus, the base for the turbine torque is NP_b/ω_b.

In our subsequent developments, we assume a two-pole machine. Thus, for the design and the analysis purposes, the machine actual number of poles is accounted for as a part of the gearbox ratio. Based on the per-unit system introduced above, assuming a single-mass model for the drive train, the per-unit turbine speed is equal to the per-unit machine rotor speed, that is, $\omega_{turpu} = \omega_{rpu}$. Therefore, throughout the rest of this chapter we use ω_{rpu} and ω_{turpu} interchangeably.

Figure 13.4 illustrates the per-unit power–speed characteristic of a wind turbine, for different wind speeds, at zero pitch angle. Since each characteristic curve in Figure 13.4 corresponds to a particular wind speed, the turbine power, P_{turpu}, becomes a function of only the rotor speed, ω_{rpu}. Figure 13.4 shows that for a given wind speed the turbine power is insignificant at small rotor speeds. However, the power increases as the rotor speed increases, until the power reaches a peak value corresponding to the peak of $C_p(\lambda)$. The power peak occurs at a rotor speed corresponding to λ_{opt}, that is, the tip-speed ratio at which $C_p(\lambda)$ reaches its peak value. Thereafter, a further increase in ω_{rpu} results in a monotonic drop

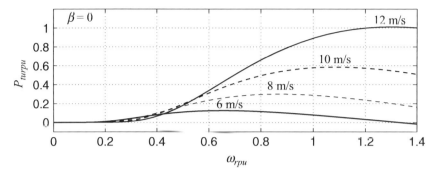

FIGURE 13.4 Power–speed characteristic curves of the wind turbine whose performance coefficient is characterized by Figure 13.3.

in turbine power until the power crosses the zero axis at a certain, relatively high, rotor speed.

Figure 13.4 also shows that the rotor speed corresponding to the maximum power shifts toward a higher value as the wind speed increases. The reason is that, for a constant β, λ_{opt} is a constant parameter ($\lambda_{opt} = 6.85$ for the turbine characterized by Fig. 13.3). Therefore, as the wind speed increases, based on (13.2) the turbine must run at a proportionally higher speed to keep λ constant at λ_{opt}.

The power-speed characteristic of Figure 13.4 reveals the main shortcoming of the constant-speed wind-power system of Figure 13.1 and the advantage of a variable-speed system. Assume that the rotor speed is fixed at 1.3 pu corresponding to the maximum power for $V_W = 12$ m/s. Now if the wind speed drops to 6.0 m/s, the turbine power drops to almost zero, as Figure 13.4 indicates. However, if the rotor speed was decreased to 0.65 pu when V_W dropped to 6.0 m/s, then a power of about 0.15 pu could still be extracted. Thus, a variable-speed wind-power system generates the maximum possible power at any given wind speed and, therefore, offers a relatively larger capacity factor. In the next section, we present a control strategy to automatically find the maximum power point of a turbine.

13.4 MAXIMUM POWER EXTRACTION FROM A VARIABLE-SPEED WIND-POWER SYSTEM

As discussed in Section 13.3, it is often desirable in a variable-speed wind-power system to produce the maximum possible electrical power, at a given wind speed that is lower than the rated wind speed. Once the wind speed exceeds its rated value, the pitch angle is increased by a feedback mechanism to limit the turbine/machine power. In this section, we present a control strategy for maximum power-point tracking.

Below the rated power, the turbine blade pitch angle is set to zero by a pitch angle control scheme. Thus, to maximize the turbine power, λ must be adjusted to λ_{opt} such that $C_p(\lambda, \beta = 0)$ becomes equal to C_{pmax}, where C_{pmax} is the peak value of

$C_p(\lambda)$, that is, $C_{pmax} = C_p(\lambda_{opt})$. Under this condition, based on (13.2), the following equation holds:

$$V_W = \frac{r\omega_{turopt}}{\lambda_{opt}}, \tag{13.3}$$

where ω_{turopt} is the turbine speed corresponding to λ_{opt}. Substituting for V_W from (13.3) in (13.1), we obtain

$$P_{turopt} = \left(\frac{0.5\rho A r^3 C_{pmax}}{\lambda_{opt}^3}\right) \omega_{turopt}^3. \tag{13.4}$$

Dividing both sides of (13.4) by ω_{turopt}, the turbine torque is given by

$$T_{turopt} = \left(\frac{0.5\rho A r^3 C_{pmax}}{\lambda_{opt}^3}\right) \omega_{turopt}^2. \tag{13.5}$$

Equations (13.4) and (13.5) can be expressed in terms of per-unit values as

$$P_{turopt\text{-}pu} = k_{opt}\omega_{ropt\text{-}pu}^3, \tag{13.6}$$

$$T_{turopt\text{-}pu} = k_{opt}\omega_{ropt\text{-}pu}^2, \tag{13.7}$$

where $\omega_{turopt\text{-}pu}$ is replaced by $\omega_{ropt\text{-}pu}$, and the constant k_{opt} is

$$k_{opt} = \frac{0.5\rho A r^3 \omega_b^3 C_{pmax}}{N^3 P_b \lambda_{opt}^3}. \tag{13.8}$$

Equation (13.6) indicates that under the constant-λ variable-speed regime, the maximum attainable turbine power is proportional to the cube of the turbine speed [86, 88, 108]. Equation (13.7) also shows that under the constant-λ variable-speed regime, the turbine torque must be proportional to the square of the turbine speed [86]. Thus, an algorithm to find the maximum power point (i) must force the turbine torque to change in proportion to the square of the rotor speed and (ii) must ensure that the relationship between the turbine speed and the wind speed is such that $\lambda = \lambda_{opt}$. These two objectives are accomplished if the following relationship is imposed on the machine [88]:

$$T_{epu} = -k_{opt}\omega_{rpu}^2, \tag{13.9}$$

where T_{epu} is machine electrical torque.

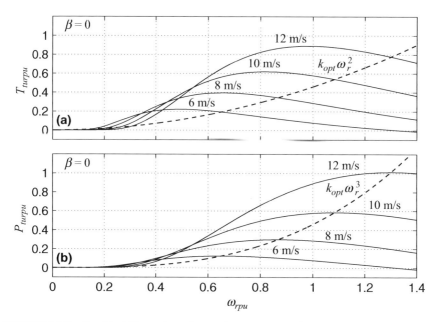

FIGURE 13.5   Graphical illustration of a maximum power point tracking strategy in which (a) the machine torque T_e is changed proportionally to ω_r^2, and (b) in steady state, when the machine torque balances the turbine torque, maximum possible power will be delivered by the turbine.

Mechanical dynamics are described by[4]

$$2H\frac{d\omega_{rpu}}{dt} = \sum T = T_{turpu} + T_{epu} = T_{turpu} - \left(k_{opt}\omega_{rpu}^2\right), \qquad (13.10)$$

where H is the inertia constant in s and defined as

$$H = \left(\frac{1}{2}J\omega_b^2\right)/P_b, \qquad (13.11)$$

where J is the moment of inertia in kgm^2, ω_b is the base value for the angular speed in rad/s, and P_b is the power base in W. A graphical visualization of the right-hand side of (13.10) is given in Figure 13.5(a), illustrating the torque–speed characteristic of a wind turbine superimposed on the curve describing $-T_{epu} = k_{opt}\omega_{rpu}^2$, that is, (13.9). As Figure 13.5(a) indicates, if $\omega_{rpu} < \omega_{ropt-pu}$, then $T_{turpu} > k_{opt}\omega_{rpu}^2$. Consequently, $d\omega_{rpu}/dt$ is positive based on (13.10), the turbine-generator set accelerates, and ω_{rpu} increases. Similarly, if $\omega_{rpu} > \omega_{ropt-pu}$, then $T_{turpu} < k_{opt}\omega_{rpu}^2$, $d\omega_{rpu}/dt$ is negative,

---

[4]To be able to directly utilize the developments of Chapter 10, we adopt the motoring convention for the machine. Thus, the positive rotor and stator currents are considered to be entering the machine, and the machine torque and all external torques are considered as driving torques, that is, they accelerate the machine.

and ω_{rpu} decreases. Therefore, the optimum operating point is a stable one and a steady state can be reached. In the steady state $d\omega_{rpu}/dt = 0$, ω_{rpu} remains constant at $\omega_{ropt-pu}$, and based on (13.10) $T_{turpu} = k_{opt}\omega_{ropt-pu}^2$. This outcome agrees with (13.7), which was developed based on the assumption of a constant λ and a maximized power.

The machine electrical power under the foregoing control strategy can be obtained by multiplying both sides of (13.9) by ω_{rpu}. Thus,

$$P_{epu} = -k_{opt}\omega_{rpu}^3. \tag{13.12}$$

Figure 13.5(b) illustrates the turbine power–speed characteristic that is superimposed on $-P_{epu} = k_{opt}\omega_{rpu}^3$, (13.12). As Figure 13.5(b) shows, for any wind speed, the intersection point of the two curves corresponds to the maximum power for that particular wind speed. Note that, as (13.12) indicates, P_{epu} is negative since a motoring convention has been adopted here.

In practice, wind speed is subject to rapid fluctuations, so is the turbine power. However, due to the inertia of the drive train, ω_{rpu} cannot assume rapid fluctuations. Therefore, the machine power formulated by (13.12) is considerably smoother than the turbine power. Although a steady-state condition is seldom reached in practice, the foregoing algorithm facilitates maximization of the electrical power in an average sense.

13.5   VARIABLE-SPEED WIND-POWER SYSTEM BASED ON DOUBLY-FED ASYNCHRONOUS MACHINE

In this section, we discuss the grid-connected variable-speed wind-power system of Figure 13.2(b) that utilizes the doubly-fed asynchronous machine and a power-electronic converter system as the main electrical building blocks. The converter system is composed of a variable-frequency VSC system (see Chapter 10) and a controlled DC-voltage power port (see Chapter 8).

13.5.1   Structure of the Doubly-Fed Asynchronous Machine-Based Wind-Power System

Figure 13.6 illustrates a more detailed schematic diagram of a wind-power system based on the doubly-fed asynchronous machine. The wind-power system utilizes a wind turbine that is mechanically coupled through a gearbox to the doubly-fed asynchronous machine. The machine stator is directly connected to the grid at the point of common coupling (PCC). The machine rotor circuit is interfaced with the PCC through an AC/DC/AC converter system. The machine side of the converter is a variable-frequency VSC system, whereas the PCC side of the converter is a controlled DC-voltage power port (see Section 8.6).[5] The two VSC systems are interfaced at

---

[5] In the technical literature, these two VSC systems are commonly referred to as *rotor-side converter (RSC)* and *grid-side converter (GSC)*, respectively.

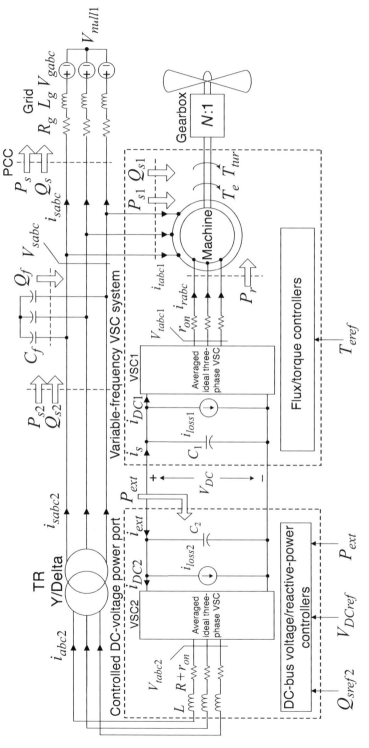

FIGURE 13.6 More detailed schematic diagram of the variable-speed wind-power system of Figure 13.2(b).

their DC sides. It should be noted that, practically, only one capacitor (bank) exists at the common DC bus. However, for the sake of reference to the developments of Chapters 8 and 10, the DC-bus capacitor is illustrated in Figure 13.6 as two separate capacitors, C_1 and C_2, corresponding to the variable-frequency VSC system and the controlled DC-voltage power port, respectively.

In the wind-power system of Figure 13.6, the variable-frequency VSC system controls the machine torque based on the algorithm presented in Section 13.4. Since the machine flux is fairly regulated due to the direct connection between the machine stator and the grid, the variable-frequency VSC system primarily controls the machine torque for the wind-power system and imposes the DC power P_{ext} on the controlled DC-voltage power port. Figure 13.6 also shows that the controlled DC-voltage power port is interfaced with the PCC through a voltage-matching transformer. Moreover, a three-phase set of shunt capacitors, C_f, is connected to the PCC. The capacitors provide a low-impedance path for the switching-frequency harmonics that are generated by both the variable-frequency VSC system and the controlled DC-voltage power port. The capacitors are required to ensure a distortion-free voltage at the PCC, especially under nonstiff grid conditions.

The function of the controlled DC-voltage power port is to regulate the DC-bus voltage irrespective of the power flow direction in the AC/DC/AC converter system, that is, the flow of power from the rotor circuit to the grid, or vice versa. The controlled DC-voltage power port regulates the DC-bus voltage through the control of the real power P_{s2}, through a feedback control mechanism. To enhance the transient response of the DC-bus voltage, a measure of P_{ext} is also included in the voltage regulation loop, as a feed-forward signal. The controlled DC-voltage power port can also independently control the reactive power component Q_{s2}. For example, Q_{s2} can be set to (partly) compensate for the machine magnetizing reactive power. Alternatively, it is possible to regulate the PCC voltage through dynamic control of Q_{s2}, in a closed-loop fashion. This option was extensively discussed in Chapter 11.

If the rotor speed is smaller than the synchronous speed, the power flows from the grid to the rotor circuit (see Section 10.3.2). Consequently, the stator power flowing to the grid is larger than the machine total electrical power. This implies that the power component that is drawn from the grid by the AC/DC/AC converter system enters the rotor and is redirected to the grid through the stator circuit. However, if the rotor speed is higher than the synchronous speed, a component of power flows from the rotor to the grid, and thus the stator power becomes smaller than the machine total power.

13.5.2   Machine Torque Control by Variable-Frequency VSC System

The control of the variable-speed wind-power system of Figure 13.6 consists of two main tasks. The first task is the machine torque control by means of the variable-frequency VSC system, under the assumption that the DC-bus voltage is regulated by the controlled DC-voltage power port. The second task is the DC-bus voltage regulation by means of the controlled DC-voltage power port. The two control tasks are independent of each other and thus treated separately. This section deals with the first task. The second task is treated in the next section.

As discussed in Section 13.4, to maximize the turbine power, the machine torque T_e must be controlled proportionally to the square of the rotor speed, and the corresponding control law is formulated by (13.9). However, the machine torque is controllable only through the torque reference command, T_{eref} (Fig. 13.6). Thus, to implement the power maximization strategy, we specify $T_{eref-pu}$ based on (13.9), as

$$T_{eref-pu} = -k_{opt}\omega_{rpu}^2. \tag{13.13}$$

The underlying assumptions here are that (i) ω_{rpu} cannot have rapid changes due to the machine inertia and therefore $T_{eref-pu}$ varies relatively slowly, and (ii) the machine torque control scheme has a fast dynamic response. The two foregoing assumptions enable us to assume that the machine torque instantly tracks its reference value, that is, $T_{epu} = T_{eref-pu}$, and therefore (13.13) and (13.9) are equivalent. Detailed analysis and design of the variable-frequency VSC system for the doubly-fed asynchronous machine were extensively discussed in Section 10.3.2.

Figure 13.7 illustrates a schematic diagram focusing on the variable-frequency VSC system part of the wind-power system of Figure 13.6. In the system of Figure 13.7, the controlled DC-voltage power port is represented by a voltage source. This virtual voltage source must provide a voltage whose minimum (transient) value still permits the operation of the variable-frequency VSC system and prevents the

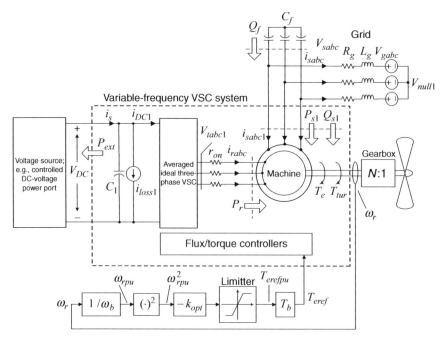

FIGURE 13.7   Simplified schematic diagram of the wind-power system of Figure 13.6 showing the torque-command generator in detail, but the controlled DC-voltage power port as a voltage source.

overmodulation, and whose maximum value is safely below the breakdown voltage of the switch cells of the two VSCs. In order not to require an excessively high DC-bus voltage, the rotor-to-stator windings turns ratio should also be optimized. Note that in the doubly-fed asynchronous machine, the rotor voltage is the largest at zero rotor speed and is zero at a rotor speed equal to the synchronous speed. Therefore, the choice of the DC voltage range and the rotor-to-stator turns ratio must take into consideration the machine parameters and permissible rotor speed variations.

Figure 13.7 also shows that the three shunt capacitors at the PCC (Fig. 13.6) are considered as part of the variable-frequency VSC system. This is done arbitrarily to emphasize that the machine reactive power can be partly compensated by these capacitors. It should be remembered that the shunt capacitors are primarily to suppress the switching-frequency harmonics generated by the two VSC systems of the wind-power system.

Figure 13.7 indicates that the per-unit torque command is determined based on (13.13). The torque command is saturated to an upper limit, to ensure that the machine is not overloaded if ω_{rpu} exceeds 1.0 pu, for example, under extreme wind conditions.

13.5.3 DC-Bus Voltage Regulation by Controlled DC-Voltage Power Port

In the wind-power system of Figure 13.6, the DC-bus voltage regulation is carried out by the controlled DC-voltage power port. The controlled DC-voltage power port also enables bidirectional power exchange between the grid and the variable-frequency VSC system or, more specifically, between the grid and the machine rotor circuit.

As noted earlier, the controlled DC-voltage power port and the variable-frequency VSC system can be considered as two independent entities in terms of their assigned control tasks. The variable-frequency VSC system (Fig. 13.7) is designed and optimized based on the machine torque control requirements, the range of rotor speed variations, and the machine parameters. The design assumes that the DC-bus voltage is within the range that ensures sound operation of the variable-frequency VSC system. This is achieved by the controlled DC-voltage power port for which the variable-frequency VSC system appears as an exogenous (DC) system, that is, an energy sink or source. In terms of operation and control, the type of the exogenous system, whether a variable-frequency VSC system or else, is of no significance to the controlled DC-voltage power port; all that matters is the power that is exchanged between the exogenous system and the controlled DC-voltage power port, that is, P_{ext}. Thus, in Figure 13.8 the variable-frequency VSC system part of the wind-power system of Figure 13.6 is represented by a power source.

As indicated in Figure 13.8, the variable-frequency VSC system imposes the power P_{ext} on the DC capacitor of the controlled DC-voltage power port, C_2. Depending on the operating condition, P_{ext} can be positive or negative in both transients and steady state. The controlled DC-voltage power port regulates the DC-bus voltage by adjusting the real-power component P_{s2}. If V_{DC} is larger than its respective reference value, P_{s2} is increased, and vice versa. In the controlled DC-voltage power port, the reactive-power component Q_{s2} can be controlled independently. Q_{s2} can be set to an

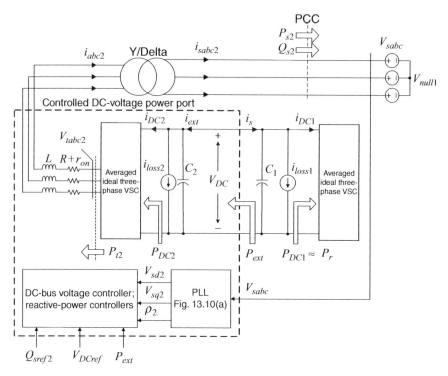

FIGURE 13.8 Simplified schematic diagram of the wind-power system of Figure 13.6 showing the controlled DC-voltage power port in detail, but the variable-frequency VSC system as an external device exchanging power with the controlled DC-voltage power port.

arbitrary value within the rating of the VSC. For example, it can be set to compensate for the machine magnetizing power. Alternatively, Q_{s2} can be dynamically controlled in a closed-loop manner to regulate the PCC voltage or the power factor of the wind-power system. The principles of operation and control of the controlled DC-voltage power port were discussed in Section 8.6.

As illustrated in Figure 13.8, the PCC voltage for the controlled DC-voltage power port is represented by an ideal three-phase voltage source, V_{sabc}, whose neutral point voltage is denoted by V_{null1}. This representation is made since the transmission line inductance and resistance, L_g and R_g, are already considered as part of the variable-frequency VSC system of Figure 13.7. However, it should be remembered that the amplitude and phase angle of V_{sabc} are functions of the real and reactive power that flow through the line, especially if the line inductance is significant.[6] The impacts of transient and steady-state excursions of V_{sabc} on the performance of the controlled DC-voltage power port are mitigated through the feed-forward actions included in the current-control scheme of the controlled DC-voltage power port (see Section 8.6). Figure 13.8 shows that the feed-forward signals are the d- and q-axis components of

---

[6]To see how such dynamics are formulated and analyzed, the reader may refer to Section 11.4.

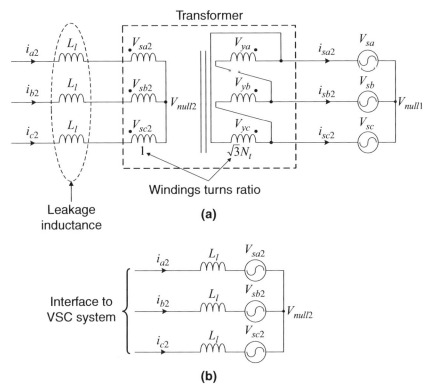

FIGURE 13.9 (a) Schematic diagram of the interface transformer of the controlled DC-voltage power port. (b) Equivalent circuit with the leakage inductance referred to the low-voltage side of the interface transformer.

V_{sabc} that are provided by a phase-locked loop (PLL). The PLL also provides the angle required for *abc*- to *dq*-frame and *dq*- to *abc*-frame transformations for the control of the controlled DC-voltage power port.

13.5.3.1 Interface Transformer and Phase-Locked Loop (PLL) To calculate the parameters of the *d*- and *q*-axis current controllers of a controlled DC-voltage power port, the inductance between the VSC AC-side terminals and the PCC must be known (see Section 8.6). In the controlled DC-voltage power port of Figure 13.8, the PLL is synchronized to V_{sabc}. Hence, the leakage inductance of the interface transformer (Fig. 13.9(a)) effectively adds to the inductance of the VSC system interface reactor L, as shown in Figure 13.9(b). Thus, the effective inductance between the VSC AC-side terminals and the PCC is $L + L_l$. Figure 13.9(b) also indicates that the VSC system is effectively synchronized to the voltage V_{sabc2}, which is phase shifted and scaled with respect to V_{sabc}. Consequently, to analyze the modified model of the controlled DC-voltage power port in which the interface transformer of Figure 13.9(a) is replaced by the equivalent circuit of Figure 13.9(b), we also need to find an equivalent model for the PLL.

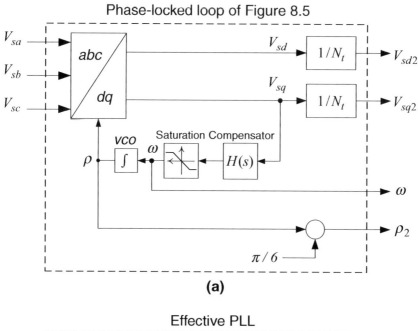

(a)

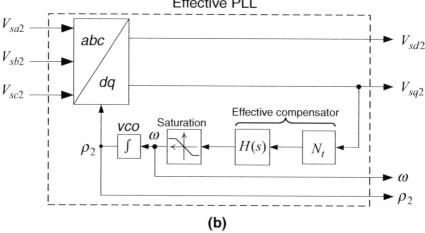

(b)

FIGURE 13.10   Block diagrams of (a) actual implementation of PLL for the VSC system of Figure 13.8 and (b) equivalent model of PLL for the case where the interface transformer of Figure 13.9(a) is replaced by the equivalent circuit of Figure 13.9(b).

Figure 13.10(a) shows a schematic diagram of the PLL employed in the controlled DC-voltage power port of Figure 13.8. This schematic diagram was introduced in Figure 8.5, and is duplicated here for ease of reference. Figure 13.10(a) illustrates that the PLL receives a measure of V_{sabc} and provides the frequency ω, the phase angle ρ_2, and

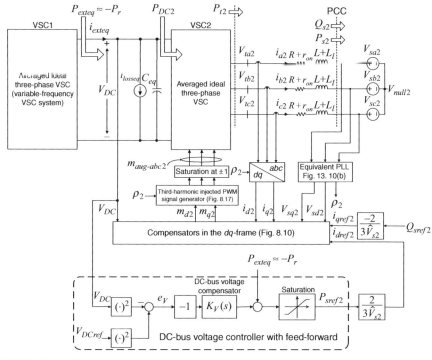

FIGURE 13.11 Schematic diagram of the controlled DC-voltage power port of the wind-power system of Figure 13.6.

the dq-frame components V_{sd2} and V_{sq2}.[7] ρ_2 is obtained by advancing the phase angle of V_{sabc} by $\pi/6$ rad, to compensate for the transformer 30° phase delay. Moreover, V_{sdq2} is calculated by multiplying V_{sdq} by $1/N_t$, where N_t is the transformer voltage ratio. Figure 13.10(b) illustrates the equivalent PLL model corresponding to the modified model of the controlled DC-voltage power port in which the interface transformer of Figure 13.9(a) is replaced by the equivalent circuit of Figure 13.9(b). For the equivalent PLL of Figure 13.10(b), V_{sabc2} is the input while ω, ρ_2, and V_{sdq2} are the outputs.

13.5.4   Compensator Design for Controlled DC-Voltage Power Port

Figure 13.11, which is similar to Figure 8.21, illustrates an equivalent model for the controlled DC-voltage power port of Figure 13.8. Figure 13.11 identifies circuit parameters based upon which the compensator for the DC-bus voltage regulator can be designed and optimized. As outlined in Section 8.6, the compensator design mainly involves (i) the d- and q-axis current decoupling and control, and (ii) the DC-bus voltage regulation and the associated feed-forward scheme.

---

[7]The principles of operation and compensator design for the PLL are extensively discussed in Section 8.3.4.

Compared to Figure 13.8, Figure 13.11 shows that the interface transformer is re-placed by the equivalent circuit of Figure 13.9(b), and therefore, the effective interface inductance of the VSC system is $L + L_l$. Moreover, the resistance of the interface reactors R also includes the transformer ohmic loss. r_{on} represents the typical on-state resistance of one switch cell of the VSC.[8] Note that since the VSC AC-side terminals view V_{sabc2} rather than V_{sabc}, the equivalent PLL model of Figure 13.10(b) is consid-ered for analysis and controller design for the controlled DC-voltage power port of Figure 13.11.

13.5.4.1 *DC-Bus Effective Capacitance and Power Loss*   Since we developed a model for the back-to-back AC/DC/AC converter of the wind-power system based on the models of a variable-frequency VSC system and a controlled DC-voltage power port, the converter DC-bus capacitor was shown as two separate capacitors, C_1 and C_2 in Figures 13.6 and 13.8, respectively; C_1 and C_2 corresponded to the variable-frequency VSC system and the controlled DC-voltage power port, respectively. Sim-ilarly, two parallel current sources were shown to represent the corresponding power losses of the two VSC systems. However, in practice only one DC-bus capacitor (bank) is employed for the AC/DC/AC converter. Moreover, for dynamic analysis and control design purposes, it is convenient to lump the capacitors and the current sources such that their effective values, $C_{eq} = C_1 + C_2$ and $i_{losseq} = i_{loss1} + i_{loss2}$, are attributed exclusively to the controlled DC-voltage power port, as shown in Figure 13.11. The reason is as follows.

In the system of Figure 13.8, consider the part designated as controlled DC-voltage power port; the following equation holds:

$$\frac{d}{dt}\left(\frac{1}{2}C_2 V_{DC}^2\right) = P_{ext} - \underbrace{P_{DC2}}_{P_{t2}} - V_{DC}i_{loss2}, \tag{13.14}$$

where P_{DC2} is the real power delivered to the DC-side terminals of VSC2. Since the averaged ideal model of the VSC is dealt with, P_{DC2} is equal to the power leaving the AC-side terminals of VSC2, that is, P_{t2}.[9] Equation (13.14) represents a power-balance equation and describes the dynamic behavior of V_{DC}, in response to the control variable P_{t2} and two exogenous inputs P_{ext} and $P_{loss2} = V_{DC}i_{loss2}$. Now consider the part of the system that encompasses C_1 and i_{loss1}; the following equation can be written:

$$\frac{d}{dt}\left(\frac{1}{2}C_1 V_{DC}^2\right) = - \underbrace{P_{DC1}}_{\approx P_r} - V_{DC}i_{loss1} - P_{ext}, \tag{13.15}$$

where P_{DC1} is the DC-side terminal power of VSC1 and is equal to the AC-side power of VSC1 since an averaged ideal model is considered. On the other hand, in

---

[8] A two-level VSC is considered.
[9] It was demonstrated in Section 5.3.1 that a nonideal two-level VSC can be considered as the combination of an averaged ideal VSC, a resistor in series with each AC-side terminal of the ideal VSC, and a current source in parallel with the DC-side terminals of the ideal VSC.

TABLE 13.1   Turbine Parameters; Example 13.1

Quantity	Value	Comments
r	35.25 m	Rotor radius
A	3904 m^2	Rotor swept area
C_{pmax}	0.421	
λ_{opt}	6.85	
k_{opt}	0.473	
N	210	Gearbox ratio including the machine number of poles
H	0.5 s	Inertia constant
ρ	1.225 kg/m^3	Air density

view of Figure 13.7, one understands that the AC-side terminal power is equal to the machine rotor power, P_r, if the power loss associated with r_{on} is ignored. Thus, P_{DC1} is approximately equal to P_r. Adding both sides of (13.14) and (13.15), we obtain

$$\frac{d}{dt}\left\{\underbrace{\frac{1}{2}(C_1 + C_2)V_{DC}^2}_{C_{eq}}\right\} = -P_{t2} - V_{DC}\underbrace{(i_{loss1} + i_{loss2})}_{i_{losseq}} + \underbrace{(-P_r)}_{P_{exteq}}. \quad (13.16)$$

Equation (13.16) describes a power-balance equation for the capacitance $C_{eq} = C_1 + C_2$, which is subjected to the (widely variable) charging power P_{exteq}, the (relatively small) discharging power $P_{losseq} = V_{DC}i_{losseq}$, and the (controllable) discharging power P_{t2}. This model is illustrated in Figure 13.11. Based on the model of Figure 13.11, the DC-bus voltage can be regulated by the control of P_{t2}, while P_{losseq} and P_{exteq} are the disturbance inputs to the control system. The transient impact of P_{exteq} on V_{DC} can be substantially mitigated if a measure of P_{exteq} is included in the control system, in the form of a feed-forward compensation. In the model of Figure 13.11, P_{exteq} is approximately equal to the negative of the machine rotor power, that is, $P_{exteq} \approx -P_r$. Thus, the feed-forward strategy can be readily implemented, as P_r can be calculated based on the machine rotor speed, synchronous speed, and the total electrical power, (10.114).[10] Guidelines for the compensator design in the controlled DC-voltage power port were extensively discussed in Section 8.6.2.

EXAMPLE 13.1   1.5 MW Doubly-Fed Asynchronous Machine Based Wind-Power System

Consider the wind-power system of Figure 13.6 with parameters given in Tables 13.1–13.3, designed for an output capacity of 1.5 MW. The wind-power system is interfaced, at the PCC, with a 13.8 kV grid via a 1.5 MVA, 13.8/2.3 kV transformer (not shown in Fig. 13.6). The inductance and resistance of the transmission line referred to the low-voltage side of the transformer are 0.175 mH

---

[10]The machine total power is the product of the machine rotor speed and torque. The machine torque in turn can be calculated from i_{rq}, for example, based on (10.76).

TABLE 13.2   Power-Electronic Converter Parameters; Example 13.1

Quantity	Value	Comments
Transformer TR nominal power	400 kVA	
Transformer TR voltage ratio	2.3/0.6 kV	Delta/Y, $N_t = 3.83$
Transformer TR leakage reactance	239 µH	L_ℓ at 0.6 kV side
Transformer TR ohmic resistance loss	9.0 mΩ	R_ℓ at 0.6 kV side
L	525 µH	Reactor inductance
R	19 mΩ	Reactor resistance including R_ℓ
r_{on}	3.0 mΩ	Switches on-state resistance
$L + L_\ell$	764 µH	Effective interface inductance
$R + r_{on}$	22 mΩ	Effective interface resistance
VSC units switching frequency	2340 Hz	39×60 Hz
C_1, C_2 (DC-bus capacitors)	2000 µF	$C_{eq} = 4000$ µF
C_f (filter capacitors)	25 µF	$Q_f = 49.8$ kVAr

and 66 mΩ, respectively. In addition, the transformer leakage inductance and winding resistance, also transferred to the transformer low-voltage side, are 0.936 mH and 35 mΩ, respectively. Thus, the effective transmission line inductance and resistance are $L_g = 1.11$ mH and $R_g = 101$ $m\Omega$, respectively.

As discussed earlier in this chapter, one of the building blocks of the wind-power system of Figure 13.6 is the variable-frequency VSC system of Figure 13.7. The variable-frequency VSC system uses the flux observer of Figure 10.15

TABLE 13.3   Machine Parameters; Example 13.1

Quantity	Value	Per-Unit Value	Comments
Base power	1.678 MW	1.0	DFIG rated power
Base voltage	1878 V	1.0	Line-to-neutral peak value
Base current	596 A	1.0	Peak value
Base frequency	377 rad/s	1.0	ω_0, ω_b
Base torque	4.451 kN m	1.0	Electrical torque
Rotor/stator turns ratio	1.0		
R_s	29 mΩ	0.00920	
R_r	26 mΩ	0.00825	Includes switches on-state resistance
L_m	34.52 mH	4.130	
L_s	35.12 mH	4.202	
L_r	35.12 mH	4.202	
σ_s	0.01736		
σ_r	0.01736		
σ	0.03384		
τ_s	1.21 s		
τ_r	1.35 s		

for which the parameter τ is 0.066 s. The explanations of the flux observer and its parameters are given in Section 10.3.2. The variable-frequency VSC system also has an embedded dq-frame current controller (Fig. 10.17) whose compensator $k(s)$ has the transfer function

$$k(s) = 15.23\frac{s + 21.86}{s}.$$

For the controlled DC-voltage power port of the wind-power system, that is, Figure 13.11, we have

$$K_V(s) = 299.66\frac{(s + 19.18)}{s(s + 2083)} \quad [\Omega^{-1}],$$

$$\frac{I_{dref2}(s)}{P_{sref2}(s)} = \frac{2}{3\widehat{V}_{s2}} = 1.361 \quad [(kV)^{-1}],$$

$$\frac{I_{qref2}(s)}{Q_{sref2}(s)} = \frac{-2}{3\widehat{V}_{s2}} = -1.361 \quad [(kV)^{-1}].$$

The commands i_{dref2} and i_{qref2} are handed to the dq-frame current-control scheme of Figure 8.10, which employs the compensators $k_d(s)$ and $k_q(s)$, as

$$k_d(s) = k_q(s) = 0.764\frac{s + 28.84}{s} \quad [\Omega].$$

Moreover, the transfer function of the feed-forward filters of the current controller is

$$G_{ff}(s) = \frac{1}{8 \times 10^6 s + 1}.$$

The transfer function of the compensator $H(s)$ of the PLL (Fig. 13.10(a)) is

$$H(s) = \frac{142,680(s^2 + 568,516)(s^2 + 166s + 6889)}{s(s^2 + 1508s + 568,516)(s^2 + 964s + 232,324)} \quad [(rad/s)/kV].$$

Under the nominal grid voltage and with i_{rd} set to zero, the machine reactive power is $Q_{s1} = 400$ kVAr, based on (10.116). The filter capacitor C_f delivers a reactive power of $Q_f = 49.8$ kVAr. Therefore, to fulfill the operation of the wind-power system at unity power factor, the AC/DC/AC converter system is required to deliver a reactive power of 350.2 kVAr to the grid. Since the variable-frequency VSC system and the controlled DC-voltage power port necessarily handle equal amounts of real power, the reactive power of 350.2 kVAr should

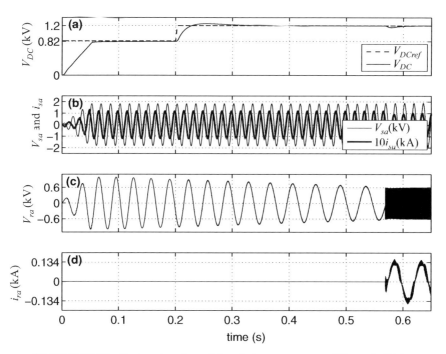

FIGURE 13.12   Start-up transient of the wind-power system of Example 13.1.

also be equally divided between them. This is achieved by setting Q_{sref2} and i_{rdref} at 175.1 kVAr and 0.063 kA, respectively, and ensures that the apparent powers of VSC1 and VSC2 are the same.

Figure 13.12 shows the start-up response of the wind-power system of Figure 13.6. Initially, all controllers are disabled, the gating pulses of both VSC1 and VSC2 are blocked, and the AC-side terminals of VSC1 are disconnected from the machine rotor.[11] However, the AC-side terminals of VSC2 are connected to the corresponding phases of the PCC, through the interface reactors and the transformer TR, Figure 13.6. Each interface reactor is also connected in series with a start-up resistor (not shown in Fig. 13.6) to limit the inrush current. Thus, as Figure 13.12(a) illustrates, the DC-bus capacitor is slowly charged to about 820 V, through the parallel diodes of the VSC2 switch cells. Figure 13.12(b) shows that although the wind-power system is in the start-up mode and exchanges a small amount of real power just to charge the DC-bus capacitor, the peak value of the grid current, i_{sabc}, is about 120 A. This is mainly due to the magnetizing reactive power of the machine of which a small fraction is supplied by C_f while the rest is drawn from the grid. Figure 13.12(b) also

[11] By means of a three-phase circuit breaker or switch, which is not shown in Figure 13.6 or Figure 13.7. This provision is necessary to eliminate the need for an excessively large DC-bus voltage for the scenario where the rotor is at the standstill or spins slowly and therefore the voltage induced in the rotor windings is relatively large.

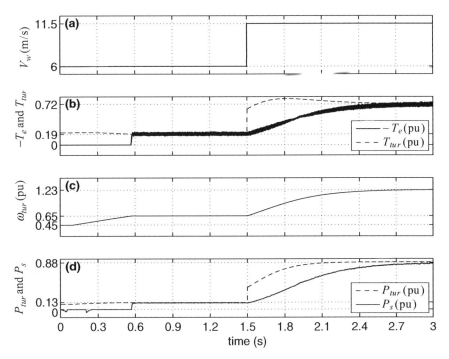

FIGURE 13.13   Response of the wind-power system of Example 13.1 to a sudden increase in wind speed.

reveals a 90° phase shift between the PCC phase voltage and the corresponding grid line current. At $t = 0.2$ s, the start-up resistors are bypassed,[12] the VSC2 gating pulses are unblocked, and the controllers of the controlled DC-voltage power port are activated. Moreover, the DC-bus voltage reference is ramped up to 1200 V. Consequently, as Figure 13.12(a) illustrates, V_{DC} tracks its reference command and settles at 1200 V subsequent to a small overshoot.

Figure 13.12(c) shows that, until $t = 0.569$ s, the rotor voltage is a sinusoidal waveform with a declining amplitude and frequency. The reason is that since the rotor terminals are disconnected from those of VSC1, the rotor assumes an open-circuit voltage due to induction from the stator windings. On the other hand, the rotor speed rises due to wind power and the resultant turbine torque. Consequently, the amplitude and the frequency of the induced voltage decrease as the rotor speed approaches the synchronous speed. During this interval, the rotor current is zero, as Figure 13.12(d) shows, and the machine produces no torque.

At $t = 0.569$ s, that is, when the rotor speed is adequately large and thus the rotor open-circuit voltage is sufficiently low, the AC-side terminals of VSC1 are

---

[12]This is usually done by closing a three-phase switch, the contacts of which are correspondingly parallel to the start-up resistors.

connected to the corresponding phases of the rotor, controllers are enabled, and gating pulses of VSC1 are unblocked. Thus, the rotor phase voltage assumes a pulse-width modulated switched waveform (Fig. 13.12(c)), and a rotor current with an amplitude of 0.134 kA develops (Fig. 13.12(d)). The amplitude of the rotor current in steady state is calculated from $\hat{i}_r = \sqrt{i^2_{rdref} + i^2_{rqref}}$ ($i_{rdref} = 0.063$ kA), and i_{rqref} is determined based on (13.13), the fact that $\omega_{rpu} = 0.65$ pu at $t = 0.569$ s, and based on (10.76).

Figure 13.13 illustrates the overall performance of the wind-power system in a 3s period composed of three subintervals, that is, (i) from $t = 0$ to $t = 0.569$ s, during which the variable-frequency VSC system is disabled and only the controlled DC-voltage power port operates to establish the DC-bus voltage, (ii) from $t = 0.569$ s to $t = 1.5$ s, during which the variable-frequency VSC system is also enabled and the wind speed is constant at 6.0 m/s, and (iii) from $t = 1.5$ s to $t = 3.0$ s, during which the wind speed increases from 6.0 to 11.5 m/s. At the beginning of the first period, the turbine has an initial speed of 0.45 pu. Since the wind speed is rather low, 6.0 m/s, the turbine has a nonzero (but small) torque; however, the variable-frequency VSC system is inactive and therefore the machine torque is zero (Fig. 13.13(b)). Consequently, the turbine speed increases as Figure 13.13(c) illustrates. Figure 13.13(d) shows that despite the fact that the turbine power is positive, no power is delivered to the grid. The reason is that the machine torque is zero over this period. Moreover, as Figure 13.13(d) shows, at two incidents real power is drawn from the grid, that is, P_s becomes negative. These instants correspond to the DC-bus capacitor initial precharging process, from 0 to 820 V, and active charging process, from 820 V to 1200 V.

At $t = 0.569$ s, the turbine and machine speeds reach 0.65 pu, the variable-frequency VSC system is enabled, and the machine rotor circuit is current controlled. Therefore, the machine torque increases and catches up with the turbine torque, as Figure 13.13(b) illustrates. Figure 13.13(c) shows that from $t = 0.569$ s until $t = 1.5$ s, the turbine speed remains relatively constant due to the torque balance. Figure 13.13(d) illustrates that in steady state, the turbine power and the power delivered to the grid become equal to about 0.13 pu, that is, 218 kW.

At $t = 1.5$ s, as Figure 13.13(a) shows, the wind speed assumes a step change from 6.0 to 11.5 m/s. Consequently, the turbine torque and power increase rapidly, as shown in Figure 13.13(b) and (d), respectively. However, unlike the turbine torque, the machine torque increases relatively slowly as Figure 13.13(b) shows. The reason is that the machine torque is changed proportionally to the square of the machine (turbine) speed, based on (13.13), and the turbine speed cannot change rapidly due to its inertia (Fig. 13.13(c)). Due to the gradual increases in the machine torque and speed, the power delivered to the grid also increases slowly (Fig. 13.13(d)). In the steady state, the turbine and machine torques become equal to about 0.72 pu (Fig. 13.13(b)), the turbine (rotor) speed settles at 1.23 pu (Fig. 13.13(c)), and the turbine power reaches a value of about

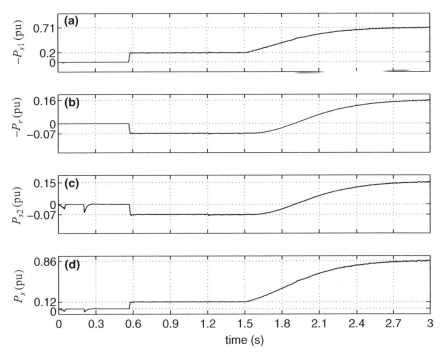

FIGURE 13.14 Real-power flow in the wind-power system of Example 13.1.

0.88 pu, that is, 1477 kW (Fig. 13.13(d)). It should be noted that due to the power losses, the power delivered to the grid is slightly smaller than the turbine power, as Figure 13.13(d) indicates.

Figure 13.14 illustrates the flow of real power through various components of the wind-power system of Figure 13.6, from the start-up to $t = 3.0$ s. Figures 13.14(a)–(d), respectively, show (a) the outgoing stator power, that is, $-P_{s1}$, (b) the outgoing rotor power, that is, $-P_r$, (c) the power delivered by the AC/DC/AC power-electronic converter, that is, P_{s2}, and (d) the net power delivered by the wind-power system to the grid, that is, $P_s = -P_{s1} + P_{s2}$. Figure 13.14(a) shows that the stator power varies between zero and about 0.71 pu, that is, 1191 kW. However, depending on whether the machine speed is below or above the synchronous speed, the rotor power can be either negative or positive, such that it varies between -0.07 pu (-117 kW) and 0.15 pu (254 kW), as Figure 13.14(b) shows. This is an interesting technical feature, further highlighted by Figure 13.14(c), as the real power that the AC/DC/AC-converter must handle is limited to only 0.15 pu. Figure 13.14(d) depicts the waveforms of the net power delivered to the grid, that is, $P_s = -P_{s1} + P_{s2}$. The two steady-state values of 0.12 pu (201 kW) and 0.86 pu (1443 kW) correspond to the wind speeds of 6.0 and 11.5 m/s, respectively; it is noted that the net power holds a cubic proportionality to the wind speed.

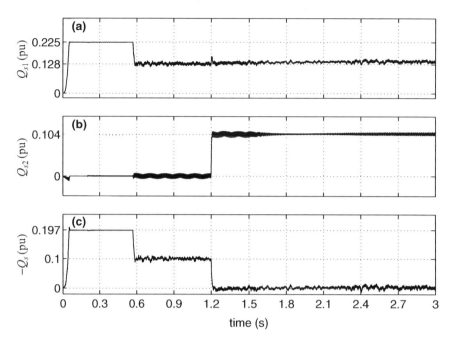

FIGURE 13.15   Reactive-power flow in the wind-power system of Example 13.1

Figure 13.15 illustrates the flow of reactive power among the components of the wind-power system of Figure 13.6, from the start-up to $t = 3.0$ s. For the time period from $t = 0$ to $t = 0.569$ s, the stator absorbs a reactive power of about 0.225 pu (377 kVAr), as Figure 13.15(a) shows. This value is about 95% of the value that we calculated based on the assumption of nominal PCC voltage and $i_{rd} = 0$. The reason for the discrepancy is that, due to the line nonzero impedance, the PCC voltage drops as the machine stator draws reactive power from the grid. This, in turn, results in a drop in the reactive power, proportional to the voltage reduction. As shown in Figure 13.15(b), Q_{s2} is zero from $t = 0$ to $t = 0.569$ s, that is, VSC2 operates at unity power factor. On the other hand, the reactive power supplied by the filter capacitors is about 0.028 pu (47 kVAr) at the reduced PCC voltage. Thus, the resultant reactive power demanded from the grid amounts to 0.197 pu (330 kVAr), as Figure 13.15(c) illustrates.

Figures 13.15(a) and (c) show that, at $t = 0.569$ s, the stator and the grid reactive power reduce to 0.128 pu (215 kVAr) and 0.1 pu (167.8 kVAr), respectively. This is due to the activation of the variable-frequency VSC system and imposition of $i_{rd} = 0.063$ kA. It should be noted that $i_{rd} = 0.063$ kA is adequate to compensate for 0.104 pu (175 kVAr) of the stator reactive power demand, under the nominal PCC voltage. However, due to the PCC voltage drop, the effective compensation is only about 0.097 pu. At $t = 1.2$ s, Q_{sref2} is subjected to a step change from zero to 0.104 pu, which is rapidly tracked by Q_{s2} as Figure 13.15(b) shows. Consequently, the reactive-power demand from

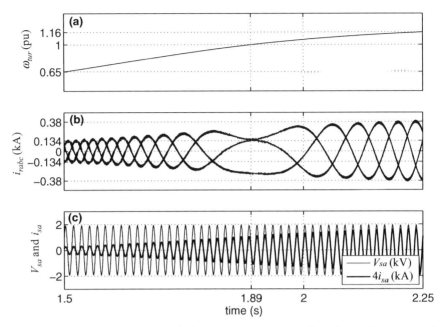

FIGURE 13.16   A close-up of the wind-power system response about the synchronous speed.

the grid reduces to about zero (Fig. 13.15(c)), and thereafter the wind-power system operates at unity power factor.

Figure 13.16 illustrates the waveforms of the rotor current, grid current, and PCC voltage, when the rotor speed is about the synchronous speed (or equivalently 1.0 pu). Figure 13.16(a) shows that, for the time period concerned, the turbine (rotor) speed increases from 0.65 to 1.16 pu. Figure 13.16(b) shows that the amplitude of the rotor current increases as the turbine speed increases. The d-axis component of the rotor current, i_{rd}, is constant at 0.063 kA to compensate for a percentage of the stator reactive power. However, the q-axis component, i_{rq}, is proportional to the square of the turbine speed and changes from 0.118 to 0.376 kA. Therefore, the amplitude of the rotor current changes from 0.134 kA (at $\omega_{turpu} = 0.65$ pu) to 0.380 kA (at $\omega_{turpu} = 1.16$ pu). Based on (10.77), the frequency of the rotor current is equal to the difference between the synchronous (grid) frequency and the rotor speed. Hence, as the rotor speed increases, the frequency of the rotor current decreases. As Figure 13.16(b) shows, at $t = 1.89$ s when the rotor (turbine) speed becomes equal to the synchronous speed, that is, $\omega_{turpu} = 1.0$ pu, the rotor current freezes momentarily. Thereafter, as the turbine speed surpasses the synchronous speed, the frequency of the rotor current becomes negative, and the sequence of the rotor current phases is reversed.

Figure 13.16(c) illustrates the grid phase-a current and the corresponding PCC phase voltage of the wind-power system, for the period of $t = 1.5$ s to $t = 2.25$ s. As indicated earlier, after $t = 1.2$ s the wind-power system operates at unity power factor. Figure 13.16(c) confirms this by showing that the grid current and voltage waveforms are in phase. Figure 13.16(c) also shows that the amplitude of the grid current increases with the increase in the turbine (rotor) speed. Ignoring the system losses, one expects that the amplitude of the grid current is proportional to the cube of the turbine speed. Figure 13.16(c) shows that the PCC voltage is almost constant.

APPENDIX A
Space-Phasor Representation of Symmetrical Three-Phase Electric Machines

A.1 INTRODUCTION

This appendix presents a space-phasor domain dynamic model for a symmetrical three-phase electric machine. The developments mainly concern the (squirrel-cage) asynchronous machine, the doubly-fed asynchronous machine[1], and the nonsalient-pole permanent-magnet synchronous machine (PMSM).

A.2 STRUCTURE OF SYMMETRICAL THREE-PHASE MACHINE

Figure A.1 illustrates a simplified electrical structure of a symmetrical three-phase machine. The rotor and the stator each has three star-connected windings. The stator windings are interfaced with either a three-phase voltage source or a three-phase current source. However, the rotor windings can be either short circuited or excited by a three-phase voltage/current source. In the former case, the machine is called the asynchronous machine (or the squirrel-cage asynchronous machine), whereas in the latter case, the machine is the doubly-fed asynchronous machine. For each phase, the positive current enters the corresponding winding (motoring convention). For each winding set, we assume that the voltages (currents) of phases b and c lag the voltage (current) of phase a, respectively, by $-120°$ and $-240°$.

With reference to Figure A.1, the mechanical angular position is defined with respect to the magnetic axis of the stator phase a winding. Thus, the windings of phases b and c are, respectively, located at positions $120°$ and $240°$. We then define the rotor angle, θ_r, as the angle between the magnetic axes of the rotor and stator phase a

---

[1]In the technical literature, the term *induction machine* is more common than the term *asynchronous machine*. However, the term *asynchronous machine* is more precise, as the induction phenomenon is not exclusive to the asynchronous machine.

Voltage-Sourced Converters in Power Systems, by Amirnaser Yazdani and Reza Iravani
Copyright © 2010 John Wiley & Sons, Inc.

413

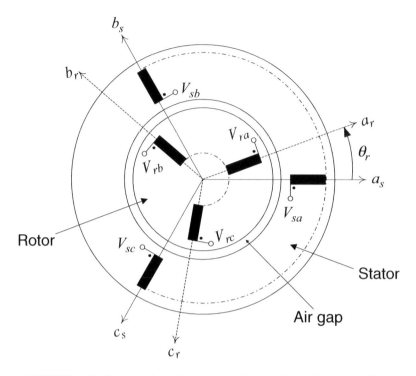

FIGURE A.1   Cross section of a symmetrical three-phase electric machine.

windings. Thus, as Figure A.1 illustrates, the rotor phase a, phase b, and phase c windings are, respectively, located at θ_r, $\theta_r + 120°$, and $\theta_r + 240°$.

A.3   MACHINE ELECTRICAL MODEL

The machine electrical model is developed based on the concept of mutually coupled inductors. For the sake of clarity, we introduce the stator and rotor variables with subscripts s and r, respectively. The following assumptions are also made:

- Neither the rotor nor the stator has saliency, and thus the air gap is uniform.
- The flux density is radial in the machine air gap.
- The flux density due to each winding in the air gap, when no other winding carries a current, is a sinusoidal function of the mechanical angular position.
- The three stator windings are identical, each with an ohmic resistance of R_s. Similarly, the three rotor windings are identical, each with an ohmic resistance of R_r.
- Both the stator and the rotor have infinite permeabilities.
- Magnetic saturation is not considered.

A.3.1   Terminal Voltage/Current Equations

Faraday's law requires that

$$\frac{d\lambda_{sa}}{dt} - V_{sa} - R_s i_{sa},$$

$$\frac{d\lambda_{sb}}{dt} = V_{sb} - R_s i_{sb},$$

$$\frac{d\lambda_{sc}}{dt} = V_{sc} - R_s i_{sc}, \qquad (A.1)$$

and

$$\frac{d\lambda_{ra}}{dt} = V_{ra} - R_r i_{ra},$$

$$\frac{d\lambda_{rb}}{dt} = V_{rb} - R_r i_{rb},$$

$$\frac{d\lambda_{rc}}{dt} = V_{rc} - R_r i_{rc}, \qquad (A.2)$$

where λ_{sabc} and λ_{rabc} are the flux linkages of the stator and rotor windings, respectively. Based on the developments of Chapter 4, (A.1) and (A.2) are equivalent to the following space-phasor equations:

$$\frac{d\vec{\lambda}_s}{dt} = \vec{V}_s - R_s \vec{i}_s, \qquad (A.3)$$

$$\frac{d\vec{\lambda}_r}{dt} = \vec{V}_r - R_r \vec{i}_r. \qquad (A.4)$$

Equations (A.3) and (A.4) are coupled through $\vec{\lambda}_s$ and $\vec{\lambda}_r$. In the subsequent sections, $\vec{\lambda}_s$ and $\vec{\lambda}_r$ are expressed in terms of the machine current.

A.3.2   Stator Flux Space Phasor

The flux linked by each stator winding is a linear function of the stator and rotor currents. For example, the flux linkage of the stator phase a winding is expressed as

$$\lambda_{sa} = L_{ss} i_{sa} + M_{ss} i_{sb} + M_{ss} i_{sc} \qquad (A.5)$$

$$+ M_1(\theta_r) i_{ra} + M_2(\theta_r) i_{rb} + M_3(\theta_r) i_{rc},$$

where L_{ss} is the self-inductance of the stator phase a winding and M_{ss} is the mutual inductance between the stator phase a winding and the stator windings for phases b

and c. M_1, M_2, and M_3 are mutual inductances between the stator phase a winding and the rotor windings for phases a, b, and c, respectively. Due to the uniformity of the air gap and the symmetry of the magnetic structure, L_{ss} and M_{ss} are not functions of the rotor position, θ_r. However, depending on the rotor position, the rotor windings can be aligned differently with respect to the stator phase a winding. Consequently, M_1, M_2, and M_3 are functions of θ_r.

When θ_r is equal to either zero or π, the axes of the stator and rotor phase a windings are aligned. When $\theta_r = 0$, $M_1(\theta_r)$ assumes its maximum (positive) value, and when $\theta_r = \pi$, $M_1(\theta_r)$ assumes its minimum (maximum negative) value. Moreover, since the flux distribution is assumed to be sinusoidal, M_1 is a sinusoidal function of θ_r. Thus,

$$M_1(\theta_r) = M_{sr} \cos \theta_r, \tag{A.6}$$

where M_{sr} is the maximum mutual inductance between a stator winding and a rotor winding [53]. It can be deduced that when $\theta_r = -2\pi/3$, the axes of the rotor phase b and stator phase a windings coincide, and thus the mutual inductance between the two windings is maximized. Similarly, when $\theta_r = 2\pi/3$, the axis of the rotor phase-c winding coincides with that of the stator phase a winding, and thus the mutual inductance between the two windings is maximized. Thus, we deduce

$$M_2(\theta_r) = M_{sr} \cos \left(\theta_r + \frac{2\pi}{3} \right), \tag{A.7}$$

$$M_3(\theta_r) = M_{sr} \cos \left(\theta_r - \frac{2\pi}{3} \right). \tag{A.8}$$

Substituting for M_1, M_2, and M_3 in (A.5), respectively, from (A.6) to (A.8), we obtain

$$\lambda_{sa} = L_{ss}i_{sa} + M_{ss}i_{sb} + M_{ss}i_{sc}$$
$$+ M_{sr} \cos (\theta_r) i_{ra} + M_{sr} \cos \left(\theta_r + \frac{2\pi}{3} \right) i_{rb} + M_{sr} \cos \left(\theta_r - \frac{2\pi}{3} \right) i_{rc}. \tag{A.9}$$

Based on the procedure that was followed to derive (A.9), similar expressions can be derived for λ_{sb} and λ_{sc}. Thus,

$$\lambda_{sb} = M_{ss}i_{sa} + L_{ss}i_{sb} + M_{ss}i_{sc}$$
$$+ M_{sr} \cos \left(\theta_r - \frac{2\pi}{3} \right) i_{ra} + M_{sr} \cos (\theta_r) i_{rb} + M_{sr} \cos \left(\theta_r + \frac{2\pi}{3} \right) i_{rc}, \tag{A.10}$$

$$\lambda_{sc} = M_{ss}i_{sa} + M_{ss}i_{sb} + L_{ss}i_{sc}$$
$$+ M_{sr} \cos \left(\theta_r + \frac{2\pi}{3} \right) i_{ra} + M_{sr} \cos \left(\theta_r - \frac{2\pi}{3} \right) i_{rb} + M_{sr} \cos (\theta_r) i_{rc}. \tag{A.11}$$

Multiplying both sides of (A.9), (A.10), and (A.11), respectively, by $(2/3)e^{j0}$, $(2/3)e^{j2\pi/3}$, and $(2/3)e^{j4\pi/3}$, adding the resultants, and employing the definition of the space phasor based on (4.2), we deduce

$$\vec{\lambda_s} = L_s \vec{i_s} + L_m e^{j\theta_r} \vec{i_r}, \tag{A.12}$$

where

$$L_s = L_{ss} - M_{ss},$$

$$L_m = \left(\frac{2}{3}\right) M_{sr}. \tag{A.13}$$

A.3.3   Rotor Flux Space Phasor

Similarly, the flux linked by the rotor windings can be formulated as

$$\lambda_{ra} = L_{rr}i_{ra} + M_{rr}i_{rb} + M_{rr}i_{rc}$$
$$+ M_{sr} \cos(\theta_r) i_{sa} + M_{sr} \cos\left(\theta_r - \frac{2\pi}{3}\right) i_{sb} + M_{sr} \cos\left(\theta_r + \frac{2\pi}{3}\right) i_{sc}, \tag{A.14}$$

$$\lambda_{rb} = M_{rr}i_{ra} + L_{rr}i_{rb} + M_{rr}i_{rc}$$
$$+ M_{sr} \cos\left(\theta_r + \frac{2\pi}{3}\right) i_{sa} + M_{sr} \cos(\theta_r) i_{sb} + M_{sr} \cos\left(\theta_r - \frac{2\pi}{3}\right) i_{sc}, \tag{A.15}$$

$$\lambda_{sc} = M_{rr}i_{ra} + M_{rr}i_{rb} + L_{rr}i_{rc}$$
$$+ M_{sr} \cos\left(\theta_r - \frac{2\pi}{3}\right) i_{sa} + M_{sr} \cos\left(\theta_r + \frac{2\pi}{3}\right) i_{sb} + M_{sr} \cos(\theta_r) i_{sc}, \tag{A.16}$$

where L_{rr} and M_{rr} are the self- and mutual inductances, respectively. Due to the symmetry of the magnetic structure, L_{rr} and M_{rr} are constant parameters. However, the mutual inductance between a rotor winding and a stator winding is a function of the rotor angle θ_r, as discussed in Section A.3.2. Multiplying both sides of (A.14), (A.15), and (A.16), respectively, by $(2/3)e^{j0}$, $(2/3)e^{j2\pi/3}$, and $(2/3)e^{j4\pi/3}$, adding the resultants, and employing the definition of the space phasor based on (4.2), we deduce

$$\vec{\lambda_r} = L_r \vec{i_r} + L_m e^{-j\theta_r} \vec{i_s}, \tag{A.17}$$

where

$$L_r = L_{rr} - M_{rr}, \tag{A.18}$$

and L_m is defined by (A.13).

A.3.4   Machine Electrical Torque

An expression for the machine electrical torque can be derived based on the principle of power balance [43, 53, 54]. The development yields the compact expression

$$
T_e = \left(\frac{3}{2} L_m\right) Im \left\{ \left(\vec{i_s} \, e^{-j\theta_r}\right) \vec{i_r}^* \right\}
$$

$$
= \left(\frac{3}{2} L_m\right) Im \left\{ \vec{i_s} \left(\vec{i_r} \, e^{j\theta_r}\right)^* \right\}. \tag{A.19}
$$

Equations (A.3), (A.4), (A.12), (A.17), and (A.19) describe the machine dynamics in the space-phasor form. The equations can be expressed in the $\alpha\beta$-frame or in an arbitrary dq-frame. For example, Chapter 10 introduced a dq-frame that renders the machine model suitable for analysis and control design purposes.

A.4   MACHINE EQUIVALENT CIRCUIT

A.4.1   Machine Dynamic Equivalent Circuit

Equations (A.3), (A.4), (A.12), and (A.17) can also be used as a basis to develop an equivalent circuit for the machine. To realize an equivalent circuit, we eliminate the terms $e^{j\theta_r}$ and $e^{-j\theta_r}$ in (A.12) and (A.17), using the following transformations:

$$
\vec{f_r'} = \vec{f_r} \, e^{j\theta_r}, \tag{A.20}
$$

or equivalently

$$
\vec{f_r} = \vec{f_r'} \, e^{-j\theta_r}. \tag{A.21}
$$

In the technical literature, (A.21) is known as *referring the rotor circuit to the stator side*. Based on (A.21), replacing $\vec{V_r}$, $\vec{i_r}$, and $\vec{\lambda_r}$ in (A.4), (A.12), (A.17), and (A.19), respectively, by $\vec{V_r'}$, $\vec{i_r'}$, and $\vec{\lambda_r'}$, we obtain

$$
\frac{d\vec{\lambda_r'}}{dt} = \vec{V_r'} - R_r \vec{i_r'} + \underbrace{j\omega_r \vec{\lambda_r'}}_{rotor\ EMF}, \tag{A.22}
$$

$$
\vec{\lambda_s} = L_s \vec{i_s} + L_m \vec{i_r'}, \tag{A.23}
$$

$$
\vec{\lambda_r'} = L_r \vec{i_r'} + L_m \vec{i_s}, \tag{A.24}
$$

$$T_e = \left(\frac{3}{2} L_m\right) Im \left\{\overrightarrow{i_s}\, \overrightarrow{i_r}^*\right\},\qquad\qquad (A.25)$$

where $\omega_r = d\theta_r/dt$ is the rotor angular velocity. The term $j\omega_r \overrightarrow{\lambda_r'}$ in (A.22) represents a voltage component, proportional to the rotor speed, which can be regarded as the rotor back EMF.

Let us define the rotor and stator leakage factors as

$$\sigma_s = \frac{L_s}{L_m} - 1, \qquad\qquad (A.26)$$

$$\sigma_r = \frac{L_r}{L_m} - 1. \qquad\qquad (A.27)$$

Then, (A.23) and (A.24) can be rewritten as

$$\overrightarrow{\lambda_s} = \sigma_s L_m \overrightarrow{i_s} + L_m \underbrace{(\overrightarrow{i_r'} + \overrightarrow{i_s})}_{\overrightarrow{i_m}}, \qquad\qquad (A.28)$$

$$\overrightarrow{\lambda_r'} = \sigma_r L_m \overrightarrow{i_r'} + L_m \underbrace{(\overrightarrow{i_r'} + \overrightarrow{i_s})}_{\overrightarrow{i_m}}, \qquad\qquad (A.29)$$

Based on (A.3), (A.22), (A.28), and (A.29), Figure A.2 presents an equivalent circuit for the machine. The equivalent circuit of Figure A.2 is known as the *air-gap flux model* or the *T-form model* of the machine [109]. The equivalent circuit of Figure A.2 represents the squirrel-cage asynchronous machine, if $\overrightarrow{V_r'}$ is zero. In the doubly-fed asynchronous machine, in addition to $\overrightarrow{V_s}$ the rotor, voltage vector $\overrightarrow{V_r'}$ is also controllable. The equivalent circuit of Figure A.2 is valid for both dynamic and steady-state conditions.

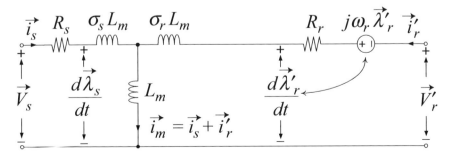

FIGURE A.2 Space-phasor domain equivalent circuit of the symmetrical three-phase machine.

A.4.2 Machine Steady-State Equivalent Circuit

In this section, we simplify the equivalent circuit of Figure A.2 to represent the steady-state behavior of the machine. If the machine rotor is short circuited, that is, $\overrightarrow{V_r'} = 0$, and the stator is excited by a balanced three-phase voltage with angular frequency ω_s, we obtain

$$\overrightarrow{i_s} = \underline{i_s} e^{j\omega_s t}, \tag{A.30}$$

$$\overrightarrow{i_r'} = \underline{i_r'} e^{j\omega_s t}, \tag{A.31}$$

$$\overrightarrow{i_m} = \underline{i_m} e^{j\omega_s t} = \left(\underline{i_s} + \underline{i_r'} \right) e^{j\omega_s t}, \tag{A.32}$$

$$\overrightarrow{V_s} = \underline{V_s} e^{j\omega_s t}, \tag{A.33}$$

where $\underline{f} = \widehat{f} e^{j\theta}$ is a complex number. Substituting for $\overrightarrow{i_s}$, $\overrightarrow{i_r'}$, and $\overrightarrow{i_m}$ in (A.28) and (A.29), from (A.30) to (A.32), we deduce

$$\overrightarrow{\lambda_s} = \underline{\lambda_s} e^{j\omega_s t}, \tag{A.34}$$

$$\overrightarrow{\lambda_r'} = \underline{\lambda_r'} e^{j\omega_s t}, \tag{A.35}$$

where

$$\underline{\lambda_s} = \sigma_s L_m \underline{i_s} + L_m \underline{i_m}, \tag{A.36}$$

$$\underline{\lambda_r'} = \sigma_r L_m \underline{i_r'} + L_m \underline{i_m}. \tag{A.37}$$

Substituting for $\overrightarrow{i_s}$, $\overrightarrow{V_s}$, and $\overrightarrow{\lambda_s}$ in (A.3), respectively, from (A.30), (A.33), and (A.34), calculating the derivative, and eliminating $e^{j\omega_s t}$ from both sides of the resultant, we obtain

$$j\omega_s \underline{\lambda_s} = \underline{V_s} - R_s \underline{i_s}. \tag{A.38}$$

Similarly, considering $\overrightarrow{V_r'} = 0$ and substituting for $\overrightarrow{i_r'}$ and $\overrightarrow{\lambda_r'}$ in (A.22), respectively, from (A.31) and (A.35), calculating the derivative, and eliminating $e^{j\omega_s t}$ from both sides of the resultant, we obtain

$$j\omega_s \underline{\lambda_r'} = -R_r \underline{i_r'} + j\omega_r \underline{\lambda_r'}. \tag{A.39}$$

Equation (A.39) can be rewritten as

$$j\omega_s \underline{\lambda_r'} = -\frac{R_r}{\left(\frac{\omega_s - \omega_r}{\omega_s} \right)} \underline{i_r'}. \tag{A.40}$$

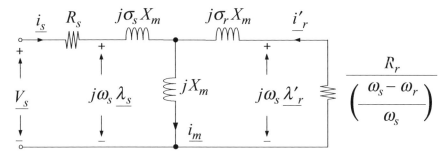

FIGURE A.3   Phasor-domain (sinusoidal steady-state) equivalent circuit of the symmetrical three-phase machine with short-circuited rotor.

Multiplying both sides of (A.36) and (A.37) by $j\omega_s$, we obtain

$$j\omega_s\underline{\lambda}_s = j\sigma_s X_m \underline{i}_s + jX_m \underline{i}_m, \tag{A.41}$$

$$j\omega_s\underline{\lambda}'_r = j\sigma_r X_m \underline{i}'_r + jX_m \underline{i}_m, \tag{A.42}$$

where

$$X_m = L_m\omega_s. \tag{A.43}$$

Equations (A.39) and (A.42) can be represented by the equivalent circuit of Figure A.3, which is the classical steady-state equivalent circuit of the asynchronous machine. Based on the equivalent circuit of Figure A.3, *blocked-rotor* and *no-load* tests, which are carried out to obtain the machine parameters, can be readily described. The term $(\omega_s - \omega_r)/\omega_s$ appearing in (A.40) and in the equivalent circuit of A.3 is referred to as *rotor slip* in the technical literature.

A.5   PERMANENT-MAGNET SYNCHRONOUS MACHINE (PMSM)

The model of the three-phase AC machine, that is, (A.3), (A.4), (A.12), (A.17), and (A.19), can be modified to represent the PMSM. The modification mainly involves the rotor structure; in the PMSM, no physical rotor windings exist and instead a permanent magnet is employed for flux generation. Figure A.4 illustrates a simplified electrical structure of the PMSM.

A.5.1   PMSM Electrical Model

To model the PMSM, we assume that there is no equivalent damper winding effect introduced by the rotor. Moreover, we neglect the rotor saliency and assume a uniform air gap [43]. This model approximately represents the surface-magnet PMSM [54], where the magnets are installed on the rotor surface in which the rotor and stator structures remain the same as those of the symmetrical three-phase AC machine of

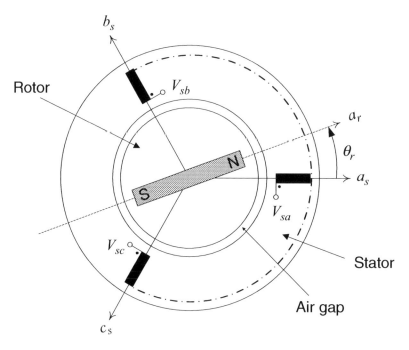

FIGURE A.4 Cross section of the three-phase PMSM.

Section A.2.[2] The stator windings constitute a set of mutually coupled inductors. Thus, the flux linked by each stator winding is a linear function of the winding current and the currents of the other two windings as if the magnet did not exist. The net flux linked by each stator winding also has a component due to the rotor magnet. However, this component is a function of the rotor angle. When the magnetic axis of a stator winding is aligned with that of the rotor permanent magnet, the flux is maximized in absolute value. Since rotor saliency is ignored, the mutual inductance between any two stator windings is constant. The fluxes linked by the stator windings can be formulated as

$$\lambda_{sa} = L_{ss}i_{sa} + M_{ss}i_{sb} + M_{ss}i_{sc} + \lambda_m \cos(\theta_r), \tag{A.44}$$

$$\lambda_{sb} = M_{ss}i_{sa} + L_{ss}i_{sb} + M_{ss}i_{sc} + \lambda_m \cos\left(\theta_r - \frac{2\pi}{3}\right), \tag{A.45}$$

$$\lambda_{sc} = M_{ss}i_{sa} + M_{ss}i_{sb} + L_{ss}i_{sc} + \lambda_m \cos\left(\theta_r + \frac{2\pi}{3}\right), \tag{A.46}$$

where λ_m is the rotor maximum flux. Multiplying both sides of (A.44), (A.45), and (A.46), respectively, by $(2/3)e^{j0}$, $(2/3)e^{j2\pi/3}$, and $(2/3)e^{j4\pi/3}$, adding the

---

[2]A PMSM with interior (buried) magnets is often attributed a nonuniform air gap and thus a salient rotor [54]. A model for the salient-rotor PMSM is presented in Example 4.10.

corresponding sides of the resultants, and employing the definition of the space phasor based on (4.2), we deduce

$$\vec{\lambda}_s = L_s \vec{i}_s + \lambda_m e^{j\theta_r}. \tag{A.47}$$

where L_s is defined by (A.13). Comparing the stator flux equation for the PMSM, that is, (A.47), with its counterpart for the symmetrical three-phase AC machine, that is, (A.12), we realize that $\lambda_m e^{j\theta_r}$ is equivalent to $L_m e^{j\theta_r} \vec{i}_r$. Thus, the electrical torque of the PMSM can be calculated by substituting for $L_m e^{j\theta_r} \vec{i}_r = \lambda_m e^{j\theta_r}$ in (A.19), and we obtain

$$T_e = \left(\frac{3}{2}\lambda_m\right) Im \left\{\vec{i}_s e^{-j\theta_r}\right\}. \tag{A.48}$$

The stator voltage and current are related through (A.1) or its equivalent, (A.3). Equations (A.3), (A.47), and (A.48) describe the dynamics of the PMSM in the space-phasor form. The equations can be expressed in $\alpha\beta$-frame by decomposing the space phasors into their real and imaginary components. The equations also provide a basis for machine control, which is usually performed in a dq-frame. In case of the PMSM, a suitable dq-frame is the one that is synchronized to the rotor angle, θ_r, as already discussed in Chapter 10.

To develop an equivalent circuit for the PMSM, we substitute for $\vec{\lambda}_s$ from (A.47) in (A.3). Thus,

$$L_s \frac{d\vec{i}_s}{dt} = \vec{V}_s - R_s \vec{i}_s - \underbrace{j\lambda_m \omega_r e^{j\theta_r}}_{back\ EMF}. \tag{A.49}$$

Based on (A.49), the equivalent circuit of Figure A.5 can be sketched for the PMSM. This equivalent circuit is valid for both steady-state and dynamic conditions.

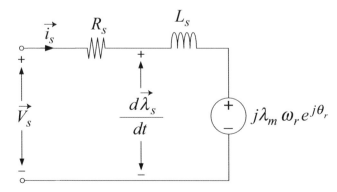

FIGURE A.5   Space-phasor domain equivalent circuit of the PMSM.

A.5.2   PMSM Steady-State Equivalent Circuit

To develop a steady-state equivalent circuit for the PMSM, we assume that the machine stator is excited by a balanced three-phase sinusoidal voltage, and the rotor speed is equal to the angular frequency of the stator voltage. Thus, the machine current is also a balanced three-phase waveform. The foregoing conditions can be expressed as

$$\overrightarrow{V_s} = \underline{V_s} e^{j\omega_s t}, \tag{A.50}$$

$$\overrightarrow{i_s} = \underline{i_s} e^{j\omega_s t}, \tag{A.51}$$

$$\theta_r = \omega_s t + \theta_{r0}, \tag{A.52}$$

where ω_s is the stator excitation frequency. Substituting for $\overrightarrow{i_s}$ and θ_r in (A.47), from (A.51) and (A.52), we deduce

$$\overrightarrow{\lambda_s} = \underline{\lambda_s} e^{j\omega_s t}, \tag{A.53}$$

where

$$\underline{\lambda_s} = L_s \underline{i_s} + \underline{\lambda_m} \implies j\omega_s \underline{\lambda_s} = j(\omega_s L_s)\underline{i_s} + j\omega_s \underline{\lambda_m}, \tag{A.54}$$

$$\underline{\lambda_m} = \lambda_m e^{j\theta_{r0}}. \tag{A.55}$$

Substituting for $\overrightarrow{V_s}$, $\overrightarrow{i_s}$, and $\overrightarrow{\lambda_s}$ in (A.3), from (A.50), (A.52), and (A.53), calculating the derivative, and eliminating the term $e^{j\omega_s t}$ from both sides of the resultant, we conclude

$$j\omega_s \underline{\lambda_s} = \underline{V_s} - R_s \underline{i_s}. \tag{A.56}$$

Equations (A.54) and (A.56) correspond to the equivalent circuit of Figure A.6, which is the classical steady-state equivalent circuit of the PMSM.

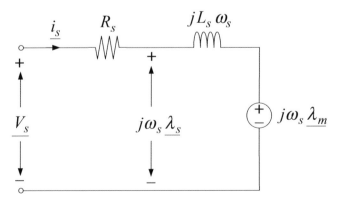

FIGURE A.6   Phasor-domain (steady-state) equivalent circuit of the PMSM.

The machine steady-state electrical torque can be calculated by substituting for $\vec{i_s} = \underline{i_s}e^{j\omega_s t}$ and $\theta_r = \omega_s t + \theta_{r0}$ in (A.48):

$$T_e = \left(\frac{3}{2}\lambda_m\right) Im\left\{\underline{i_s}e^{-j\theta_{r0}}\right\}, \tag{A.57}$$

Let $\underline{i_s} = \hat{i_s}e^{j\theta_i}$. Then, (A.57) can be written as

$$T_e = \left(\frac{3}{2}\lambda_m\right) Im\left\{\hat{i_s}e^{j(\theta_i-\theta_{r0})}\right\}$$

$$= (\frac{3}{2}\lambda_m)\hat{i_s}\sin\underbrace{(\theta_i - \theta_{r0})}_{\delta}. \tag{A.58}$$

The phase difference between the machine current and the machine internal EMF, denoted as $\delta = (\theta_i - \theta_{r0})$ in (A.58), is known as *load angle* in the technical literature. For a given torque, the machine current is minimum when $\delta = \pi/2$ rad.

APPENDIX B
Per-Unit Values for VSC Systems

B.1 INTRODUCTION

It is often more convenient to express a power-electronic converter system in a per-unit term. This can be achieved based on the following per-unitization system.

B.1.1 Base Values for AC-Side Quantities

The base values for a VSC system AC-side quantities are given in Table B.1. As shown in the table, the base voltage for a VSC system is chosen as the peak value of the line-to-neutral voltage of the point of common coupling (PCC); this is in contrast to the conventional power system for which the rms line-to-neutral voltage represents the base voltage. The rated three-phase power is selected as the base power.

B.1.2 Base Values for DC-Side Quantities

The DC-side base values are determined based on those of the AC side. The base power is the same for both DC and AC sides. However, the DC-side base voltage is defined to be two times the AC-side base voltage. This is to obtain the AC-side voltage of 1.0 pu from the DC-side voltage of 1.0 pu, at unity modulation index. The base values for DC-side quantities are summarized in Table B.2.

EXAMPLE B.1   Model of Three-Phase VSC-Based Rectifier

Figure B.1 illustrates a schematic diagram of a three-phase VSC system that operates in the rectifying mode of operation and supplies a DC RL load. The open-loop model of the VSC system is described by

$$L\frac{di_d}{dt} = -Ri_d + L\omega i_q + \frac{1}{2}m_d V_{DC} - v_{sd},$$

$$L\frac{di_q}{dt} = -Ri_q - L\omega i_d + \frac{1}{2}m_q V_{DC} - v_{sq},$$

Voltage-Sourced Converters in Power Systems,   by Amirnaser Yazdani and Reza Iravani
Copyright © 2010 John Wiley & Sons, Inc.

TABLE B.1   Based Values for VSC AC-Side Quantities

Quantity	Symbol and Expression	Description
Power	$P_b = \dfrac{3}{2} V_b I_b$	VA rating of the VSC
Voltage	$V_b = \widehat{V}_s$	Amplitude of the line-to-neutral nominal voltage
Current	$I_b = \dfrac{2P_b}{3V_b}$	Amplitude of the nominal line current
Impedance	$Z_b = \dfrac{V_b}{I_b}$	
Capacitance	$C_b = \dfrac{1}{Z_b \omega_b}$	
Inductance	$L_b = \dfrac{Z_b}{\omega_b}$	
Frequency	$\omega_b = \omega_0$	Usually the power system nominal frequency

TABLE B.2   Based Values for VSC DC-Side Quantities

Quantity	Symbol and Expression	Description
Power	$P_{b-dc} = V_{b-dc} I_{b-dc} = P_b$	Same as the AC-side base power
Voltage	$V_{b-dc} = 2V_b$	
Current	$I_{b-dc} = \dfrac{3}{4} I_b$	
Impedance	$R_{b-dc} = \dfrac{8}{3} Z_b$	
Capacitance	$C_{b-dc} = \dfrac{3}{8} C_b$	
Inductance	$L_{b-dc} = \dfrac{8}{3} L_b$	

$$C\frac{dV_{DC}}{dt} = -i_l - i_{DC} = -i_l - \frac{3}{4}\left(m_d i_d + m_q i_q\right),$$

$$L_l \frac{di_l}{dt} = -R_l i_l + V_{DC}. \tag{B.1}$$

Let us signify a per-unitized value by the underline. Thus, for the AC-side quantities we have

$$L = L_b \underline{L},$$

$$R = Z_b \underline{R},$$

$$v_{sd} = V_b \underline{v_{sd}},$$

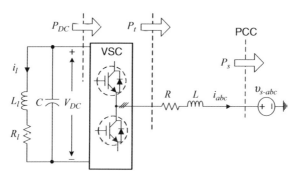

FIGURE B.1   Schematic diagram of a three-phase VSC-based rectifier.

$$v_{sq} = V_b\underline{v_{sq}},$$
$$i_d = I_b\underline{i_d},$$
$$i_q = I_b\underline{i_q},$$
$$\omega = \omega_b\underline{\omega}, \tag{B.2}$$

and for the DC-side quantities, we obtain

$$C = \frac{3}{8}C_b\underline{C},$$
$$L_l = \frac{8}{3}L_b\underline{L_l},$$
$$R_l = \frac{8}{3}Z_b\underline{R_l},$$
$$V_{DC} = 2V_b\underline{V_{DC}},$$
$$i_l = \frac{3}{4}I_b\underline{i_l}. \tag{B.3}$$

Substituting from (B.2) and (B.3) in (B.1), and using the relationships between the base values as given in Table B.1, one deduces the following per-unitized set of equations:

$$\frac{1}{\omega_b}\underline{L}\frac{d\underline{i_d}}{dt} = -\underline{R}\underline{i_d} + \underline{L}\underline{\omega}\underline{i_q} + m_d\underline{V_{DC}} - \underline{v_{sd}},$$

$$\frac{1}{\omega_b}\underline{L}\frac{d\underline{i_q}}{dt} = -\underline{R}\underline{i_q} - \underline{L}\underline{\omega}\underline{i_d} + m_q\underline{V_{DC}} - \underline{v_{sq}},$$

$$\frac{1}{\omega_b}\underline{C}\frac{d\underline{V_{DC}}}{dt} = -\underline{i_l} - (m_d\underline{i_d} + m_q\underline{i_q}),$$

$$\frac{1}{\omega_b} L_l \frac{di_l}{dt} = -R_l i_l + V_{DC}. \tag{B.4}$$

It should be noted that based on the foregoing per-unit system, we do not express the modulating signals m_d and m_q in per unit terms. The reason is that the absolute values of the modulating signals are between zero and unity, and thus expressing them in per-unit terms does not yield more insight. Equation (B.4) indicates that each derivative term of an original equation is premultiplied by the factor $1/\omega_b$ in the corresponding per-unit counterpart. This factor can be avoided if the time is also expressed in per-unit terms using the base value $t_b = 1/\omega_b$. Based on such per-unitization of time, (B.4) assumes the form

$$L\frac{di_d}{dt} = -Ri_d + L\omega i_q + m_d V_{DC} - v_{sd},$$

$$L\frac{di_q}{dt} = -Ri_q - L\omega i_d + m_q V_{DC} - v_{sq},$$

$$C\frac{dV_{DC}}{dt} = -i_l - \left(m_d i_d + m_q i_q\right),$$

$$L_l\frac{di_l}{dt} = -R_l i_l + V_{DC}. \tag{B.5}$$

REFERENCES

1. N. Hingorani and L. Gyugyi, *Understanding FACTS: Concepts and Technology of Flexible AC Transmission Systems*, IEEE Press, 2000.

2. Y. H. Song and A. T. Johns, *Flexible AC Transmission Systems (FACTS)*, IEE, 1999.

3. E. Acha, C. R. Fuerte-Esquivel, H. Ambriz-Perez, and C. Angeles-Camacho, *FACTS: Modelling and Simulation in Power Networks*, Wiley, 2004.

4. E. Acha, V. G. Agelidis, O. Anaya-Lara, and T. J. E. Miller, *Power Electronic Control in Electrical Systems*, Newnes, 2002.

5. R. M. Mathur and R. Varma, *Thyristor-Based FACTS Controllers for Electrical Transmission Systems*, Wiley/IEEE, 2002.

6. X. P. Zhang, C. Rehtanz, and B. Pal, *Flexible AC Transmission Systems: Modelling and Control*, Springer-Verlag, 2006.

7. N. Hatziargyriou, H. Asano, R. Iravani, and C. Marnay, "Microgrids," *IEEE Power and Energy Magazine*, vol. 5, no. 4, pp. 78–94, July/August 2007.

8. H. Akagi, E. H. Watanabe, and M. Arades, *Instantaneous Power Theory and Applications to Power Conditioning*, Wiley/IEEE, 2007.

9. T. Larsson, A. Edris, D. Kidd, and F. Aboytes, "Eagle Pass Back-to-Back Tie: A Dual Purpose Application of Voltage Source Converter Technology," *IEEE Power Engineering Society Summer Meeting*, vol. 3, pp. 1686–1691, July 2001.

10. J. Arillaga, *High Voltage Direct Current Transmission*, IEE Power Engineering Series 6, Peter Peregrinus Ltd., 1983.

11. V. Sood, *HVDC and FACTS Controllers: Applications of Static Converters in Power Systems*, Kluwer Academic Publishers, 2004.

12. O. Wasynczuk and N. A. Anwah, "Modeling and Dynamic Performance of a Self-Commutated Photovoltaic Inverter System," *IEEE Transactions on Energy Conversion*, vol. 4, no. 3, pp. 322–328, September 1989.

13. M. N. Marwali and A. Keyhani, "Control of Distributed Generation Systems. Part I. Voltages and Currents Control," *IEEE Transactions on Power Electronics*, vol. 19, no. 6, pp. 1541–1550, November 2004.

Voltage-Sourced Converters in Power Systems,　by Amirnaser Yazdani and Reza Iravani
Copyright © 2010 John Wiley & Sons, Inc.

14. K. Satoh and M. Yamamoto, "The Present State of the Art in High Power Semiconductor Devices," *Proceedings of the IEEE*, vol. 89, no. 6, pp. 813–821, July 2001.

15. B. J. Baliga, "The Future of Power Semiconductor Device Technology," *Proceedings of the IEEE*, vol. 89, no. 6, pp. 822–832, July 2001.

16. N. Mohan, T. M. Undeland, and W. P. Robbins, *Power Electronics, Converters, Applications, and Design*, 3rd edition, Wiley, 2003.

17. B. Wu, *High-Power Converters and AC Drives*, Wiley/IEEE, 2006.

18. A. Alesina and M. G. B. Venturini, "Analysis and Design of Optimum-Amplitude Nine-Switch Direct AC–AC Converters," *IEEE Transactions on Power Electronics*, vol. 4, no. 1, pp. 101–112, January 1989.

19. S. B. Dewan and A. Straughn, *Power Semiconductor Circuits*, Wiley, 1974.

20. D. G. Holmes and T. A. Lipo, *Pulse Width Modulation for Power Converters: Principles and Practice*, Wiley/IEEE, 2003.

21. M. Saeedifard, H. Nikkhajoei, R. Iravani, and A. Bakhshai, "A Space Vector Modulation Approach for a Multimodule HVDC Converter System," *IEEE Transactions on Power Delivery*, vol. 22, no. 3, pp. 1643–1654, July 2007.

22. M. Hagiwara, H. Fujita, and H. Akagi, "Performance of a Self-Commutated BTB HVDC Link System Under a Single-Line-to-Ground Fault Condition," *IEEE Transactions on Power Electronics*, vol. 18, no. 1, pp. 278–285, January 2003.

23. C. Schauder, M. Gernhardt, E. Stacey, T. Lemak, L. Gyugyi, T.W. Cease, and A. Edris, "Development of ±100 MVAR Static Condenser for Voltage Control of Transmission Systems," *IEEE Transactions on Power Delivery*, vol. 10, no. 3, pp. 1486–1493, July 1995.

24. J. Holtz, "Pulsewidth modulation: A Survey," *IEEE Transactions on Industrial Electronics*, vol. 39, no. 5, pp. 410–420, December 1992.

25. H. W. Van Der Broeck, H. Skudelny, and G. V. Stanke, "Analysis and Realization of a Pulsewidth Modulator Based on Voltage Space Vectors," *IEEE Transactions on Industry Applications*, vol. 24, no. 1, pp. 142–150, January/February 1988.

26. R. Wu, S. B. Dewan, and G. R. Slemon, "Analysis of an AC-to-DC Voltage Source Converter Using PWM with Phase and Amplitude Control," *IEEE Transactions on Industry Applications*, vol. 27, pp. 355–364, March/April 1991.

27. A. Nabavi Niaki and M. R. Iravani, "Steady-State and Dynamic Models of Unified Power Flow Controller (UPFC) for Power System Studies," *IEEE Transactions on Power Systems*, vol. 11, pp. 1937–1942, November 1996.

28. J. A. Sanders and F. Verhulst, *Averaging Methods in Nonlinear Dynamic Systems*, Springer-Verlag, 1985.

29. H. A. Khalil, *Nonlinear Systems*, 3rd edition, Prentice-Hall, 2002.

30. J. G. Kassakian, M. F. Schlecht, and G. C. Verghese, *Principles of Power Electronics*, Addison-Wesley, 1991.

31. P. T. Krein, J. Bentsman, R. M. Bass, and B. L. Lesieutre, "On the Use of Averaging for the Analysis of Power Electronic Systems," *IEEE Transactions on Power Electronics*, vol. 5, pp. 182–190, April 1990.

32. R. W. Erickson and D. Maksimovic, *Fundamentals of Power Electronics*, 2nd edition, Kluwer Academic Publishers, 2001

33. E. Davison, "The Robust Control of a Servomechanism Problem for Linear Time-Invariant Multivariable Systems," *IEEE Transactions on Automatic Control*, vol. AC-21, no. 1, pp. 25–34, February 1976.

34. W. M. Wonham, "Towards an Abstract Internal Model Principle," *IEEE Transactions on Systems, Man, and Cybernetics*, vol. SMC-6, no. 11, pp. 735–740, November 1976.

35. P. J. Antsaklis and O. R. Gonzalez, "Compensator Structure and Internal Models in Tracking and Regulation," *Proceedings of 23rd Conference on Decision and Control*, Las Vegas, NV, pp. 634–635, December 1984.

36. G. F. Franklin and A. E. Naeini, "Design of Ripple-Free Multivariable Robust Servomechanisms," *Proceedings of 23rd Conference on Decision and Control*, Las Vegas, NV, pp. 1709–1714, December 1984.

37. J. J. D'Azzo and C. H. Houpis, *Linear Control System Analysis and Design: Conventional and Modern*, 4th edition, McGraw-Hill, 1995.

38. K. Ogata, *Modern Control Engineering*, 4th edition, Prentice-Hall, 2001.

39. X. Yuan, W. Merk, H. Stemler, and J. Allmeling, "Stationary-Frame Generalized Integrators for Current Control of Active Power Filters with Zero Steady-State Error for Current Harmonics of Concern Under Unbalanced and Distorted Operating Conditions," *IEEE Transactions on Industry Applications*, vol. 38, no. 2, pp. 523–532, March/April 2002.

40. D. N. Zmood and D. G. Holmes, "Stationary Frame Current Regulation of PWM Inverters with Zero Steady-State Error," *IEEE Transactions on Power Electronics*, vol. 18, no. 3, pp. 814–822, May 2003.

41. H. Akagi, Y. Kanazawa, and A. Nabae, "Instantaneous Reactive Power Compensators Comprising Switching Devices Without Energy Storage Components," *IEEE Transactions on Industry Applications*, vol. IA-20, no. 3, pp. 625–630, May/June 1984.

42. P. Kundur, *Power System Stability and Control*, McGraw-Hill, 1994.

43. W. Leonhard, *Control of Electrical Drives*, 3rd edition, Springer-Verlag, 2001.

44. L. Angquist and L. Lindberg, "Inner Phase Angle Control of Voltage Source Converter in High Power Applications," *IEEE Power Electronics Specialists Conference PESC 91*, pp. 293–298, June 1991.

45. L. Xu, V. G. Agelidis, and E. Acha, "Development Considerations of DSP-Controlled PWM VSC-Based STATCOM," *IEE Proceedings: Electric Power Application*, vol. 148, no. 5, pp. 449–455, September 2001.

46. A. R. Bergen, *Power System Analysis*, Prentice-Hall, 1986.

47. M. C. Chandorkar, D. M. Divan, and R. Adapa, "Control of Parallel Connected Inverters in Standalone AC Supply Systems," *IEEE Transactions on Industry Applications*, vol. 29, no. 1, pp. 136–143, January/February 1993.

48. M. H. Rashid, *Power Electronics, Circuits, Devices, and Applications*, 3rd edition, Pearson Prentice-Hall, 2003.

49. S. Chung, "A Phase Tracking System for Three Phase Utility Interface Inverters," *IEEE Transactions on Power Electronics*, vol. 15, pp. 431–438, May 2000.

50. A. B. Plunkett and F. G. Turnbull, "Load-Commutated Inverter/Synchronous Motor Drive Without a Shaft Position Sensor," *IEEE Transactions on Industry Applications*, vol. IA-15, no. 1, pp. 63–71, January/February 1979.

51. R. Wu and G. R. Slemon, "A Permanent Magnet Motor Drive Without a Shaft Sensor," *IEEE Transactions on Industry Applications*, vol. 27, no. 5, pp. 1005–1011, September/October 1991.

52. T. Noguchi, K. Yamada, S. Kondo, and I. Takahashi, "Initial Rotor Position Estimation Method of Sensorless PM Synchronous Motor with No Sensitivity to Armature Resistance," *IEEE Transactions on Industrial Electronics*, vol. 45, no. 1, pp. 118–125, February 1998.

53. P. C. Krause, O. Wasynczuk, and S. D. Sudhoff, *Analysis of Electric Machinery*, IEEE Press, 1995.

54. P. Vas, *Vector Control of AC Machines*, Oxford University Press, 1990.

55. K. Thorborg, *Power Electronics*, Prentice-Hall, 1988.

56. J. S. Lai and F. Z. Peng, "Multilevel Converters: A New Breed of Power Converters," *IEEE Transactions on Industry Applications*, vol. 32, pp. 509–517, May/June 1996.

57. J. Rodriguez, J. Pontt, G. Alzamora, N. Becker, O. Einenkel, and A. Weinstein, "Novel 20-MW Downhill Conveyor System Using Three-Level Converters," *IEEE Transactions on Industrial Electronics*, vol. 49, pp. 1093–1100, October 2002.

58. J. Rodriguez, J. S. Lai, and F. Z. Peng, "Multilevel Inverters: A Survey of Topologies, Control, and Applications," *IEEE Transactions on Industrial Electronics*, vol. 49, no. 4, pp. 724–738, August 2002.

59. A. Nabae, I. Takahashi, and H. Akagi, "A New Neutral-Point-Clamped PWM Inverter," *IEEE Transactions on Industry Applications*, vol. IA-17, pp. 518–523, September/October 1981.

60. R. Sommer, A. Mertens, C. Brunotte, and G. Trauth, "Medium Voltage Drive System with NPC Three-Level Inverter Using IGBTs," *IEEE PWM Medium Voltage Drives Seminar*, pp. 3/1–3/5, May 11, 2000.

61. A. Yazdani and R. Iravani, "A Generalized State-Space Averaged Model of the Three-Level NPC Converter for Systematic DC-Voltage-Balancer and Current-Controller Design," *IEEE Transactions on Power Delivery*, vol. 20, no. 2, pp. 1105–1114, April 2005.

62. D. H. Lee, S. R. Lee, and F. C. Lee, "An Analysis of Midpoint Balance for the Neutral-Point-Clamped Three-Level VSI," *IEEE Power Electronics Specialists Conference*, PESC98, vol. 1, pp. 193–199, May 17–22, 1998.

63. C. Newton and M. Sumner, "A Novel Arrangement for Balancing the Capacitor Voltages of a Five-Level Diode Clamped Inverter," *IEE Power Electronics and Variable Speed Drives*, no. 456, pp. 465–470, September 21–23, 1998.

64. M. K. Mishra, A. Joshi, and A. Ghosh, "Control Schemes for Equalization of Capacitor Voltages in Neutral Clamped Shunt Compensator," *IEEE Transactions on Power Delivery*, vol. 18, pp. 538–544, April 2003.

65. C. Newton and M. Sumner, "Neutral Point Control for Multi-Level inverters: Theory, Design and Operational Limitations," *IEEE Industry Application Society Annual Meeting*, pp. 1336–1343, October 5–9, 1997.

66. G. Scheuer and H. Stemmler, "Analysis of a 3-Level-VSI Neutral-Point-Control for Fundamental Frequency Modulated SVC-Applications," *IEE AC and DC Power Transmission*, no. 423, pp. 303–310, April 29–May 3, 1996.

67. C. Osawa, Y. Matsumoto, T. Mizukami, and S. Ozaki, "A State-Space Modeling and a Neutral-Point Voltage Control for an NPC Power Converter," *Power Conversion Conference*, vol. 1, pp. 225–230, August 3–6, 1997.

68. M. P. Kazmierkowski and L. Malesani, "Current-Control Techniques for Three-Phase Voltage-Source PWM Converters: A Survey," *IEEE Transactions on Industrial Electronics*, vol. 45, no. 5, pp. 691–703, October 1998.

69. C. D. Schauder and R. Caddy, "Current Control of Voltage-Source Inverters for Fast Four-Quadrant Drive Performance," *IEEE Transactions on Industrial Electronics*, vol. IA-18, pp. 163–171, March/April 1982.

70. J. A. Houldsworth and D. A. Grant, "The Use of Harmonic Distortion to Increase the Output Voltage of a Three-Phase PWM Inverter," *IEEE Transactions on Industry Applications*, vol. IA-20, pp. 1224–1228, September/October 1984.

71. M. Mohaddes, D. P. Brandt, and K. Sadek, "Analysis and Elimination of Third Harmonic Oscillations in Capacitor Voltages of 3-Level Voltage Source converters," *IEEE PES Summer Meeting*, vol. 2, pp. 737–741, July 16–20, 2000.

72. A. Yazdani and R. Iravani, "An Accurate Model for the DC-Side Voltage Control of the Neutral Point Diode Clamped Converter," *IEEE Transactions on Power Delivery*, vol. 21, no. 1, pp. 185–193, January 2006.

73. R. Pena, R. Cardenas, R. Blasco, G. Asher, and J. Clare, "A Cage Induction Generator Using Back to Back PWM Converters for Variable Speed Grid Connected Wind Energy System," *IEEE Industrial Electronics Conference*, IECON'01, vol. 2, pp. 1376–1381, 2001.

74. C. K. Sao, P. W. Lehn, M. R. Iravani, and J. A. Martinez, "A Benchmark System for Digital Time-Domain Simulation of a Pulse-Width-Modulated D-STATCOM," *IEEE Transactions on Power Delivery*, vol. 17, pp. 1113–1120, October 2002.

75. Y. Ye, M. Kazerani, and V. H. Quintana, "Modeling, Control, and Implementation of Three-Phase PWM Converters," *IEEE Transactions on Power Electronics*, vol. 18, pp. 857–864, May 2003.

76. P. W. Lehn and M. R. Iravani, "Experimental Evaluation of STATCOM Closed-Loop Dynamics," *IEEE Transactions on Power Delivery*, vol. 13, pp. 1378–1384, October 1998.

77. T. M. Rowan and R. J. Kerkman, "A New Synchronous Current Regulator and an Analysis of Current-Regulated PWM Inverters," *IEEE Transactions on Industry Applications*, vol. IA-22, no. 4, pp. 678–690, March/April 1986.

78. V. Kaura and V. Blasko, "Operation of a Phase Locked Loop System Under Distorted Utility Conditions," *IEEE Transactions on Industry Applications*, vol. 33, no. 1, pp. 58–63, January/February 1997.

79. J. Svensson, "Synchronization Methods for Grid-Connected Voltage Source Converters," *IEE Proceedings: Generation, Transmission, and Distribution*, vol. 148, no. 3, pp. 229–235, May 2001.

80. L. G. B. Rolim, D. R. da Costa, and M. Aredes, "Analysis and Software Implementation of a Robust Synchronizing PLL Circuit Based on *pq* Theory," *IEEE Transactions on Industrial Electronics*, vol. 53, no. 6, pp. 1919–1926, December 2006.

81. D. A. Paice, *Power Electronics Converter Harmonics: Multipulse Methods for Clean Power*, Wiley/IEEE Press, 1999.

82. C. Schauder and H. Mehta, "Vector Analysis and Control of Advanced Static VAR Compensators," *IEE Proceedings C*, vol. 140, pp. 299–306, July 1993.

83. A. Yazdani, "Control of an Islanded Distributed Energy Resource Unit with Load Compensating Feed-Forward," *IEEE Power Engineering Society General Meeting*, 7 pp. July 20–24, 2008.

84. M. B. Delghavi and A. Yazdani, "A Control Strategy for Islanded Operation of a Distributed Resource (DR) Unit," *IEEE Power and Energy Society General Meeting*, 8 pp. July 26–30, 2009.

85. H. Karimi, A. Yazdani, and R. Iravani, "Negative Sequence Current Injection for Fast Islanding Detection of a Distributed Resource Unit," *IEEE Transactions on Power Electronics*, vol. 23, no. 1, pp. 298–307, January 2008.

86. R. Pena, J. C. Clare, and G. M. Asher, "A Doubly-Fed Induction Generator Using Back-to-Back PWM Converters Supplying an Isolated Load from a Variable Speed Wind Turbine," *IEE Proceedings on Power Applications*, vol. 143, pp. 380–387, September 1996.

87. S. Muller, M. Deicke, and R. W. De Donker, "Adjustable Speed Generators for Wind Turbines Based on Doubly-Fed Induction Machines and 4-Quadrant IGBT Converters Linked to the Rotor," *IEEE Industry Applications Magazine*, vol. 8, no. 3, pp. 26–33, May/June 2002.

88. R. Datta and V. T. Ranganathan, "Variable-Speed Wind-Power Generation Using Doubly-Fed Wound-Rotor Induction Machine: A Comparison with

Alternative Schemes," *IEEE Transactions on Energy Conversion*, vol. 17, no. 3, pp. 414–421, September 2002.

89. D. W. Novotny and T. A. Lipo, *Vector Control and Dynamics of AC Drives*, Oxford University Press, 1996.

90. B. K. Bose, *Power Electronics and Variable Frequency Drives*, IEEE Press, 1997.

91. R. Cardenas, R. Pena, G. M. Asher, J. Clare, and R. Blasco-Gimenez, "Control Strategies for Power Smoothing Using a Flywheel Driven by a Sensorless Vector-Controlled Induction Machine Operating in a Wide Speed Range," *IEEE Transactions on Industrial Electronics*, vol. 51, no. 3, pp. 603–614, June 2004.

92. T. M. Jahns, G. B. Kliman, and T. W. Neumann, "Interior PM Synchronous Motors for Adjustable-Speed Drives," *IEEE Transactions on Industry Applications*, vol. 22, no. 4, pp. 738–747, July/August 1986.

93. B. K. Bose, "A High-Performance Inverter-Fed Drive System of an Interior Permanent Magnet Synchronous Machine," *IEEE Transactions on Industry Applications*, vol. 24, no. 6, pp. 987–997, November/December 1988.

94. S. Y. Morimoto, Y. Takeda, T. Hirasa, and K. Taniguchi, "Expansion of Operating Limits for Permanent Magnet Motor by Current Vector Control Considering Inverter Capacity," *IEEE Transactions on Industry Applications*, vol. 26, no. 5, pp. 866–871, September/October 1990.

95. R. Mihalic, P. Zunko, I. Papic, and D. Povh, "Improvement of Transient Stability by Insertion of FACTS Devices," *IEEE/NTUA Proceedings of Athens Power Tech Conference, APT 93*, vol. 2, pp. 521–525, September 1993.

96. J. F. Gronquist, W. A. Sethares, F. L. Alvarado, and R. H. Lasseter, "Power Oscillation Damping Control Strategies for FACTS Devices Using Locally Measurable Quantities," *IEEE Transactions on Power Systems*, vol. 10, no. 3, pp. 1598–1605, August 1995.

97. E. Stacey, T. Lemak, L. Gyugyi, T. W. Cease, and A. Edris, "Operation of −100 MVAr TVA STATCON," *IEEE Transactions on Power Delivery*, vol. 12, no. 4, pp. 1805–1811, October 1997.

98. M. Noroozian, A. Edris, D. Kidd, and A. J. F. Keri, "The Potential Use of Voltage-Sourced Converter-Based Back-to-Back Tie in Load Restoration," *IEEE Transactions on Power Delivery*, vol. 18, pp. 1416–1421, October 2003.

99. G. C. Paap, "Symmetrical Components in the Time-Domain and Their Applications to Power Network Calculations," *IEEE Transactions on Power Systems*, vol. 15, pp. 522–528, May 2000.

100. L. Moran, P. D. Ziogas, and G. Joos, "Design Aspects of Synchronous PWM Rectifier-Inverter Systems Under Unbalanced Input Voltage Conditions," *IEEE Transactions on Industrial Electronics*, vol. 28, no. 6, pp. 1286–1293, November/December 1992.

101. T. Sun, Z. Chen, and F. Blaabjerg, "Flicker Study on Variable Speed Wind Turbines with Doubly-Fed Induction Generators," *IEEE Transactions on Energy Conversion*, vol. 20, no. 4, pp. 896–905, December 2005.

102. A. Yazdani and R. Iravani, "A Neutral-Point Clamped Converter System for Direct-Drive Variable-Speed Wind Power Unit," *IEEE Transactions on Energy Conversion*, vol. 21, no. 2, pp. 596–607, June 2006.

103. M. Chinchilla, S. Arnaltes, and J. C. Burgos, "Control of Permanent-Magnet Generators Applied to Variable-Speed Wind-Energy Systems Connected to the Grid," *IEEE Transactions on Energy Conversion*, vol. 21, no. 1, pp. 130–135, March 2006.

104. P. M. Anderson and A. Bose, "Stability Simulations of Wind Turbine Systems," *IEEE Transactions on Power Apparatus and Systems*, vol. PAS-102, pp. 3791–3795, December 1983.

105. M. P. Kazmierkowski, R. Krishnan, and F. Blaabjerg, *Control in Power Electronics, Selected Problems*, Academic Press, 2002.

106. S. Heier, *Grid Integration of Wind Energy Conversion Systems*, 2nd edition, Wiley, 2006.

107. J. G. Slootweg, H. Polinder, and W. L. Kling, "Representing Wind Turbine Electrical Generating Systems in Fundamental Frequency Simulations," *IEEE Transactions on Energy Conversion*, vol. 18, no. 4, pp. 516–524, December 2003.

108. Y. D. Song and B. Dhinakaran, "Nonlinear Variable Speed Control of Wind Turbines," *IEEE Proceedings of International Conference on Control Applications*, pp. 814–819, August 1999.

109. G. R. Slemon, "Modelling of Induction Machines for Electric Drives," *IEEE Industry Application Society Annual Meeting*, pp. 111–115, 1988.

INDEX

Voltage-Sourced Converters in Power Systems, by Amirnaser Yazdani and Reza Iravani
Copyright © 2010 John Wiley & Sons, Inc.

Printed in Poland
by Amazon Fulfillment
Poland Sp. z o.o., Wrocław
30 October 2020

3f71b041-5fbe-4bc6-b8f0-843779572eb7R01